ETHICS EXPERTISE

Philosophy and Medicine

VOLUME 87

The titles published in this series are listed at the end of this volume

ETHICS EXPERTISE

HISTORY, CONTEMPORARY PERSPECTIVES, AND APPLICATIONS

Edited by

LISA RASMUSSEN

University of Alabama, Birmingham, AL, U.S.A.

Springer

A C.I.P. Catalogue record for this book is available from the Library of Congress.

ISBN-10 1-4020-3819-4 (HB)
ISBN-13 978-1-4020-3819-8 (HB)
ISBN-10 1-4020-3820-8 (e-book)
ISBN-13 978-1-4020-3820-4 (e-book)

Published by Springer,
P.O. Box 17, 3300 AA Dordrecht, The Netherlands.

www.springeronline.com

Printed on acid-free paper

Printed in the Netherlands.

TABLE OF CONTENTS

TABLE OF CONTENTS

ACKNOWLEDGMENTS

The authors who contributed to this volume have considerably enhanced my understanding of the issues involved in ethics expertise. I thank each of them for their thoughtful contributions and for their (expert) patience as the volume was brought to fruition.

Thanks also go to anonymous reviewers of the individual papers and an anonymous reviewer of the volume. I would especially like to thank Prof. Dr. Oliver Scholz, who reviewed the volume for Springer Publishers and offered many helpful suggestions. Reviewing is a laborious activity, and I am grateful for the time and energy spent on behalf of this volume.

The editor of the *Philosophy & Medicine* series, H. Tristram Engelhardt, Jr., has been supportive throughout the process of producing this volume. I thank him as ever for his mentorship and generosity.

Finally, thanks to my husband, Benjamin Hippen, for our ongoing philosophical conversations.

CHAPTER 1

LISA M. RASMUSSEN

INTRODUCTION: IN SEARCH OF ETHICS EXPERTISE[1]

1. INTRODUCTION

Expertise is commodified. In a busy, specialized world where one person simply cannot master all fields, we turn to experts to help us with a variety of issues. It is not surprising, then, that a variety of people from a variety of disciplines have stepped forward as bioethicists or "ethics experts."[2] It may be difficult to find many who would actually *call* themselves moral or ethical experts, but the possibility of moral expertise is the presumption behind much of the practice in bioethics consultation, expert bioethics witnessing in court, policy recommendations and the offering of media "sound bites." This volume addresses the question of whether ethics expertise exists, what training would convey such expertise, and what claims of expertise warrant.

A first response to the idea of ethics expertise might be skepticism or suspicion. One might be *skeptical* because questions of right and wrong have been debated for a significant portion of human history; why should we expect that we can answer them now? And if we cannot answer these questions, the skeptic might ask, how can someone become an ethical expert? *Suspicion* might arise because, as Stephen Wear points out in this volume (Chapter 13), the canonization of liberty, autonomy and diversity mitigates against formally recognizing or ceding authority to any such expertise even if it were conceded to exist. A proponent of the importance of individual freedom, or the fact of moral plurality, or the positive value of diversity, might resist any attempt to recognize ethical expertise.

However, two issues should give us pause before we resist ethical expertise altogether. First, the proliferation of bioethics experts suggests that people find them helpful. Although it is not immediately clear that such experts are gurus of the good, intellectual humility at least recommends that we honestly investigate the phenomenon. We may eventually conclude that such experts are charlatans at worst or misguided at best, but we may also conclude that there is indeed a thing called ethics expertise.

Second, expertise comes in many forms. So again, we may conclude that there *is* such a thing as ethics expertise, but it is an expertise that does not threaten autonomy or diversity. For example, if we interpret ethical expertise as expert

L. M. Rasmussen (Ed.), Ethics Expertise: History, Contemporary Perspectives, and Applications, 1-12.

scholarly knowledge about the content of particular moral theories, nothing practical necessarily follows. An answer to the question "Does ethics expertise exist?" might always be answered by, "It depends on what you mean by 'ethics expertise'," but this need not be a pernicious relativism.

Much of the literature on moral expertise begins with the question of whether or not it is even possible. Each author interprets ethics expertise differently, and draws conclusions about that particular interpretation. For example, Peter Singer argues that the process of deciding right and wrong involves gathering information, determining what counts as relevant information and how it fits with a particular moral position, and avoiding one's own bias. For this reason, he suggests,

> [s]omeone familiar with moral concepts and with moral arguments, who has ample time to gather information and think about it, may reasonably be expected to reach a soundly based conclusion more often than someone who is unfamiliar with moral concepts and moral arguments and has little time. So moral expertise would seem to be possible (Singer, 1972, pp. 116-117).

In contrast, Cheryl Nobel (1982) suggests that moral expertise is the *provision of moral wisdom*, and that moral philosophers are too theory-bound to provide such wisdom (the 'teaser' for the article reads: "Have philosophers substituted moral reasoning for moral wisdom?").[3] Bruce Weinstein (1994) distinguishes between expertise in descriptive ethics, expertise in metaethics, expertise in normative ethics, and expertise in living a good life. Rosemarie Tong recognizes room both for expertise as "command over specific matters of fact and/or value" and as "ability to achieve 'consensus' about what is 'true'/'false', or 'right'/'wrong'" (p. 409). Scott Yoder (1998) suggests that ethical expertise is not about making correct or true judgments, but rather it is the ability to justify coherently one's moral judgment. Not only do authors disagree on whether ethics expertise exists, they disagree on what it *is*.

Therefore, we can imagine a variety of definitions of ethics expertise, which include the type of knowledge such experts possess, the degree of fallibility permitted of such experts (and how this fallibility would be measured), and whether it is objective/scholarly knowledge or, on the other hand, substantive knowledge about the good. Among the salient questions are:

1. How does ethics expertise relate to morality and ethics in general? That is, does an ethics expert possess moral wisdom in the sense of knowing the Truth about the right and the good? Is an ethics expert also a good person, and if so, is he so by definition? That is, can we imagine an ethics expert who is a bad person, or must knowledge and behavior always cohabit in an ethics expert?
2. What kind of training gives one ethics expertise? Is it only graduate school education in philosophy or theology?
3. What kind of political or legal authority does an ethics expert possess? If an ethics expert advises on government policy, are policy makers mistaken to override that advice? Would a bioethics consultant's expert opinion be sufficient to override the wishes of others, including the patient or family member? Even if that were illegal, would it be unwise, morally speaking?
4. How (apart from credentialing dependent on question 2) can the rightness of an ethics expert's opinion be assured? On what basis can it be challenged?

Section I examines historical philosophical understandings of expertise in order to situate the current institution of bioethics. Section II focuses on philosophical analyses of the concept of expertise, asking, among other things, how it should be understood, how it can be acquired, and what such expertise warrants. Finally, section III addresses topics in bioethics and how ethics expertise should or should not be brought to bear in these areas, including expertise in the court room, in the hospital room, in the media, and in making policy.

2. A GUIDED HISTORICAL TOUR

As Scott LaBarge points out, Plato's dialogues can be viewed as an extended treatment of the concept of moral expertise, so it is fitting to begin the volume with an examination of "Socrates and Moral Expertise". Given Socrates' protestations (the Oracle at Delphi notwithstanding) that he knows nothing, LaBarge observes that it would be interesting to determine both what a Socratic theory of moral expertise might be and whether Socrates qualified as such an expert. Plato's model of moral expertise is what LaBarge calls "demonstrable expertise", which is concerned mainly with the ability to attain a goal and to explain how one did it. The problem with this account is that when one tries to solve the various problems in the model – for example, allowing that moral expertise is not an all-or-nothing skill – then one is immediately faced with the "credentials problem".

As LaBarge puts it, "...since we have acknowledged that expertise comes in degrees, even moral expertise, and thus that even genuine moral experts can disagree from time to time, the crucial task that remains is to explain just how moral experts can be recognized" (p. 24). But how can non-experts recognize and judge the credentials of such experts? In the case of car repair, LaBarge explains, a layperson might be able to judge a mechanic's expertise by observing the results of his work: if a problem is solved after the mechanic's work, the layperson is justified in believing that the mechanic knew what he was talking about. Moral expertise, however, is not so straightforward: "Confronted by competing camps of experts who each offer different and incompatible explanations of moral issues, the non-expert is not well-situated to judge which of the competing camps possesses genuine expertise, if any" (p. 25). And in this case, unlike the case of auto repair, there is no objective criterion by which the non-expert can make a reasonable assessment of expertise: "...on the issues where we feel the greatest need for the guidance of a moral expert...the shape of a successful solution to our moral problems is typically the very thing at the heart of the controversy" (p. 26). The unsatisfying answer to this dilemma is that "[s]ince the nature of success in moral expertise is itself under dispute, and this dispute characteristically does not produce a clear winner recognizable to all, ultimately the non-expert cannot confidently assert that any of the claimants to moral expertise is a genuine expert" (p. 27).

It makes sense to take Socrates at his word when he protests that he is not a moral expert, because according to the six Platonic criteria LaBarge offers, Socrates does in fact fail to measure up:

> Thus Socrates turns out to deny he is a moral expert for much the same reasons that we found make moral experts so hard for non-experts to identify. He is unable to resolve the confusion and disagreement even of intelligent, well-meaning ... and he acknowledges that he is unable reliably to produce conviction even on matters where he is confident he has the truth (p. 30).

Instead, Socrates has what LaBarge calls "dialectical expertise". In his activities as gadfly, he constantly prods people regarding their values, attempting to make their souls more beautiful not by imbuing those souls with value, but rather by removing "ugly ignorance" (p. 32). By exposing the contradictions inherent in his interlocutors' belief sets, Socrates "deflat[es] their presumption to moral expertise and...inspir[es] a new commitment to seek such expertise" (p. 33). Thus Socrates is right in asserting his ignorance, and the Oracle is right in asserting his wisdom.

According to Aristotle, "the human good proves to be activity of the soul in accord with virtue" (quoted in Khan, p. 41). But, as Carrie-Ann Biondi Khan points out in "Aristotle's Moral Expert: The *Phronimos*", it turns out not to be easy to live in accord with virtue, so we need moral experts to help us. She first gives a brief description of Aristotle's moral theory, and then demonstrates the crucial role the *phronimos*, or moral expert, plays in this theory.

The human good, as Aristotle understands it, is based on the *telos* (purpose or end) of humans. He argues that the highest end of human life is to use the reasoning part of the soul, and to use it virtuously. Moreover, moral virtue and moral knowledge are essential parts of this end. Virtue, Khan points out, has four key features: it is a mean between two extremes; it involves decision and deliberation; it requires the use of reason; and it must be acquired through habituation. Aristotle himself believed that there would be very few people who could do all of these things, both because it is difficult to attain the requisite sort of wisdom and because it is difficult to overmaster one's passions. Accordingly, for Aristotle, moral experts must exist in order for *others* to live the good human life.

Of course, the next pressing question is how one becomes a *phronimos*. Khan discusses three problems with Aristotle's account of the method – problems that may leave one in doubt as to whether or how such expertise can really be acquired. Interestingly, the 'credentials problem' that LaBarge discusses (see above) arises again in Khan's discussion of Aristotle: if one is not oneself a moral expert, how can one recognize those who are? For the general question of moral expertise, perhaps the most intriguing question that arises in Khan's discussion is whether the existence and identification of the moral expert is indeed necessary for the rest of us to live the good human life?

In "Hume on True and False Philosophy," Christopher Tollefsen argues that some kinds of contemporary expertise are examples of 'false' philosophy, but that there is a possible kind of expertise that demonstrates 'true' philosophy. A central feature of Hume's theory was the claim that morality and the moral impulse derived from sentiments rather than reason. The impact of this, according to Tollefsen, was that Hume came to see the use of abstract reason in morality as

> not simply wrongheaded, but potentially dangerous and divisive...by attempting to transcend custom and sentiment, an impossible task, [it] could only end up seizing covertly upon some principle of common life, and transforming it into an overarching principle of reason intended to rule over common life (p. 56).

Once one starts down this path, however, if one is honest, the search will eventually result in true philosophy, in Hume's theory.

Tollefsen presents Livingston's account of the dialectic that moves the honest person from false to true philosophy. When one uses abstract reason to find an Archimedean point from which to view morality, one is subject to a variety of possible errors. In order to overcome these errors, one can break judgments down into smaller pieces – but the result is that each of those pieces are themselves subject to error, so one has in fact multiplied problems rather than solved them. One is reduced to profound skepticism. Yet we clearly are not frozen by skepticism in our daily lives, and this proves, according to Hume, that "belief is not, after all, founded on reason, but on custom" (p. 57).

But if this is the case, what is true philosophy, and how does it differ from false philosophy? According to Tollefsen, there is a negative and a positive aspect of Hume's true philosophy. The negative task of true philosophy is "the abatement of pride that drives the construction of systems and theories" – i.e., the skeptical task. The positive task, Tollefsen shows, has several components, including "refin[ing] and moderat[ing] the passions generally," investigating the principles in common life, and taking part in "public conversation on matters of societal import". In the end, Hume's account is "a portrait of the true philosopher as a kind of rhetorician" (p. 69).

It is interesting to view moral expertise in bioethics through the Humean prism of true and false philosophy. Many approaches in bioethics, Tollefsen points out, use reason to "elevat[e] a single abstract principle and impos[e] it from above" (e.g., utilitarianism), yet other approaches attempt to embrace the context of bioethics, with all its complexity, emotional aspects and tangle of motivations. Though there are challenges to be mounted against a Humean version of moral expertise, those favoring such an approach in bioethics can find sympathetic accounts, Tollefsen suggests, in the works of Edmund Pellegrino, Leon Kass, or William May.

"Moral Expertise: A Millian Perspective" by Dale Miller presents the concept of moral expertise as Mill might have interpreted it. Miller begins by pointing out the distinction between theoretical moral expertise, the ability "to evaluate the arguments for and against rival moral theories and pick out the one for which the strongest case can be made" (p. 75), and practical moral expertise, "being able to 'apply' a moral theory, in order to determine in concrete terms what it requires from us" (p. 75). Theoretical expertise is not particular to any theory, but the requirements for practical expertise, Miller points out, will vary from theory to theory.

Though there is disagreement about whether Mill himself was an act or a rule utilitarian, Miller points out that Mill seems to have affirmed that "the rules of ordinary morality are a good first approximation" (p. 80) of how to conduct ourselves morally. Clearly then, Miller argues, Mill would understand expertise in terms of moral rules. This expertise can come in higher and lower forms, "with the lower consisting merely of acquaintance with a decent set of such rules, and the higher consisting of the ability to improve upon the set that is conventionally accepted" (p. 81). Just as Aristotle held that the *phronimos* (see above) was a rare person, Mill holds that it is 'the philosopher' rather than the multitude who is able to improve on the commonly accepted rules of morality (Miller, p. 81). In Mill's case, claims to the higher form of expertise require that one have empirical knowledge (especially of human nature) and knowledge of human character. Echoing Aristotle's notion of habituation, Mill holds that for someone to have the higher form of expertise, one must know "what character traits or habits are most desirable, and how they can be instilled in people" (p. 81). Mill was dubious about his own and others' abilities to instill these traits, and likewise would have been dubious about anyone's claim to moral expertise.

What of the question of moral experts as moral exemplars? Miller observes that Mill was a moral externalist, so he would not hold any necessary connection between possessing moral expertise (even of the higher type) and being *motivated* to do the right thing. If this is so, then while we might today engage moral philosophers *cum* moral experts to help us solve the profound dilemmas in bioethics (which might require revising our commonly accepted but conflicting moral rules), we should not expect them personally to be paragons of virtue.

"The Ineffable and the Incalculable: G.E. Moore on Ethical Expertise" by Ben Eggleston neatly sets out Moore's theory of ethics and indicates what Moore's thoughts might have been regarding ethical expertise. Though Moore himself was more concerned with the good than the right, his theory can nonetheless address the question of moral rightness, and it is this course that Eggleston pursues.

As a consequentialist, Moore held that it is in virtue of the consequences brought about by an action that such an action is right; i.e., productive of the most good. But there are three aspects to Moore's defense of consequentialism that impact what he might have to say about ethical expertise. First, Moore held consequentialism to be analytically true, and would claim that any moral expert had to embrace this position (because not to recognize the truth of this claim would be an indication of faulty judgment). Second, the expert must understand that the good is ineffable yet possess the intuition to see it, and third, he must be able to make judgments about the best means to a given good end.

Moore holds that questions about right actions cannot be answered without first answering two other questions: "We must know both what degrees of intrinsic value different things have, and how these different things may be obtained" (quoted in Eggleston, p. 93). The first question concerns knowing what "the good" is, but it turns out to be difficult to do this. Moore argues that when people try to explain what the good is, for example by saying "the good is the pleasant," they commit the "naturalistic fallacy" – that is, they make the mistake of thinking that by describing a fact or feature about something (its pleasantness, for example), they are saying something about the good. According to Moore, the good is "a simple, indefinable, unanalysable object of thought", or as Eggleston summarizes it, the good is "ineffable". Therefore an ethical expert must be able "to discern, by intuition, the goodness and badness, both absolute and relative, of various possible ends" (p. 96). As Eggleston points out, there are a number of potential problems with this view, but it is a central feature of Moore's position.

The final trait an ethical expert must possess is the ability to make empirical judgments of cause and effect – that is, he must be able to determine the best means to a given end. Here we meet with what Eggleston has identified as the incalculability problem: as with other consequentialist theories, Moore's theory suffers from the fact that it is very demanding. One must know all of the possible acts he could perform and be able to calculate all the possible consequences of each. Between the difficulty of knowing the good and the best way of obtaining it, Moore's theory seems impossible to implement. However, Eggleston points out, in the end Moore recognizes common-sense morality as a reasonable approximation of the correct moral rules.

Griffin Trotter's paper, "Pragmatism and Moral Expertise", begins by pointing out the pluralistic nature of contemporary secular societies, indicating that part of the question of ethical expertise stems from the lack of a commonly-recognized moral code or moral authority. In such societies, Trotter suggests, there are four possible models of the ethical expert. The first is mainly a theoretical expert, who knows about moral theory in a scholarly sense. The second kind of expert is an inquirer; she will know how best to conduct an inquiry into what is right and wrong, applying a particular theory. Third, the 'surveyor' is someone who knows about peoples' moral customs, in an anthropological sense. Finally, the 'counselor' "is a person who helps individuals and communities become ethically good, or do ethically good things" (p. 106), however that individual or community defines "good". The first two types of expertise focus more on epistemic issues, while the latter two constitute what Trotter calls the 'therapeutic option'.

According to Trotter, with certain caveats, Peirce and Royce fall more into the epistemic camp, while Dewey and James fall into the therapeutic camp. As Royce and Dewey focused more on social ethics, Trotter spends more time investigating their views on ethical expertise. Royce "...bases his account of the moral life on the struggle of the individual to cultivate a coherent and satisfying life plan," which requires becoming conscious of one's will and turning outward, to one's community, for help in the formulation of the plan. Here, of course, is space for an ethical expert, but his authority "extends only to fellow loyalists within a particular moral community" (p. 110). That is, an ethical expert can help someone in a particular, committed community to formulate a life plan, but cannot offer the same expertise to someone outside that community.

John Dewey, on the other hand, had a more "therapeutic" view of expertise. He argued for a certain kind of life, and thought that political life and political leaders ought to be organized towards that end: "Like contemporary perfectionists..., Dewey thinks that states are obligated to discern and articulate the characteristics of good living, and then to encourage good lives through legislation" (p. 114). A Deweyan ethical expert would advise both legislators and individuals about the good life. Trotter concludes by giving reasons to prefer Royce's version of ethical expertise to Dewey's – reasons that certainly apply to bioethics today.

3. CONTEMPORARY PERSPECTIVES

Mary Ann Cutter's paper, "Expert Moral Choice in Medicine: A Study of Uncertainty and Locality" begins the section of the volume focused on contemporary examinations of ethical expertise. She argues that moral knowledge is localized, so the ethics expert (especially in bioethics) will need to be able to understand and explain to others the local terrain surrounding particular medical decisions.

She begins, as did Trotter, by acknowledging the pluralistic nature of moral beliefs. This does not imply ethical relativism, she argues, but it does force us to "begin to accept that moral decision making...is fraught with uncertainty" (p. 128). "Localization" of moral knowledge is an attention to the concrete facts of a case, allowing for flexibility in judgment. For example, Cutter explains, clinical problems are described in certain ways depending on who is looking at them: a genetics expert biologist will look at an AIDS case through the lens of genetics, while an immunologist will focus on T-cell counts, and a public health official will worry about lifestyle and public exposure. As she points out, "[a] local account of clinical problems allows for a range of judgments" (p. 130), but also "there is *much* that is shared within and across communities of medical professionals" (p. 132).

Often the way in which such consensus is reached is by negotiation: "the *process* of arranging the terms and conditions of a transaction...within the constraints of a given world" (p. 132). The bioethics expert can aid in the negotiation process by serving as

> geographers of facts, values, and socio-political influences ... They are trained to illuminate conflicting moral interests. They can rehearse the reasons for holding one moral position as opposed to another and can often provide the best arguments for their moral positions (p. 133).

Such experts will therefore need interdisciplinary training, but they will not in the end possess a privileged position relative to moral truth.

In "Societal Consensus and the Problem of Consent: Refocusing the Problem of Ethics Expertise in Liberal Democracies," Corinna Delkeskamp-Hayes argues that while the discussion of ethical expertise in the literature focuses on whether or not individuals ought to *offer* expertise, the real focus should be on whether others should *demand* it. According to her, Western society bases its moral legitimacy on a commitment to granting certain freedom rights, especially the right to autonomous decision-making. In many manifestations, ethics expertise focuses on simply enabling individuals to realize that freedom. However, she argues, in many cases this expertise "concern[s] content-rich moral solutions to value-impregnated practical problems" (p. 146). The main problem with this use of expertise is that "[i]nsofar as ethics experts guide the use of political power in concrete situations, their service....turns out to legitimize the use of force against the non-consenting" (p. 146). The irony, of course, is that such expertise "morally justifies interfering with those whose right to moral self-determination was supposed to have been respected" by these experts in the first place (p. 146). She suggests that there are two solutions to this dilemma. First, we can impose a moratorium on decisions in which not everyone involved with the decision agrees. However, practical limitations on actually obtaining everyone's agreement (especially where it concerns an institutional or government policy) render this an inadequate solution, as does the fact of moral pluralism. Second, she suggests that we can renounce the idea of common action completely, instead adopting an arrangement in which "the project of regulating...the lives of people who are divided by irredeemably different moral commitments could be trusted to different sets of regulatory agencies, or diverse kinds of regulatory, policy, or even legal frameworks" (p. 152). The alternative, she concludes, is to continue to allow ethics expertise as it is usually conceived "merely [to secure] an ethical veneer for the continuing self-destruction of [the state's] liberal commitments" (p. 153).

Lisa Parker takes a novel perspective on the subject in "Ethical Expertise, Maternal Thinking, and the Work of Clinical Ethicists." She begins with a very helpful review of some of the major work in the area of ethical expertise, in particular Bruce Weinstein's concept of "performative ethics experts". She then offers an account of performative expertise, specifically as it applies in the medical arena, grounded in Sara Ruddick's conception of 'maternal' thinking.

Clinical ethics, Parker points out, is responsible for creating in medical students and consultees (i.e., patients or members of the health care team) the "capacity for seeing the world in a way that makes particular demands upon them, including demands for care, justice, and humility" (p. 181). Like 'mothering' work, the goal of this kind of activity is to create independent agents who possess their own moral expertise. If clinical ethical expertise is construed in this way, she points out, it escapes many of the criticisms against it. For example, one criticism of clinical ethics is that it abuses the autonomy of patients by substituting the expert's judgment for the patient's. Instead, Parker argues, clinical experts might actually help to foster the exercise of autonomy by pointing out to patients the pitfalls of various arguments, or the existence of choices that have not occurred to the patient. "So long as the ethics expert supplies her reasons, she provides the opportunity to enhance her consultee's moral and decisional autonomy" (p. 193).

4. APPLICATIONS

Insofar as we accept the existence of some sort of ethical expertise, how might it be used in bioethics today? Robert Veatch begins this section with "The Roles of Scientific and Normative Expertise in Public Policy Formation: The Anthrax Vaccine Case." One of the central points of his paper is that when experts are gathered to make public policy recommendations, invariably they are asked to provide expertise that they do not necessarily possess. He takes recent policy work on the anthrax vaccine as an illustrative (but not unique) example of this claim.

As Veatch points out, even claims that appear scientific and straightforward (e.g., that a reaction is "adverse" or that a vaccine is "safe") involve value judgments (what is adverse to one person may not be adverse to another; "safe" vaccines may have side effects acceptable to some and not to others). In fact, he argues, "…*every* public policy decision involves assessments of potential risks and benefits. No interventions can be labeled 'safe' independent of the judgment that it is 'safe enough for the objective being pursued'" (p. 215).

This is not to imply anything untoward about the members of such committees, only that they may not have some of the expertise required. There are ethics experts "skilled in moral analysis, in the history of ethics, and in what conclusions have been reached by past participants in moral disputes" (p. 217). But, Veatch observes, "I know of no evidence that the consensus view of ethicists is superior or more correct on moral and other normative questions than the consensus view in the general public" (p. 218). His conclusion is to suggest three strategies "for protecting against this inevitable impinging of values in the scientific consensus of experts": the appointment of multiple panels, establishing political criteria for choosing panelists and the discounting or adjusting of expert opinions to accord with society's values.

A new application of ethical expertise is the focus of Kenneth Cust's "Philosophers Return to the *Agora*." Philosophical counseling takes knowledge of and experience in philosophical reasoning and uses it to help those who experience some tension or difficulty in their lives. Of course the first question that arises in the mind of the layperson is how philosophical counseling differs from psychotherapy, but as Cust points out, psychotherapists have a difficult time defining their own field. Overlap between the two fields may be something to expect, and need not gainsay either field.

Yet there are other criticisms of the novel field. Cust discusses the charge that philosophical counselors are sophists – those "learned men" who sold their intellectual services to wealthy ancient Athenians. First, Cust argues, simply accepting a fee for what one does in philosophical counseling does not render it an unacceptable activity. It is the taking of money on false pretenses that makes the sophist a charlatan, and Cust points out that if real services are delivered in philosophical counseling, no deception is involved. It seems reasonable to assume that some problems are philosophical (for example, beliefs based on false assumptions which cause the believer problems) and could benefit from philosophical counseling. As Cust puts it, "[w]hat philosophical counselors can bring to the metaphorical couch of the client is … the education of the client on *how* to think, not *what* to think" (p. 232). This observation also addresses the criticism that philosophical counselors have unjustifiably entered the legitimate domain of psychotherapy – because if philosophical problems exist and are different than psychological problems, then different expertise is required to address them. Cust notes that philosophical counselors must be careful to familiarize themselves with a variety of psychological disorders and psychoactive drugs, so as to recognize when a client really needs another kind of care or therapy in lieu of philosophical counseling. He concludes with helpful examples of how philosophical counseling can proceed.

Clinical consultation is the subject of Stephen Wear's "Ethical Expertise in the Clinical Setting." This is a particularly relevant area for the investigation of ethical expertise, as the field examines the wisdom and practicability of accreditation in the field. Wear argues that ethical expertise in the clinical setting is real, and that such expertise includes both procedural and substantive knowledge. He also takes a stand against ethical relativism, arguing that "there are right and wrong answers to certain ethical issues, as well as better and worse ways of responding to ethical conflicts and disputes, and ... certain people are much more capable of formulating such answers or responses" (p. 243).

There seems to be agreement on the claim that clinical ethics consultants are (or should be) experts in procedural skills, though the degree to which an individual possesses it may be open to question. There is also general agreement on the more substantive expertise that these consultants should have regarding areas such as codes of ethics, health law, clinical terms, etc. Wear points out that while mastering this set of information may be impossible for an individual, ethics consultation services use a variety of committee methods to handle the task.

A deeper question at the heart of clinical ethics consultation is whether experts in this area can know more about right and wrong than others. Wear argues that they can, despite the millennia of arguing over moral philosophy. Unlike the theoretical world of academia, Wear points out, features of the clinical setting allow for a substantive form of ethical expertise. The clinical issue requires resolution, not ongoing theoretical debate. This moves us, as he points out, from relativism to the search for the best option given the circumstances. And this best option is not merely achieved by the bioethics consultant facilitating consensus. Wear argues that there is a canon of clinical ethics, including accepted moral truths, that guide the practice of clinical ethics consultation. As a result, it is possible that sometimes the job of the clinical ethicist is simply the political task of getting others to see and agree to the right answer. As he sums up this argument, "...clinical ethics experts are legitimately seen as having special standing to frame, guide and, at times, declare the end result of an ethics dispute at the bedside" (p. 256).

Ana Smith Iltis notes, in "Bioethical Expertise in Health Care Organizations," that there are multiple possible, legitimate accounts of bioethical expertise. These accounts can depend on the roles experts play, such as expert witnesses in the court, policy analysts, etc. They can also depend on the content which is the focus of a particular expert's attention, such as genetics or medical research.

It becomes clear, as Iltis points out, that the knowledge base required of bioethics experts will vary depending on the roles they play, their area of specialty, and the context in which they act. For example, a religious institution which takes its moral commitments seriously will want to hire a bioethics consultant who, if she does not actually hold those values, will at least know them thoroughly and act and consult based upon them. As she sums up, "[i]nstitutional mission and identity can justifiably frame the scope of bioethics consultation within particular institutions, just as an organization's mission and identity should shape the activities of all employees" (p. 265).

In "The Expert Ethics Witness as Teacher," Kenneth Kipnis reflects on more than a decade of experience in this particular application of ethics expertise. He begins by reflecting on the path that led him from graduate study in philosophy to serving as an expert witness in court. He became interested in the application of philosophy, particularly in the hospital, which eventually involved him in a variety of bioethics activities. He came to think that the classroom methods he had developed would also serve in the courtroom.

As Kipnis presents it, both teaching and expert witnessing

> draw from six key concepts: 1) the fact of ethical pluralism; 2) the need for professional consensus; 3) the practice of ethical justification; 4) the varieties of ethical competence; 5) the role of professional statespersons; 6) the renegotiation of the social contract (p. 272).

An expert witness can uses these ideas "...to illuminate for the court the scope, the justification, and the import of pertinent ethical standards applicable to doctors" (p. 269). Through this activity, the expert witness can help "...to identify and achieve a stable and just accommodation between those involved in the delivery of health care and the members of the communities they serve" (p. 278).

5. CONCLUSION

Given the diverse interpretations of ethics expertise, the pertinent question is not whether ethics experts exist: on some definitions, they clearly do. Rather, what we are most concerned with is what such expertise warrants. Should individuals (or policy-making bodies, or institutions) cede moral authority and moral responsibility to ethics experts? If one identifies ethics experts as experts in the intricacies of moral arguments, nothing follows about their moral authority or responsibility. Alternatively, if ethics experts are held to possess knowledge of the good or the right, then individuals may choose to follow their advice as authoritative. If both kinds of expertise are possible, the next course of investigation is what kind of training conveys that expertise, and how we may and should utilize such experts.

NOTES

1. This paper and the volume as a whole take no stand on the difference between "moral" and "ethical" expertise (though individual authors may have reasons for choosing one or the other). I use the term "ethics expertise" instead of "ethical expertise" to make clear that the subject matter is not expertise conducted ethically, but rather, expertise in ethics.

2. In an early, well-known article on this subject, Cheryl Noble points out that "To an age beginning to be fed up with its own skepticism and relativism about 'values', the philosophers' claim that they know how to think rationally and objectively about values has great appeal. *The claim also suits an era accustomed to solving its problems by delivering them to the correct set of problem solvers*" (1982, p. 7, emphasis added).

3. Peter Singer provides a response to Noble's article immediately following it (1982). He takes particular issue with the implication in Noble's article that philosophers should withdraw from attempting to cast light on moral questions, especially in the public arena. After all, he suggests, "[g]reater clarity, the avoidance of logical fallacies, an understanding of moral concepts, and a critical examination of arguments may not be sufficient to remedy the situation, but what else can we try? Applying these skills is what applied moral philosophy is all about" (1982, p. 11).

REFERENCES

Burch, Robert (1974). 'Are There Moral Experts?' *The Monist* 58, 646-658.
Noble, Cheryl (1982). 'Ethics and Experts,' *The Hastings Center Report* (June), 7-9.
Singer, Peter (1972). 'Moral Experts,' *Analysis* 32, 115-117.
Singer, Peter (1982). 'How Do We Decide?' *The Hastings Center Report* (June), 9-11.
Tong, Rosemarie (1991). 'The Epistemology and Ethics of Consensus: Uses and Misuses of "Ethical" Expertise,' *Journal of Medicine and Philosophy* 16(4), 409-426.
Weinstein, Bruce (1994). 'The Possibility of Ethical Expertise,' *Theoretical Medicine* 15, 61-75.
Yoder, Scot (1998). 'The Nature of Ethical Expertise,' *The Hastings Center Report* 28(6), 11-19.

SECTION I: A GUIDED HISTORICAL TOUR

CHAPTER 2

SCOTT LABARGE

SOCRATES AND MORAL EXPERTISE

Where the nature of moral expertise is concerned, the Platonic Socrates is of special interest for at least two reasons. First, the Socratic dialogues[1] can all be understood as episodes in a larger project exploring the concept of expertise (*technê* in Greek) and its application to morality. In them we find Socrates and his interlocutors perpetually involved with questions of what it means to be an expert, why it is important to be led by experts in certain areas (morality above all) or to become an expert oneself, and how normal people who claim no moral expertise can determine to whom we can safely entrust ourselves and our loved ones as experts. Indeed, this last problem — call it the "credentials problem" — might fairly be described as the central problem of the Socratic dialogues; over and over we find Socrates concerned to verify the credentials of those who claim a moral expertise that makes them authorities in the education of children and the management of states. Socrates' famous discussions of substantive moral questions almost invariably occur within the context of an examination of some would-be expert's bid for special moral authority, and must be understood in that light. Second, Plato's Socrates consistently denies that he is a moral expert, but paradoxically he has been taken ever since as one of the world's paradigms of moral expertise. This state of affairs compels us to ask whether we and Socrates can both be right about his status as an expert.

My goals are accordingly double: first, to explicate and assess the theory of expertise that Plato constructs in these dialogues, and second, to examine how this theory of expertise helps to explain Socrates' understanding of his own successes and failures as a person on a lifelong quest for moral expertise. Along the way, I will argue that: (1) Plato's straightforward model of moral expertise, which chiefly involves reliably getting the right answers about questions of moral value in such a way that the answers are *demonstrably* right to any reasonable interlocutor — inescapably falls prey to the credentials problem. If Plato's explicitly offered model of moral expertise is the right one, a genuine moral expert cannot offer non-experts any credential recognizable to non-experts as genuine. (2) Plato's Socrates was not and did not believe himself to be that kind of moral expert. (3) Plato's Socrates was and believed himself to be a different kind of expert — what I will describe below as a "dialectical expert". And

L. M. Rasmussen (Ed.), Ethics Expertise: History, Contemporary Perspectives, and Applications, 15-38.

finally (4) this "dialectical expertise" offers us an attractive alternative model of moral expertise which could be usefully adopted today.

1. PLATO'S MODEL OF EXPERTISE

I'll begin with an exposition of Plato's analysis of the concept of expertise in general, to be followed by, first, a characterization of moral expertise in particular, and second, by a critical assessment of Plato's analysis. For Plato, to be an expert is to possess a *technê*, a word which can also be translated in some contexts as "skill", "art",[2] or "craft". All of the Socratic dialogues make extensive use of this concept; in fact, it would not be an exaggeration to say that the Socratic dialogues are an extended meditation on the nature of expertise generally, the relationships between different specific expertises, and the importance of expertise (especially moral expertise) for human life. When we approach the dialogues from that perspective, a remarkably consistent picture of what Plato thought expertise consists in emerges. For Plato, to be an expert is to meet the following criteria:[3]

(1) *The expert must have a holistic grasp of a well-defined subject matter.* No one is ever simply an expert, full stop; an expert is always expert at something. Thus a mathematician and a potter may both be experts, but their different subject matters make them different sorts of expert. This is no doubt an unremarkable point, and really it bears more on the nature of expertise than on the criteria for being an expert, but it plays an important role in some of the Socratic dialogues. For instance, in the *Ion* Socrates forces Ion, a professional reciter of Homer's poems, to admit that he does not really possess an expertise because he cannot specify what his expertise is an expertise in (*Ion* 537a ff.); and in the *Protagoras*, Socrates first gets his young friend Hippocrates to temper his enthusiasm for Protagoras by showing him that he really has no idea what specific subject matter it is that Protagoras is expert in (*Prot.* 311b-312e; cf. *Soph.* 232b-235a).

Furthermore, Socrates always assumes not only that the subject matter of an expertise is well-defined, but that the expert's knowledge covers the full range of the relevant subject matter. Thus he says, for instance, that Ion cannot be an expert about Homer without being an expert about poetry generally (*Ion* 532c). This might strike us as problematic; doctors, for instance, often specialize in particular areas of the body such that one could be an expert on the heart but not the body as a whole. If Socrates thinks the proper subject matter of medicine is health as a whole, then he will not be content with this way of proceeding. I cannot take the time here to discuss this issue fully, but note that the intuition driving Socrates here is simply that we expect experts not merely to know isolated pieces of information but to grasp a discipline as some sort of whole, though we may carve up the available disciplines differently than Socrates does.

(2) *Experts will successfully achieve the goals of their expertise.* Plato seems consistently willing to link the possession of expertise with the achievement of success, at times pushing the link so hard that he risks implausibility. For instance, in the *Euthydemus* Socrates claims that wisdom (understood on the model of expertise) is the same thing as good luck: "So wisdom makes men fortunate in every case, since I don't

suppose she would ever make any sort of mistake but must necessarily do right and be lucky — otherwise she would no longer be wisdom" (*Euthyd.* 280a).[4] There are also straightforward statements of this criterion in the *Laches* and the *Charmides*. At one point Laches argues that there is some reason to think that the so-called experts in weapons training are not experts at all because none of them is particularly successful as a soldier (*Lach.* 183c), while Socrates claims shortly thereafter that if an expert cannot prove his credentials by producing his teachers or students (see criterion 6 below), he can instead prove them by displaying the products of his craft, which presumably demonstrate his ability to achieve the goals of the expertise (*Lach.* 185e). And Socrates admits in the *Charmides* that if everything were done as knowledge directs, everything would be done successfully (*Chrm.* 172a).

It is clear enough from these passages that success is a critical element in expertise, but there are nevertheless some senses in which the success criterion is vague. The *Hippias Minor*, for instance, discusses a puzzling ambiguity in our discussions of success; it seems that the expert runner, for instance, can do both what we commonly take to be successful running — running quickly — and what we take to be unsuccessful running — running slowly — better than the non-expert. This suggests that we should not understand success as simply the attainment of what we intuitively consider the goal of the expertise to be, but rather the ability to achieve whatever result the expert may happen to desire in his area of expertise. The expert doctor cures more successfully than the non-expert, but he also poisons more successfully. Thus we might say that the expert will always be successful because he will always achieve the well-considered goal which he sets himself in his area of expertise.

(3) *Experts can explain why the things they say and do are correct and why the mistakes of others were mistakes.* In *Alcibiades*, Socrates chastises Alcibiades for claiming to be an expert statesman and yet being unable to answer questions about the expertise he has claimed (*Alc.* 109a). Likewise, that which distinguishes a mere knack from genuine expertise in the *Gorgias* is the rational understanding which the expert has of the nature of his subject matter, an understanding which enables him to explain why things happen as they do (*G.* 465a and 500e-501a). Also, in the *Meno*'s "Road to Larissa" passage, Socrates suggests that the thing which separates true belief from actual knowledge (understood again on the model of expertise) is the fact that the former tends to escape us while the latter is "tied down by an account of the reason why" (*M.* 98a). Though both true belief and knowledge can achieve the same results in a particular instance — i.e., they satisfy the success criterion equally — the fact that the knowledgeable person understood and can explain why the right result happened shows that he is more reliable when his beliefs are put to the test the next time around.

The other criteria are really corollaries of the first three, especially of the success and explanation criteria. These remaining criteria are:

(4) *Experts in a given field will agree with one another concerning questions in the domain of their expertise.* Plato apparently thinks that if a particular subject area is an area of expertise, all of the practicing experts will always agree with one another on the facts within their expertise. Socrates says in the *Meno* that the fact that the supposed teachers of virtue themselves cannot agree on whether their subject matter is teachable is grounds for thinking that virtue is not an expertise at all (*M.* 95b). We see a similar

sentiment expressed in *Republic* I, where Socrates argues that, "In any branch of knowledge or ignorance, do you think that a knowledgeable person would intentionally try to outdo other knowledgeable people or say something better or different than they do, rather than doing and saying the very same thing as those like him?" (*Rep*. 350a). The emphasis here is on motives, but it is nonetheless clear that Socrates expects all the experts in a given field to speak and act univocally where that expertise is concerned. The criterion is most plainly stated, however, in *Alcibiades* I, where Socrates and Alcibiades agree that if those who claim to be experts disagree about anything in that area of expertise we will claim that they are not really experts (*Alc*. 111b and ff.)[5]

(5) *Experts in a given area will recognize each other as such.* In the *Ion*, Socrates spends a good deal of time trying to convince a somewhat dense Ion that it is the expert who can best judge whether what is said regarding his area of expertise is rightly said and whether what is done is rightly done (*Ion* 531d-532a and 538a). But if any expert can not only speak and act rightly in her area of expertise but establish whether other people's statements and actions in that area are right or wrong, then it follows that any expert ought to be able to recognize a fellow expert through his words and deeds.

(6) *Experts can teach their expertise to others.* To the extent that possessing an expertise consists in understanding a subject matter intellectually, the material of the expertise can be discursively conveyed to others.[6] Plato claims this most clearly in the *Meno*, where he asks, "...(I)s it plain to anyone that men cannot be taught anything but knowledge?....But if virtue is a kind of knowledge, it is clear that it could be taught" (*M.* 87c). In the *Gorgias* Socrates tells Callicles that proof that he has learned an expertise from distinguished experts would itself be sufficient to establish his credentials as an expert, as would being able to point to students he himself had taught his expertise (*G.* 514a-515a). Similarly, Socrates tells Alcibiades that the ability to make others wise is a necessary accompaniment to wisdom (*Alc*. 118c-d) and his fellow speakers in the *Laches* that they should seek a teacher who himself had been taught by the best teachers and who could point to his own students (*Lach*. 185a-b and 186a-b). Given the reappearance of teachability in so many dialogues, it is certain that it is a central element in Plato's concept of expertise.

Putting all of these criteria together, we get a concept of expertise which emphasizes two main features of the expert: her ability to get results (criterion 2), and her ability to explain how she did it (criterion 3). The other criteria (apart from criterion 1, which is merely formal) fall more or less directly out of these two; criterion 6 follows directly from the explanation criterion, while criteria 4 and 5 assume that the experts' shared intellectual understanding of how to achieve their goals make agreement and mutual recognition both possible and necessary.

2. PLATONIC MORAL EXPERTISE

Consider now moral expertise in particular: given Plato's concept of expertise, what should we expect moral expertise to consist in?

I will take the criteria for expertise one by one. First, what would the specific subject matter of moral expertise be? One can answer this question in different ways, but let us begin with what seems the most obvious way to answer it, namely that the subject matter of moral expertise is, quite simply, morality. And I am confident that if we asked most laypeople what it means to be an expert in morality, they would say it means knowing what is right and wrong, not just in any specific area of activity but generally. Now, the ancients understood this subject matter somewhat differently than we do in ways that have been discussed by many authors,[7] but there is a deep similarity between ancient and present-day moral thought insofar as both focus on what we might call "ultimate values". That is, both ancients and moderns claim to discuss values that (a) ought to trump all other values in our decision-making and (b) are relevant to every area of human life. In the language of Plato's *Euthydemus*, moral expertise is the "ruling wisdom" that directs and coordinates the activity of all the other kinds of expert (*Euthyd.* 291c ff).

If ultimate values are the subject matter of moral expertise, then Plato would expect moral experts (criterion 4) to agree with each other about what the ultimate values we should pursue are and how they relate to one another, and he would expect them (criterion 5) to recognize each other on the basis of this agreement (and, presumably, on the basis of their ability to explain the grounds for the agreement.) Moreover, given the subject matter we have described for moral expertise, the moral expert's success must consist in recognizing and living by the right set of ultimate values (criterion 2), whether this involves embodying the nature of human flourishing which preoccupied the ancients or enacting the proper solution to the "hard cases" that occupy so much modern ethical discussion.

Finally, criterion 3 of course demands that moral expertise should also include the ability to explain our reasons for weighting values as we do, and criterion 6 demands that the moral expert should be able to teach others to understand these reasons and endorse the conclusions that follow from them, thereby turning them into moral experts as well.

3. HOW PLAUSIBLE IS PLATO'S MODEL OF EXPERTISE?

Now that the outline of Plato's model of expertise (and moral expertise in particular) is clear, let us pause to ask how reasonable this model is. I imagine that many readers will have doubts about the model by now, and rightly so.

One aspect of Plato's model that is bound to draw the modern reader up short is his emphasis on the explanation criterion. For those of us schooled in the epistemological distinction between "knowing how" and "knowing that",[8] Socrates' unrelenting emphasis on the intellectual, discursive element in expertise can be off-putting. (This is perhaps particularly true where moral expertise is concerned, but more on that below.) No doubt we expect results from, for instance, an expert billiards player, but if a consistently successful billiards player finds himself unable to explain to the rest of us just how he made a particular shot, we do not therefore write him off as a lucky amateur, nor do we (as the Socrates of the *Gorgias* might) write off billiards-playing as a knack rather than an expertise. These facts strongly suggest that most of us have a concept of expertise that is much less rigidly intellectualized than Socrates'.

Nevertheless, in Socrates' defense we should note first that, all other things being equal, in many areas we consider a person who can explain what she is doing to be more (or more nearly) expert than one who cannot. If we are faced with two experts who can each achieve the same results in their area of expertise, but one of them can explain the underlying mechanics of what she is doing while the other cannot, we will typically consider the one with the explanation more expert. Imagine you had to choose to receive your medical care from one of two doctors, both of whom had very similar records of medical success, but one doctor proved incapable of explaining to you why the treatment he prescribed was the correct one while the other could do so. Other things being equal, I imagine almost everyone would prefer the articulate doctor. Second, this criterion also seems to be stronger in the cases of some expertises than in others. For instance, even if a person is capable of "just seeing" what the right answer to a complicated math problem is, we only consider that person a mathematical expert if she can provide a mathematical proof that demonstrates the rightness of her intuition. Thus we should acknowledge that there is a continuum of expertises along which explanation plays a greater or lesser role; it may play no role at all in an area like billiards-playing,[9] while at the other extreme there are disciplines like mathematics where Plato is right to make the capacity to explain a necessary condition. (From this point on I will call this the "explanation continuum").

Given the close relationship between the explanation and teachability criteria, it should not be surprising that some have questioned the importance of the ability to teach along with that of the ability to explain. At one extreme, Burch has rejected Plato's position completely:

> Why should experts in a subject be even potential teachers of it? There is every difference between possessing a skill and possessing the ability to communicate that skill to others. Having a know-how, even an intellectual know-how, does not necessarily involve being able to formulate its canons, rules, and criteria for the purpose of communicating them to others. A person who can expertly make a carom need not be able to explain to anyone else how to do it, nor even show by example how to do it. Bridge, musical composition, mathematics, and literature involve one set of skills, whereas pedagogy and didactics involve another (Burch (1974) p. 653).

But this is only partly right. In the case of expertises like billiards where success is the most important criterion of expertise, we will not hold an inability to teach against an expert any more than we do the inability to explain what he is doing. But where the ability to give an explanation is essential to the possession of expertise, as in mathematics, the ability to teach others is likewise essential. Now, this does not imply that every expert mathematician will be charming, sympathetic, patient, or inspiring. And to the extent that the best teachers combine the ability to lay open the fundamentals of their discipline with these features that actually motivate their students to learn, expert mathematicians may not be especially *good* teachers. (I suspect that is really what makes Burch hesitate to insist that at least some kinds of expert must be able to teach — he thinks being able to teach entails being able to motivate students to learn.) But the criteria for being a good teacher are different from the criteria for being a teacher at all. In order to be a teacher at all, one need only be capable of explaining her subject matter to a willing and capable student, and this is all that Plato's teachability criterion requires. Ultimately, then, the the ability to teach will track the ability to explain across the continuum of expertises; the ability to teach is essential to expertise just so far as explanation is.

Moving on, perhaps the most problematic aspect of Plato's model is that it is utterly unforgiving of imperfection or failure of any kind. The Platonic expert understands her entire discipline completely, in such a way that she is infallible in the affairs of her discipline and never encounters a soluble problem in the discipline that she cannot successfully resolve. At his most extreme, Plato even goes so far in the *Euthydemus* as to imply that luck (as we normally understand it) has no grip on the expert's activities; what we normally call bad luck is really a person's failure to adequately grasp the situation around her. If a person were truly an expert, she would understand her practical situation so completely that she would control every factor that bears on the success of her project. But while this standard may perhaps be acceptable for an abstract discipline like mathematics (and it is disputable even there), it is of course unattainable for any expertise that has to deal with the unpredictable world of practical activity. Thus, minimally, we need to revise Plato's success criterion so that success does not always require attaining some positive goal so much as avoiding error in that pursuit. If success were in every case a matter of achieving the ultimate goal of the expertise then no expert general could ever lose a battle, but that surely seems too strong a requirement. We should rather understand expertise as the ability to do the best that one possibly could do given the facts of the expert's situation and the resources at hand. For example, an expert general like Chief Joseph may lose a battle, but because he knows that the battle could not be won he retreats from the field toward the strongest possible position with the least possible losses. Similarly a non-expert general might win a victory and so appear to have achieved success, but to have paid so dearly for the victory that his army will not be able to survive the next attack.

Even altered in this way, however, the success criterion still seems too strong because it disregards the limits on our knowledge that beset any area of expertise worthy of study. This is largely a product of Plato's first criterion of expertise, that an expert must grasp the entirety of the discipline's subject matter. When this criterion is combined with the success criterion, the result is that one could never claim to be a medical expert so long as there was a single curable disease she did not know how to cure, nor an expert in physics so long as there was a single aspect of the physical universe that had not been plumbed to its depths. In other words, for Plato, expertise is an all-or-nothing proposition; there are no degrees of expertise.

This feature of Plato's model is also largely responsible for the agreement criterion. Plato can insist that all experts will agree with each other about matters of their shared expertise only because he stipulates that all true experts in a given discipline will understand exactly the same body of information in exactly the same way. If there were any areas of the discipline that were still understood so poorly by some participants that a dispute about the matter were possible, Plato would deny that all the disputants were really experts to begin with. But this will not do. Medical experts will disagree about the nature of a new disease and its proper treatment, and scientists will offer competing theories about their objects of study, and they may all yet be experts.

So imagine that we alter Plato's model to acknowledge these objections: we insist that expertise comes in degrees, and that one can be an expert without being infallible in his area of expertise. What results?

We would first have to make a corresponding change to the mutual recognition criterion. Plato thought that experts would necessarily recognize each other because he thought that they would share the same understanding of the same range of material, but once we allow for disagreements among experts that common ground can no longer be taken for granted. In response we could give up the mutual recognition and agreement criteria entirely, but that seems too extreme a reaction. It is reasonable to acknowledge that experts can disagree with each other to some degree, but surely people cannot disagree with each other about *every aspect* of a subject matter and still all be experts on it. For that to be possible, genuine experts would have to be able to disagree about (at least) which phenomena they are dealing with, what the basic problems confronting them are, what the fundamental facts of their discipline are, what counts as evidence in the discipline, and when a problem has been solved. But if two people disagree about that much, either at least one of them is going to be getting things wrong to the point that they no longer achieve success in their endeavors, or, where both still achieve success, their understandings and explanations of what they are doing will be so divergent that we might well wonder whether they are still talking about the same thing. This suggests that the agreement criterion is not intrinsically flawed, but rather is made too extreme by Plato's insistence that experts be omniscient in their discipline. We should thus expect experts to share *some* common ground.

The question that then faces us is just how much common ground all the experts must share, and whether this will always suffice to make mutual recognition of experts possible. Perhaps at this point a concrete case might be helpful. Imagine a typical Western doctor confronted with, for instance, a New Age healer who claims to cure diseases and ease pain by manipulating patients' auras. Both healers may share some rudimentary area of agreement here — they are both dealing with human bodies and trying to cure the same conditions, and they are both satisfied when the condition disappears, so there are rough similarities between them — but they start from such different assumptions and employ such different methodologies that it is difficult to think of them both as experts in the same way. Now, if one of the two consistently has little success in helping patients, we would have excellent reason to doubt that that person is an expert. But let us imagine also that both practitioners have substantial and roughly equal degrees of success in helping their patients and that both have internally consistent theories they could offer to explain this success; what would we say of them? We might grant that both are experts of a sort — perhaps we could imagine going to either for help — and each might even (grudgingly?) acknowledge the other's accomplishments. We might even imagine them recognizing each other as health experts in a broad sense, in the way that some Western doctors have begun to take alternative therapies like acupuncture seriously. Nevertheless, I do not think we would expect them to recognize each other as the *same kind* of expert.

This example suggests that, once we acknowledge the existence of different theoretical paradigms competing for adoption in a given area, different layers of agreement become relevant to the assessment of expertise. Successful healers of all paradigms are able to recognize each other as health experts broadly speaking because they all share a broad notion of what healing is. They all work on and talk about human bodies, and they all assess results in roughly the same way. But they are not all the same kind of health expert because they understand what they are doing in fundamentally different ways, such that they do not share a common vocabulary and theoretical ontology which allows them to talk to each other in terms that all can understand. Their understandings of the human body may be so divergent that they are talking about the same thing only in the weakest sense. It would take some kind of *unified* explanation, one that accounts for the practical success of both, to bring them close enough together that they could conceivably recognize each other as experts of the same kind in the narrow sense.

This leaves us with the result that the agreement and mutual recognition criteria will manifest themselves differently in different areas of expertise. As we saw above, in some areas of expertise the explanation criterion has a much higher profile than in others. Where the role of explanation is minimal, the only agreement we will really insist that experts share will involve standards of success. Billiards experts will have to agree about criteria for success in billiards, and then the experts will recognize each other through their repeated successes. But in areas where explanation plays a larger role, we have to make room for both a broad kind of agreement and a narrow kind of agreement, and different kinds of recognition that track them. When people agree in a general way about the set of problems they are grappling with and about what success in dealing with these problems consists in, then all those who reliably and repeatedly achieve solutions by that standard must recognize each other as experts in a broad sense. But to recognize each other as experts in the narrow and complete sense, they would have to share a vocabulary and ontology in such a way that they could, for instance, share a research project in their area. But this also gives us an answer to our earlier question: whatever the discipline, we can probably tolerate a wide area of disagreement at the narrow, theoretical level, provided that all the parties to the dispute have met a certain agreed-upon standard of success in the discipline. This also suggests a limit to the disagreement we can tolerate: if the different parties have such different concepts of their discipline that they cannot even agree upon a standard for success, we must be skeptical that they are all indeed experts, or at least experts of the same kind who are talking about the same thing in even a broad sense.

When we finally return to moral expertise, this critique of Plato's general account of expertise suggests that we need to revise our earlier concept of expertise in at least a couple of ways. First, we have to acknowledge that moral expertise, like other forms of expertise, is not an all-or-nothing proposition; moral experts will not always get everything in their discipline right, and accordingly it is to be expected that genuine experts will disagree on some matters. Second, we need to place moral expertise on the explanation continuum. To what extent is the ability to give an explanation essential to moral expertise?

Interestingly, we could make an argument to place moral expertise at either end of the continuum. We could place it at one extreme (with billiards) and simply require the moral expert to be a "good person" who is committed to the right values and makes the right moral decisions, whether she can explain why they are right or not. If we choose this option, we basically treat the moral expert as a moral exemplar, a person who reliably makes good moral choices in such a way that we could safely use her as a sort of moral barometer, without worrying about how she got her results. Or we could place it at the other extreme (with mathematics) and emphasize that we require the moral expert to be able to discuss these values with others in such a way as to demonstrate her moral system's correctness. Or, finally, we could place moral expertise somewhere between the two extremes (with medicine), *valuing* the ability to give an explanation but *requiring* only success. Which option should we take?

I sympathize with those who want to assert the value of moral exemplars, but it seems to me that such a person is not really a moral expert. Her moral success may make her capable of leading an extraordinary life — a feature we might envy — and those who decide to emulate her may see their own lives improve in kind — a feature we might admire. But however virtuously and conscientiously she may lead her life, and however good an example she is to the rest of us, I do not think we should call such a person a moral expert unless she can speak articulately about her moral convictions. For better or worse, those who claim moral expertise in our society typically claim a wisdom that bears not just on their own lives but on the lives of people in general. However righteous the person, she cannot simply say "Follow me" and expect others to acknowledge her moral authority without further justification, especially given the full roster of competitors. She must defend her commitments in the public forum, and that entails giving explanations. I therefore think we need to place moral expertise over with mathematics at the more cognitive end of the explanation-continuum.

This means that one of the hallmarks of Plato's model of moral expertise is still largely intact; we should hold on to the explanation criterion for moral expertise in just the way Plato suggests. But since we have acknowledged that expertise comes in degrees, even moral expertise, and thus that even genuine moral experts can disagree from time to time, the crucial task that remains is to explain just how moral experts can be recognized. Our analysis so far has suggested that the place to look for such a criterion is not in the sort of shared explanatory paradigm that would let experts agree in the narrow sense, but rather in a standard of success that all experts agree upon and achieve. I turn now to an assessment of the prospects for such a standard in the case of moral expertise.

4. THE CREDENTIALS PROBLEM AND MORAL EXPERTISE

Let us assume for now that the revised Platonic model of moral expertise is the one we should adopt. What prospect is there for such a moral expert to establish her credibility in such a way that she will be recognizable to those who lack her expertise? For instance, how could such an expert establish her credibility to the participants in a court of law — lawyers, judge, and jury — who did not claim for themselves any special moral knowledge or authority?

Consider for a moment the way that a different kind of expert — say, an expert auto mechanic — might deal with such a question; a possible customer who knows almost nothing about auto care comes into his shop and asks him to prove that he is an expert. Now, the credentials problem threatens the mechanic as well, because the person who knows almost nothing about cars is not well prepared to assess all the mechanic's evidence of his expertise. Imagine that our auto expert agrees with other experts and is recognized as an expert by his fellows. Still, since the non-expert does not share the understanding of automotive affairs in which the experts' expertise consists, he has nothing more than the experts' say-so to show that they are genuinely experts. If our non-expert is a careful consumer who has visited several autoshops, and he has heard a contradictory diagnosis of his automotive problem elsewhere, the fact that one group of mechanics agree with each other and testify to each others' expertise is no proof that they are really the experts. A group of fake auto experts could achieve the same appearance with a little coordination. Similarly, the non-expert will be hard-pressed to make much of any explanation the expert might give him; understanding automotive explanations requires understanding automobiles.

On the other hand, the non-expert has one piece of evidence by which to test the mechanic's expertise which is easy for him to assess. If the mechanic says he can fix a car, but after he has worked on it the car still does not work, that is good evidence that the mechanic is not an expert. Similarly, if he says he can fix a car, works on it, and the car works as it is supposed to, that is good evidence that the mechanic is an expert.[10] In other words, the success criterion of expertise is quite accessible to the non-expert in this case, since, even if the non-expert knows very little about cars, presumably he can at least be counted on to understand how cars are supposed to perform and recognize good and bad car performance. Thus, the best bet for the non-expert looking for an expert mechanic is to focus on results.

Consider next the case of the non-expert in search of a moral expert, someone who can offer reliable guidance in the realm of ultimate values. He faces the same difficulties the non-expert in search of a mechanic did: as a non-expert he lacks the comprehensive moral understanding to assess the foundations and adequacy of the experts' explanations, consensus, and mutual support. Confronted by competing camps of experts who each offer different and incompatible explanations of moral issues, the non-expert is not well-situated to judge which of the competing camps possesses genuine expertise, if any.

But the non-expert in search of a moral expert is worse off still, for where the search for a mechanic could focus on results, the search for a moral expert is much more precarious. The focus on results was possible in the case of mechanics because all concerned, experts and non-experts, could at least agree on what success consists in and recognize success when it was achieved. No such consensus is available in the case of moral expertise, because the ultimate goal of moral deliberation and action is itself one of the very issues that is at the heart of ethical disputes. We will no doubt find widespread agreement on some (maybe even on many) points in our moral debates—for instance, we can probably count on just about everyone denying the moral acceptability of slavery, rape, and other practices — but on the issues where we feel the greatest need for the guidance of a moral expert, i.e. the issues where we find the greatest disagreement and controversy, the shape of a successful solution to our moral problems is typically the very thing at the heart of the controversy. If, once a mechanic has fixed your car, another mechanic claims the first mechanic did not know what he was doing, the fact that you can drive away from him is good evidence that the second mechanic is wrong. But if, once you have taken a moral expert's advice and adopted certain values as your own, another moral expert claims that you have been led astray, how could you prove her wrong? You could not simply reply, "But look, I have adopted the right values and so have achieved moral success!" because the values they defend as the right ones are likely to be one of the very things the claimants to moral expertise argue about.

This does not mean that the genuine moral expert and her followers have nothing more to say for themselves. On the Platonic model of expertise we have been developing, the genuine expert has an intellectual understanding of morality to which she can turn to provide her with arguments in defense of her commitments, presumably including her standard of moral success,[11] and she can employ these against the competing claimants to expertise. Now, for the non-expert watching the competition and waiting to see whom he should follow, this is a crucial moment. If one of the claimants marshals her arguments so successfully that all of her competitors find themselves compelled to give up their own positions about the nature of moral success and take up hers, that would be strong evidence that she is a genuine moral expert.

The question that faces us, then, is whether we really think that such an achievement is possible. And much as I wish it were otherwise, I fear it is not. There may be other forms of expertise in which it is possible to marshal arguments that command the universal assent of all rational agents — mathematics and logic come instantly to mind, and to some extent the natural sciences and other empirical enterprises may qualify as well. But my experience of ethical debate, which I imagine is shared by many, is that however well one's views hold up in argument, however much one can make a single or several or indeed all one's opponents give up argumentative ground, ethical debate is just not the sort of enterprise in which one can raise one debater's hand at the end of the day and declare her the winner, not just for today but for good and all. One can always assume that, however bad a drubbing they have taken, some of one's opponents will not believe that they have been genuinely beaten, will remain unconvinced that they ought to give up their commitments to their ultimate values, and will refuse to acknowledge the (from their point of view) temporary victor as a greater moral authority.

Now, there are different ways we can diagnose this result. One way is to claim that the experts' opponents are simply too dim intellectually to understand what the argument has shown; I regrettably confess I have had students who simply seem incapable of following arguments well enough to see the problems that their views commit them to, and so remain blithely content with them. A second way to account for the continued impasse is to blame the deficient character of the moral expert's opponents. That is, we can attribute to them either a weakness of will which prevents them from committing themselves to the values which they now recognize intellectually as the true ones, or a wickedness which keeps them perversely committed to the wrong values even though they have been fairly beaten. Finally, we can say that the problem lies not with the argumentative opponents, but with the nature of the material under debate. Perhaps the best arguments available about ultimate values are just not capable of silencing once and for all those who do not already accept them.

I have no doubt that thick-headedness and failures of character play some role in explaining the interminability of moral debate. It would be surprising if there were not some thick-headed, wicked, and/or weak-willed people out there with whom moral debate is an exercise in futility. But it is very hard to believe that all cases of persistent moral disagreement can be explained in these ways. The reality of moral debate is that sometimes rational, fair-minded, charitable agents persist in their commitment to their values even in the face of powerful arguments that weigh against them.[12] Often, those who disagree with us even have arguments of their own.

If this description of moral debate is accurate, it has depressing implications for the non-expert who seeks moral guidance, for it seems to preclude his recognizing the genuine moral expert by watching the moral debate unfold. Since the nature of success in moral expertise is itself under dispute, and this dispute characteristically does not produce a clear winner recognizable to all, ultimately the non-expert cannot confidently assert that any of the claimants to moral expertise is a genuine expert.

The persistence of moral disagreement also has serious implications for the very possibility of moral expertise on the Platonic model, since it undermines the claim of the moral expert to be able to teach her expertise to others. It would be one thing if the moral expert's inability to demonstrate to others the correctness of her moral commitments was attributable only to the intellectual deficiencies or character flaws of her would-be students. But if the moral expert's failure as a teacher is a result of her inability to establish the success of her moral system in such a way that the non-expert could no longer reasonably doubt her, then Plato's model of expertise gives us grounds to deny her the title of moral expert.[13]

5. SOCRATES' DENIAL OF MORAL EXPERTISE

Now think about Socrates. In Plato's dialogues he explicitly and repeatedly denies that he is a moral expert, but he has been taken to be paradigmatic of such an expert ever since. Both of these facts require further elucidation.

Start with Socrates' assertions of his ignorance. There has been much argument about the scope of Socrates' self-declared ignorance — does he claim to be ignorant of absolutely everything, or does he confine his ignorance to the affairs of natural philosophy and/or morality? — but I will leave that larger debate aside.[14] All we need to discuss for our purposes is Socrates' disclaimer of moral expertise, and this is fairly easy to establish. In reporting a conversation with Callias, for instance, he says:

> (Socrates speaking to Callias) "Now since (your sons) are men, whom do you have in mind to supervise them? Who is an expert in this kind of excellence, the human and social kind?...Is there such a person, or is there not?"..."His name, Socrates, is Evenus, he comes from Paros, and his fee is five minas." I thought Evenus a happy man, if he really possesses this art,[15] and teaches for so moderate a fee. Certainly I would pride and preen myself if I had this knowledge, but I do not have it, gentlemen (*Ap.* 20b-c).

Here he clearly denies that he is an expert in human *aretê*, a word translated here as "excellence" but which is often translated "virtue".[16] Similarly, the Socratic dialogues are full of Socrates' assertions that he does not know how to define important moral terms, and so is as greatly in need of a teacher in these matters as his interlocutors.[17]

The pressing question for us is why he claims this kind of moral ignorance. Given that Socrates says surprisingly little about his grounds for claiming ignorance, scholars tend to focus instead on the question whether he really means it when he says he is ignorant. The different responses to this second question tend to fall into two broad camps: either his profession of ignorance reflects a genuine belief that he is ignorant, or the profession is ironic, serving some more[18] or less[19] devious argumentative purpose. A full discussion of this question would take us too far afield, so I will instead simply explain why, on Plato's account of moral expertise in the Socratic dialogues, it makes sense for Socrates genuinely to believe that he is not a moral expert.[20]

This requires us to return to Plato's criteria for expertise. One of these, the mutual recognition criterion, is irrelevant here, since Socrates has found no one capable of meeting the explanation criterion in such a way that he is even a candidate for moral expertise. We might expect the same to hold of the agreement criterion, since there is no expert around for Socrates to agree with, but that would be too hasty. Consider the following passage from the *Hippias Minor*:

> I make mistakes as to the way things are, and don't know how they are — I find it sufficient evidence of this that when I am with one of you who are highly regarded for wisdom, and to those whose wisdom all the Greeks bear witness, I show myself to know nothing. For I think pretty well none of the same things as you do; yet what greater evidence of ignorance is there than when someone disagrees with wise men?...So indeed now, I don't agree with what you are saying but disagree very strongly....However, sometimes I believe the opposite, and I go back and forth about all this — plainly because I don't know (*Hp. mi.* 372b3-e1).[21]

We must be careful here. Socrates cannot be serious when he speaks as though Hippias is a genuine moral expert; the whole of the dialogue is a demonstration of Hippias' inability to resolve the moral difficulties Socrates lays out. So to that extent Socrates is being ironic, and he does not genuinely take this to be a case of an amateur disagreeing with an expert. But his admission that he finds himself frequently changing his mind is important, because it entails that Socrates himself sometimes falls prey to the same sort of failure that plagues his interlocutors: he finds himself with an unstable belief set that leads him to pronounce at one time "*p*" and at another "*not p*". The possession of genuine expertise would require one to be consistent not only with other experts, but also with oneself. Moreover, we see this sort of shifting of positions take place in the dialogues themselves, most strikingly in the *Protagoras*, where Socrates first argues that virtue cannot be taught (*Prot.* 319a ff.) but later reverses himself and argues that it can be taught (*Prot.* 361a-c).[22]

When we add the explanation criterion to the mix, things get more complicated. For we *never* find Socrates short on explanations for his positions, even when, as in the *Protagoras*, he changes his view in the course of the single dialogue. When he argues that virtue is not teachable, he gives reasons for believing that to be so, and likewise when he argues that virtue is teachable, he gives reasons for that as well. It should be pointed out that on occasion Socrates even offers an explanation that he claims is successful, as when he says in the *Gorgias* of his proof that it is better to suffer wrong than to do it that:

> These conclusions, at which we arrived earlier in our previous discussions are, I'd say, held down and bound by arguments of iron and adamant....And if you or someone more forceful than you won't undo them, then anyone who says anything other than what I'm now saying can't be speaking well. And yet for my part, my account is ever the same: I don't know how these things are, but no one I've ever met, as in this case, can say anything else without being ridiculous (*G.* 508e-509a).

This passage warrants consideration.[23] Socrates claims not only to believe that it is better to suffer wrong than to do it, but to have "arguments of iron and adamant" that justify that belief. Yet immediately we find him denying that he has knowledge (read: "expertise") in this area. It is one thing for Socrates to deny expertise when he gives separate explanations for contradictory positions, as in the *Protagoras*; in such a situation it is sensible for him to be skeptical whether either of his explanations is satisfactory. But here, where he claims that any argument for a competing proposition comes out looking "ridiculous", how can he still deny expertise? Moreover, if success in moral matters consists in knowing how to prioritize one's ultimate values, does not this particular piece of moral reasoning give Socrates at least one instance of success?

Socrates does not explicitly say how he thinks he falls short in this case, but we can speculate. Part of the reason is perhaps that on Plato's model of expertise having a satisfactory explanation for one proposition is not sufficient to make one an expert. As I said above in laying out the first criterion for expertise, an expert grasps not just isolated pieces of knowledge but an entire discipline, and Socrates nowhere, in the *Gorgias* or elsewhere, claims he has achieved that.

More important, though, is the teachability criterion. In some sense, the whole of the Socratic project is to find someone who can reliably teach him moral truth so that he can live the best human life and teach others to do the same. For Plato's Socrates, the gold standard for moral expertise is the ability reliably to turn others into moral experts. He says in the *Meno* that:

> ...(V)irtue would be neither an inborn quality nor taught, but comes to those who possess it as a gift from the gods which is not accompanied by understanding, unless there is someone among our statesmen who can make another into a statesman. If there was one, he could be said to be among the living as Homer said Tiresias was among the dead, namely, that "he alone retained his wits while the others flitted about like shadows." In the same manner such a man would, as far as virtue is concerned, here also be the only true reality compared, as it were, with shadows (*M.* 99e-100a).

This is why in the *Protagoras* and *Meno* he expresses so many doubts about the teachability of virtue: the people like Themistocles and Pericles who have the greatest reputation for virtue seem to do a dismal job passing that virtue on to others — even to their sons, whom they would be especially motivated to improve if they could.[24] Socrates knows from personal experience that the sophists do not understand moral matters clearly enough to qualify as experts, and the great statesmen of his day have done a dismal job turning their children into men of substance. But Socrates, even when he is confident that he has the right answer to a specific moral problem, knows better than to think that he meets the teachability standard. The dialogues are full of his denials that he is a teacher,[25] and Socrates knew perfectly well that many of the people he spent the most time arguing with had turned out quite badly indeed. Critias and Charmides, for instance, became notorious for their involvement with the Thirty Tyrants, and Alcibiades, who seems to have been one of Socrates' special personal projects, turned out to live a life of epic excess. Even in the case of beliefs which he thought he could explain, like the one from the *Gorgias*, Socrates knew full well that he was not going to bring Callicles and Polus around to his point of view even with his best arguments. In the face of such failures, Socrates could not claim to meet his own standards for expertise.

To be fair, the sort of interlocutor we often see Socrates confronted with is not exactly ideal. Over and over Plato gives Socrates surly, unresponsive, dialectically unscrupulous people to talk to, and more often than not these men appear to have little interest in seeking the truth for its own sake.[26] An inability to bring people like Thrasymachus and Callicles around to one's point of view should not count as a failure of moral expertise.[27] But on the other hand, Socrates sometimes has people like Nicias or Lysis to talk to. These figures might not be as slick as Socrates, but they are not fools, they seem to be genuinely interested in the truth, and one suspects they are eager to be persuaded by a satisfactory account if Socrates can give one. If Socrates were genuinely a moral expert, some of these people seem to have been excellent candidates for training in the craft.

Thus Socrates turns out to deny he is a moral expert for much the same reasons that we found make moral experts so hard for non-experts to identify. He is unable to resolve the confusion and disagreement even of intelligent, well-meaning interlocutors — indeed, his own shifting views reflect that confusion and disagreement — and he acknowledges that he is unable reliably to produce conviction even on matters where he is confident he has the truth.[28] He would be the first to tell us that if we want to find what makes him special, we have to look somewhere other than his special moral knowledge.

6. SOCRATES' UNIQUENESS

That Socrates is special, even he agreed. Twice in the Socratic dialogues Socrates claims not only that he is unique, but that his uniqueness lets him serve a role of the utmost benefit to the Athenians. When he is asked what penalty for himself he will offer the jury as an alternative to death, he responds that what he deserves for giving his fellow citizens "the greatest benefits" is the penalty of free meals in the Prytaneum for life, the same "punishment" assigned to Olympic victors (*Ap*. 36b-e), and warns them that once he is gone, they will be hard put to find another "gift from the god" as a replacement for him (*Ap*. 30e-31b). And though he denied above in a passage from the *Meno* that he was a true statesman, in the *Gorgias* he says exactly that:

> I believe that I'm one of a few Athenians — so as not to say I'm the only one,[29] but the only one among our contemporaries — to take up the true political craft and practice the true politics. This is because the speeches I make on each occasion do not aim at gratification but at what's best. They don't aim at what's pleasant (*G*. 521d).

These are not the assertions of a man who thinks he has nothing to offer his fellows.

What does Socrates think he can do for the Athenians that makes him so valuable? In the same two dialogues, Socrates also offers some indications of the source of his uniqueness. For instance, in the *Apology* he says, "What has caused my reputation is none other than a certain kind of wisdom. What kind of wisdom? Human wisdom, perhaps. It may be that I really possess this, while those whom I mentioned just now (the sophists) are wise with a wisdom more than human" (*Ap*. 20d-e). He does not explicitly characterize the content of this "human wisdom," but the story he goes on to tell about his examinations of reputedly wise Athenians allows us to make some inferences. The thing that distinguishes him from the politicians, poets, and craftsmen is that he is fully aware of his own (and others') limitations where everyone else is not. In other words, Socrates has the ability to recognize when he and others lack wisdom, and this ability makes him of tremendous use to the city. Armed with his awareness of human ignorance, he stirs up his fellow citizens like a gadfly, "doing nothing but persuading both young and old among you not to care for your body or your wealth in preference to or as strongly as for the best possible state of your soul" (*Ap*. 30a).

Now, this last statement is provocative, because it sounds like the sort of thing a genuine moral expert might tell people. After all, the subject matter of expertise is ultimate values, and here Socrates confidently recommends a clear set of priorities in the pursuit of common values: go ahead and pursue worldly goods, but be sure to care for your soul at least as much. Socrates' claim in the *Gorgias* that he is the only true practitioner of political expertise is similarly troubling, since early in the *Gorgias* he defines that expertise as the wisdom that knows how to make the soul fit and healthy, just as gymnastic and medical expertise do for the body (*G*. 464a ff.) Thus Socrates' claim to practice political expertise seems to amount to the claim that he knows how to make human souls fit and healthy — precisely the knowledge we have seen him explicitly deny he possesses (see esp. *Ap*. 20b-c).

We need not, however, take Socrates to be confused here, especially if we take into consideration how very important he thinks it is to recognize one's own limitations, and how great a difference this can make to one's life. In the *Sophist*, a late dialogue in which Socrates is only a minor character, Plato offers us an extended passage (228d-231b) about the nature and proper treatment of ignorance that is surely meant to call the Socrates of the early dialogues to mind. The Eleatic stranger distinguishes two kinds of ignorance: the normal kind, such as ignorance of a specific subject matter like medicine, and a more general kind, "a large, difficult type of ignorance marked off from the others and overshadowing all of them" (229c). This ignorance involves "not knowing, but thinking that you know. That's what probably causes all the mistakes we make when we think" (*ibid.*). The first, discipline-specific kind of ignorance can be treated by teaching of the sort proper to each of the different disciplines (229d), but the treatment of the broader kind of ignorance is more complicated. It involves in part admonition, a "method of scolding or gently encouraging" (229e), but sometimes admonition alone will not suffice, for "if someone thinks he's wise, he'll never be willing to learn anything about what he thinks he's clever at," and with such people admonition is ineffective (230a). In such cases, the proper treatment is to:

> cross-examine someone when he thinks he's saying something though he's saying nothing. Then, since his opinions will vary inconsistently, (the teachers) will easily scrutinize them. They collect his opinions together during the discussion, put them side by side, and show that they conflict with each other at the same time on the same subjects in relation to the same things and in the same respects. The people who are being examined see this, get angry at themselves, and become calmer toward others. They lose their inflated and rigid beliefs about themselves that way, and no loss is pleasanter to hear or has a more lasting effect on them. Doctors who work on the body think it can't benefit from any food that's offered to it until what's interfering with it from inside is removed. The people who cleanse the soul...likewise think the soul, too, won't get any advantage from any learning that's

> offered to it until someone shames it by refuting it, removes the opinions that interfere with learning, and exhibits it cleansed, believing that it knows only those things that it does know, and nothing more...For all these reasons, Theaetetus, we have to say that refutation is the principal and most important kind of cleansing (230b-d).

The stranger then says that this kind of teaching characterizes the "noble sophist" (231a-b; as opposed to the common sort of sophist that peoples Plato's dialogues.)

This passage gives a strikingly accurate portrait both of what Socrates does for his interlocutors and why what he does is so valuable. When Socrates says in the *Gorgias* that he is the only true statesman in Athens because only he genuinely improves the souls of his fellow citizens, he is not claiming to make their souls positively beautiful by teaching them what ultimate values they should accept. Rather, through testing their beliefs he improves their souls by removing the chief source of their souls' ugliness — their belief that they are already beautiful, i.e., already know what values they should pursue. And once this ugly ignorance is removed, his fellows are prepared to begin the search for the values that will make their souls truly beautiful. Note as well that in the very act of showing his fellows their limitations, Socrates prods them toward adopting a provisional set of priorities that can require them to reshape their lives profoundly. Without having a final account of the proper set of ultimate values to adopt, Socrates can nonetheless quite reasonably tell others that, in the absence of genuine, positive knowledge about the priorities we ought to have, our chief priority ought to be the attainment of such knowledge. (One does not have to be a moral expert to understand how valuable moral expertise would be.[30])

7. SOCRATES THE DIALECTICAL EXPERT

If this is the contribution that makes Socrates valuable to his fellow citizens, we should ask how he is able to make such a contribution. In one sense the answer is obvious and has already been stated: Socrates is adept at conducting verbal examinations of his interlocutors in such a way that he reveals to them the contradictions lurking in their belief-sets, thereby deflating their presumption to moral expertise and (ideally) inspiring a new commitment to seek such expertise. So this is where Socrates' uniqueness clearly lies, in the conduct of conversations about what we do know and what we should most want to know.[31]

But this truth warrants further unpacking, for this ability to conduct a conversation is multi-faceted, containing elements that need to be distinguished. It certainly includes a high degree of logical sophistication; if Socrates is to detect contradictions, he must of course be able to grasp what follows from what and thus recognize which beliefs produce inconsistent consequences. Socrates also has to possess the ability to see quickly and accurately which beliefs provide the crucial assumptions at the core of his interlocutor's position. But we must always remember that Socrates is not interested in detecting contradiction purely for its own sake. That project serves the larger project of testing broader claims to expertise — especially moral expertise. And in order to accomplish this, Socrates has to have a clear grasp of what expertise consists in. That is, if Socrates is to judge who is and is not an expert, he must have an epistemological theory which defines the nature of expertise satisfactorily and which he can look to to shape his examinations and judge their results.[32]

Happily, we have already seen Plato lay out just such a theory in section I above. Over a range of dialogues Plato's Socrates puts to work a clear and comprehensive grasp of the criteria that anyone who lays claim to an expertise of any kind must meet. Moreover, the possessor of such a theory himself meets the criteria for expertise that the theory provides:[33] (1) The subject matter of the expertise is well-defined. (2) Possession of this expertise would guarantee success in judging claims to expertise in the sense that the possessor will always make the best judgment possible given the information available and never think her judgment is more reliable than it is.[34] (3) Possession of the theory of expertise would allow one to explain why she does or does not believe another to be an expert, as in fact we regularly see Socrates do in the dialogues. (4) Possessors of such expertise would agree with each other about the nature of expertise and (5) could recognize each other by testing each other's grasp of the theory.[35] And (6) this expertise can clearly be taught to (at least some) others. Socrates seems to have produced a flock of young imitators who went around testing claims to expertise in much the way Socrates did (*Ap.* 22c), and when he claims not to have taught these young men anything, we must understand him to mean that he did not *set out* to teach them this expertise and that he never received payment or sought a reputation for it.

If this is correct, then Socrates' "human wisdom", which gives him awareness of his own and others' expertise and ignorance, is more than that simple awareness alone; it is a kind of expertise itself. Taking this expert grasp of the nature of expertise together with great logical sophistication, we might call Socrates' human wisdom "dialectical expertise", expertise in how to conduct a conversation to establish whether someone is or is not an expert. Beyond that, all Socrates needs in order to test others' claims to expertise is enough rudimentary information about different subject matters to enable him to have an intelligent discussion about those matters at all. (I have in mind information like our wide-spread non-expert understanding of what counts as success in auto repair or a basic, non-expert grasp of fundamental moral concepts.)

Now, at least two problems confront the attempt to attribute such an expertise to Socrates. First, if Socrates genuinely believes himself to possess such an expertise, we may wonder how he can say that he "knows nothing worthwhile" (*Ap.* 21d4-5), that "human wisdom is worth little or nothing" (*Ap.* 23a7) or is "worthless" (*Ap.* 23b3-4). Second, we must face the uncomfortable fact that the *Charmides* contains a series of arguments meant to cast doubt on the possibility of knowledge of knowledge and ignorance (or expertise regarding expertise), or on the value of such knowledge if it were possible.

I have already dealt with the second objection elsewhere (LaBarge, 1997), so I will not cover the same ground again here. I will, however, offer a response to the first objection. Interestingly, the *Charmides*, the source of the second objection, offers the answer to the first. At one point in that dialogue, Socrates says he is willing for the sake of argument to assume the possibility of the knowledge of knowledge and ignorance, and then asks what the value of that knowledge would be. His first attempt to answer that question is:

> We'd live all our lives without making any mistakes, and not just those of us who possessed (such knowledge), but all those other people who were governed by us as well, because we'd neither try ourselves to do what we didn't know, but would find those who did and hand the matter over to them, nor trust those whom we governed to do anything except what they were likely to do properly — and that would be what they possessed knowledge of....When error has been removed and correctness leads the way, people in those circumstances must do admirably and well in their every activity, and people who do well must be happy (*Chrm.* 171d-172a).

He also says that this knowledge benefits us by making us better at learning anything at all and at conducting the conversations that lead to learning (*Chrm.* 172b).[36] So it would seem that the knowledge that lets us recognize experts and fakes would be genuinely worthwhile.

Shortly thereafter, however, Socrates reverses himself. He argues instead that we should judge the value of any expertise only in terms of the contribution it makes to happiness, and that ultimately the only expertise that makes a positive contribution to the attainment of happiness is moral expertise. Even if the ability to recognize experts allowed us to recognize the moral expert, it would be her moral expertise which we should credit for our resulting happiness, not our ability to recognize her (*Chrm.* 173d-175a).

In order to assess the value of dialectical expertise properly, we need to split the difference between these two contradictory assessments. On the one hand, we must acknowledge the contribution that dialectical expertise does indeed make, especially when we are unable to locate a genuine moral expert. As the first passage from the *Charmides* and the long passage from the *Sophist* above indicate, at a bare minimum this expertise will allow us to avoid making many mistakes simply by curing our epistemic arrogance and re-orienting us so that we spend less time pursuing things of dubious value and more time seeking a moral expert. These are genuine benefits. But on the other hand, Socrates is painfully aware of how much more beneficial moral expertise would be; without it, we can never be sure how happiness should be pursued. So compared to moral expertise, dialectical expertise really does look like it is "worth little or nothing". Thus, if Socrates is indeed a dialectical expert, by his own assessment there is no contradiction in his claiming both that he gives the Athenians a great benefit and that his wisdom is ultimately nothing much to speak of.

8. MODERN MORAL EXPERTISE AS DIALECTICAL EXPERTISE

To return to the present day, what does all of this mean for those who today call themselves "moral experts"? If we take Socrates as our model, the most important results are the following. Whether or not we continue to use the term "moral expert", we should acknowledge that the understanding that we claim to have and the success which we claim to make possible is better described under the rubric of dialectical than of moral expertise. We should not promise a permanent and complete resolution of all moral disagreement; such a promise could never be satisfied and would ultimately generate resentment. What we can promise to do is to conduct a discussion of moral issues that clarifies the true grounds behind different parties' commitments, deflates occasional claims to moral infallibility, steers the discussion away from well-known philosophical errors, and helps opponents see each others' views more clearly. We can also help others get a clearer sense of the ideal results we should be seeking: an account of the moral issues at stake that not only resolves any related disputes, but does so in such a way that all reasonable people can understand why the disputes should be resolved as they were. Where the most difficult moral disagreements are concerned, we will probably never achieve this ideal, but nonetheless the kind of explanatory account which would satisfy Plato's criteria for expertise ought to be our standard. To the extent that we fall short of that ideal, we must acknowledge our conclusions to rest on intuitions or articles of faith with which reasonable people can disagree.

This is not to say that we should avoid taking stands on individual moral issues or trying to persuade others to join us in our commitments. Socrates himself had many such commitments and certainly was not chary of encouraging others to share them. What we must avoid is acting as though the purpose of experts' involvement in the ethical decision-making process is to generate the decision that the expert believed in advance was the right one. We are not ethical compasses that can be simply looked to for authoritative direction. However right our pronouncements may be, their authority is no stronger than the cogency of the arguments we give for them. Our proper role, then, is to create an environment for discussion in which those and all other arguments can get a fair hearing and the work of persuasion can go forward with as little obstruction as possible. Here, it is not the rightness of our ethical commitments but the sophistication of our epistemic and dialectical tools that determines our level of success. When we look back to Socrates as our model, this is the chief lesson we owe him.[37]

NOTES

1. The arguments about the proper division of Plato's dialogues into groups or developmental periods are interminable, and I do not intend to enter into them here. But the group of dialogues I intend to discuss as Socratic includes, in no particular order, *Euthyphro*, *Apology*, *Crito*, *Alcibiades* I (which I assume to be genuine, though not much turns on it here), *Charmides*, *Laches*, *Lysis*, *Euthydemus*, *Ion*, and the *Hippias Major* and *Minor*. I will also treat *Protagoras*, *Gorgias*, and *Meno* as Socratic for the purposes of this paper, although these dialogues share enough features with the "middle" dialogues (e.g. *Republic*, *Phaedo*, and *Symposium*) that their position is problematic.

Moreover, I side with those scholars who do not believe that we can read Plato's dialogues — even the Socratic dialogues — as a reliable guide to the philosophical positions of the historical Socrates. I accordingly will treat the Socrates of the Socratic dialogues as a (quite consistently drawn) literary character.

All quotations from Plato are taken from the Hackett edition of his complete works, edited by John Cooper and translated by several hands.

2. "Art" should be understood in the sense of a discipline generally rather than specifically of the fine arts, although Plato would consider some of the fine arts (e.g., painting, sculpture, and music) to count as forms of *technê*.

3. Much of the material on the criteria for expertise is adapted from my treatment of the same issue in LaBarge (1997).

4. Plato also says this somewhat obliquely in *Republic* I, although the fact that the claim appears chiefly in the mouth of Thrasymachus should make us pause before we attribute the view to Plato. Nevertheless, when Thrasymachus argues:

> ...(D)o you describe someone who makes mistakes about his patients as a doctor in virtue of the fact that he makes mistakes? Or do you describe someone who makes mistakes in his calculations a mathematician, at precisely the time when he is making a mistake, and in virtue of the mistake he is making? It's true that the expression is in our language: we say

> that a doctor or a mathematician makes mistakes; but in fact, in my opinion, to the extent that each of them is what we call him, he never makes mistakes. And the consequence of this is that...no professional makes mistakes: a mistake is due to a failure of knowledge, and for as long as that lasts he is not a professional (*Rep.* I 340d-e).

We ought to note a couple of things before we discount the position as unsocratic. First, Socrates does not challenge this particular claim (though he does reject the larger argument of which it is a part.) Second, however unpleasant Thrasymachus may be, he is not stupid, and this position coheres quite nicely with the claim of the *Euthydemus* that we cease being wise when we start making mistakes.

5. Strictly speaking, Socrates has gone too far here; the disagreement of reputed experts signals not that *all* but only that *some* of those who disagree are not experts. Note also *Hp. mi.* 372b-e, in which Socrates' claim that his disagreement with experts like Hippias is evidence of his ignorance, although laced with irony about who precisely is ignorant, proves the general point about expertise.

6. This is not to say that an expert could teach absolutely anybody. The student must bring to the learning process the requisite abilities, basic knowledge, and motivations to make learning possible. Plato is fully aware of this; it is the truth behind the Noble Lie of the metals at *Rep.* 414c ff., and also informs his development of the Wax Block metaphor (*Tht.* 194c-195a).

7. See e.g. Annas (1993) 4-10, and the articles in Crisp and Slote (1997).

8. See e.g. Chapter 2 of Ryle (1949).

9. I do think, however, that for a discipline to count as an expertise the success of the experts has to be *explicable*. An expert billiards player need not be able to explain how he achieves success so regularly, but his success ought to be explicable by reference to some sort of law-like regularities. A gambler who wins regularly in games of pure chance (which does not include, for instance, poker) is lucky, not expert. (This also means that, even if a billiards expert need not himself be a good teacher of billiards, billiards ought to be *teachable by somebody*.)

10. This evidence, being inductive, is defeasible. The prudent non-expert will remember that the mechanic's success could be a product of luck (presumably that is all some ancient medical success amounted to), or that his failure could be a result of forces outside his control (even the best doctors lose patients.) Of course, repeated and varied tests of the expert would strengthen the non-expert's convictions one way or the other.

11. This is not necessarily true. We might agree with Aristotle or Mill, for instance, that there are moral first principles which are not amenable of proof and claim that the standard of moral success is among these. If the *Seventh Letter* is indeed by Plato, he himself may have held such a view. (See Plato's extended attack on Dionysius' claim to have grasped Plato's entire moral theory at 341b-345a, especially 344a-b.)

12. I should add that it seems to me the domain in which such continued moral disagreement between fair and rational agents is possible is not co-extensive with the domain of moral debate generally. Commitment to some values is evidence of irrationality, wickedness, or both. Thus I think it is no longer possible to conscientiously defend slavery (if indeed it ever was). But many of the moral dilemmas we face in, for instance, medical practice today are such that conscientious people can reasonably disagree about fundamentals even in the face of powerful arguments.

13. Note that the establishment of an agreed upon criterion for moral success would ground only the kind of broad agreement that would allow genuine experts with different explanatory paradigms to recognize each other as experts in the broad sense. There could still turn out to be incompatible moral theories that meet the standard for moral success.

14. Those who are interested in this issue and in issues about Socrates' ignorance generally can find an excellent discussion of them in Chapter 8 of Benson (2000) which I have found very helpful.

15. Translating *technê*.

16. "Excellence" is often an appropriate translation because, though the possession of *aretê* sometimes entails (where humans are concerned) specifically moral excellence, other sorts of things, including inanimate objects, can possess *aretê* as well in a way that does not involve morality at all. One of the chief disputes in ancient Greek society centered on the centrality of moral superiority to human *aretê*; that issue is crucial to, for instance, the *Gorgias*.

17. Here are some examples, drawn from the list in Benson (2000) 173: Temperance (*Chrm.* 165b4-c2); Courage (*Lach.* 200e2-5); Virtue (*M.* 71a1-7, 71b2-3, and 80d1-8); Justice (*Rep.* 337d3-4, 337e4-5, and 354b9-c3); the Fine (*Hp. mai.* 286c8-e2 and 304d4-e5); Friend (*Lys.* 223b4-8).

18. E.g., Thrasymachus' accusation in the *Republic* that Socrates' profession of ignorance is meant to give him a dialectical advantage over those who have actual views to defend; see *Rep.* 337a4-7.

19. E.g., Versenyi's assertion that Socrates claims ignorance only to encourage others to figure out the truth for themselves; see Versenyi (1963), 120-121.

20. See Benson (2000), 176-180, for a thorough treatment with which I am largely in agreement.
21. See also 376b8-c6.
22. I suppose one could chalk this shift of views up to Socratic deviousness, arguing that Socrates really knows which of the positions is true but pretends not to so as to show up Protagoras, but if so, Socrates, by disguising his knowledge, would not be abiding by his statement in the *Laches* that one ought always to help others become better (*Lach.* 200e).
23. Socrates makes several other confident statements of moral fact that require similar explication; these are usefully collected in Vlastos (1985).
24. See *M.* 93a ff. and *Prot.* 319c-320b. See also *Lach.* 179b-d.
25. See especially *Ap.* 23c-24b.
26. Beversluis (2000) makes an intriguing case that Socrates is sometimes not above the use of argumentative dirty tricks himself. I think that there is something to this, and that this has important implications for the way we understand Socrates' project and method, but I cannot devote space to that issue here.
27. Note that Socrates cannot explain his failure with these figures in as many ways as we can; where we might readily attribute the stubbornness of a Callicles or the changeability of an Alcibiades to a weakness of character, the intellectualist Socrates must chalk them up to a purely intellectual failure. This general difference may make us think Socrates is harder on himself than he should be.
28. Interestingly, this result does not seem to make him doubt the possibility of an expert who could do these things; despite his repeated failures at discovering or becoming such an expert, he never gives up the search. See Benson (2000) 180-185.
29. It would have been nice if Plato had given us a hint as to who else Socrates may have had in mind. Aristides, often nicknamed "the Just", is a decent candidate, although in the *Meno* Socrates explicitly singles Aristides out as one of the excellent men who failed to raise excellent children (*M.* 94a). Perhaps he is thinking of Solon.
30. This is also the message of the "protreptic passage" from the *Euthydemus*, a delightful dialogue which is too little read by non-specialists. Socrates argues there (*Euthyd.* 277d-282e) that wisdom is the most important good because all other goods are only beneficial to us if our use of them is guided by wisdom. I.e., the goodness of all goods other than wisdom is dependent on their use by wisdom. (See also *Ap.* 30b.) Thus it is reasonable to put off the pursuit of all other goods until we have succeeded in achieving wisdom. As in the *Sophist*, the result is that, without possessing moral expertise of a kind that tells us what our ultimate values should be, we can nonetheless reasonably adopt a set of provisional priorities which make the pursuit of moral wisdom our first priority.
31. Thus Santas on the *Laches*: "...(I)n this section Socrates is established as the expert! Not an expert in fighting in armor of course. Nor a sophistic expert or orator, one who gives the answers in a great speech, *à la* Protagoras. Not even an expert about courage. Socrates is an expert in a *search* for the nature of courage" (Santas (1971) 181).
32. See Austin (1987) 27 and Benson (2000) 3-6 and Chapter 9. Not all scholars, however, would agree that Socrates needs a sophisticated epistemological theory; see e.g. Woodruff (1989) 90, Vlastos (1991) 239 n.18, and Brickhouse and Smith (1994) 7.
33. In LaBarge (1997) I called the possession of this theory "technical expertise", i.e., expertise about the nature of *technê*.
34. We cannot expect complete success — in this case, knowledge in every case whether a claim to expertise is genuine or not — because on occasion we simply will not have all the evidence we need to know whether or not a person's claim is true. Sometimes this is because, as Plato points out in the *Charmides*, we have to already know something about the subject matter under discussion before we can evaluate claims to expertise in that area (*Chrm.* 169e-171c). In any case, we should treat such cases in the same way I suggested in section III above, by noting that the expert cannot be blamed if he lacks the materials required for the practice of his craft (good leather for the cobbler, sufficient evidence for the evaluator of expertise), and that in such cases success consists in doing the best one can possibly do under the circumstances and making no mistakes. (For instance, the expert evaluator of experts would not make the mistake of thinking he had a certain evaluation of a claim to expertise if he lacked all the evidence necessary to justify such a judgment.)
35. Even in the *Charmides*, where Socrates worries that a person who had knowledge of knowledge and ignorance would not be able to recognize experts and non-experts in any field besides that of knowledge and ignorance, he still acknowledges that the knower of knowledge and ignorance will recognize others who share that specific knowledge (171c).
36. The strong similarity between Socrates' practice as described in the *Apology* and this description of the knowledge of knowledge in the *Charmides* has been noted by many commenters, including Grube (1935) 219,

Dyson (1974) 104, Schmid (1983) 342, Austin (1987) 23, Kahn (1988) 546, Morris (1989) 49, and Benson (1995) 98, though many of them find this similarity chiefly a source of confusion.

37. Many people have offered useful criticisms and suggestions as I wrote this paper. I wish in particular to thank Jay Alberg, Robert Audi, Mane Hajdin, Philip Kain, Mike Meyer, Larry Nelson, Bill Prior, Manny Velasquez, and Brad Wilburn. Also, this paper grew out of a piece I wrote as a graduate student at the University of Arizona, and I thus owe a debt to the people who discussed that earlier version with me then: Julia Annas, Paul Bloomfield, Mark McPherran, and Dan Russell.

REFERENCES

Annas, J. (1993). *The Morality of Happiness*. Oxford: Oxford University Press.

Austin, S. (1987). 'The Paradox of Socratic Ignorance (How to Know That You Don't Know).' *Philosophical Topics*, 15, 23-34.

Benson, H., ed. (1992). *Essays on the Philosophy of Socrates*. Oxford, New York: Oxford University Press.

— (1995). 'The Dissolution of the Problem of the Elenchus.' *Oxford Studies in Ancient Philosophy*, 13, 45-112.

— (2000). *Socratic Wisdom*. Oxford: Oxford University Press.

Beversluis, J. (2000). *Cross-Examining Socrates: A Defense of the Interlocutors in Plato's Early Dialogues*. Cambridge: Cambridge University Press.

Brickhouse, T. C. & Smith, N. D. (1994). *Plato's Socrates*. Oxford, New York: Oxford University Press.

Burch, R.W. (1974). 'Are There Moral Experts?' *Monist*, 58, 646-658.

Cicero (1933, reprint 1994). *De Natura Deorum, Academica*. H. Rackham (Trans.). Cambridge, MA: Harvard University Press.

Crisp, R. and M. Slote, eds. (1997). *Virtue Ethics*. Oxford: Oxford University Press.

Dyson, M. (1974). 'Some Problems Concerning Knowledge in Plato's *Charmides*.' *Phronesis*, 23, 216-28.

Grube, G.M.A. (1935, reprinted in 1980). *Plato's Thought*. Indianapolis: Hackett Pub. Co., Inc.

Kahn, C. H. (1988). 'Plato's *Charmides* and the Proleptic Reading of the Socratic Dialogues.' *Journal of Philosophy*, 85, 541-49.

LaBarge, S. (1997). 'Socrates and the Recognition of Experts.' In: McPherran (1997), pp. 51-62.

McPherran, M., ed. (1997). *Wisdom, Ignorance, and Virtue: New Essays in Socratic Studies*. Edmonton: Academic Printing and Publishing.

Morris, T.F. (1989). 'Knowledge of Knowledge and of Lack of Knowledge in the *Charmides*.' *International Studies in Philosophy*, 21, 49-61.

Plato (1997). *Complete Works*. J. Cooper (Ed.). Indianapolis: Hackett Pub. Co., Inc.

Ryle, G. (1949). *The Concept of Mind*. Chicago: University of Chicago Press.

Santas, G. (1971). 'Socrates at Work on Virtue and Knowledge in Plato's *Laches*.' In: Vlastos (1971), pp. 177-208.

Schmid, W.T. (1983). 'Socratic Moderation and Self-Knowledge.' *Journal of the History of Philosophy*, 21, 339-48.

Versenyi, L. (1963). *Socratic Humanism*. New Haven: Yale University Press.

Vlastos, G. (1971). *The Philosophy of Socrates: A Collection of Critical Essays*. Notre Dame: University of Notre Dame Press.

— (1985). 'Socrates' Disavowal of Knowledge.' *Philosophical Quarterly*, 35, 1-31.

— (1991). *Socrates: Ironist and Moral Philosopher*. Ithaca, NY: Cornell University Press.

Woodruff, P. (1989). 'Plato's Early Theory of Knowledge.' In: Benson (1992), pp. 86-106.

CHAPTER 3

CARRIE-ANN BIONDI KHAN

ARISTOTLE'S MORAL EXPERT: THE *PHRONIMOS*[*]

"The agent they would call [the *phronimos*[1]] is the one who studies well each question about his own [good], and he is the one to whom they would entrust such questions"
—Aristotle (*Nicomachean Ethics* VI 7, 1141a26-28, trans. Irwin).[2]

1. INTRODUCTION

Despite the contemporary, Western, democratic impulse toward relying on one's own common sense or judgment in matters both personal and political, people in this sort of society often seek guidance from "moral experts". This occurs when individuals treat physicians, political analysts, astrologers, religious leaders, and so on as bearers of special knowledge whose pronouncements are to guide their choices and lead them to better lives. For example, the front cover of a recent issue of *Newsweek* queried "What Would Jesus Do?"[3] Though who is seen as a moral expert changes over time, the general practice of seeking moral expertise has a long history. One of the earliest explicit considerations of moral expertise was developed by the ancient Greek philosopher Aristotle (384-322 B.C.). Aristotle's moral expert uses his "eye of the soul"—honed by experience and made possible by good moral character—to act virtuously and prescribe right action to others. As such, moral expertise is *practical* wisdom rather than *theoretical* wisdom, the latter of which is the account advocated by Aristotle's teacher Plato. Like Aristotle's historically influential account of virtue ethics in general, his account of the moral expert, namely, the *phronimos*, has made a resurgence in the field of bioethics.[4]

Technological breakthroughs in medical science have blinded the public as well as many medical practitioners to the ethical dimension of medical practice. The relatively recent introduction of hospital ethics committees and institutional review boards is an attempt to rectify this problem. These much-needed bodies, though, have tended (at least in the United States) to focus on adopting general rules and policies that are supposed to be applied—in the name of fairness and efficiency—in the same way to all those involved in the medical decision-making process.

L. M. Rasmussen (Ed.), Ethics Expertise: History, Contemporary Perspectives, and Applications, 39-53.

Many neo-Aristotelian virtue ethicists (Foot, 2001; Hursthouse, 1991; Sherman, 1991), however, have found such an approach problematic. The main problems with such a procedure are twofold: (1) to apply general rules in the same way to all is to ignore morally relevant features of different circumstances, and (2) it is not always clear that a rule—for example, "Upon providing all relevant information, let the patient decide"—will be very helpful in guiding the actions of either doctor or patient.

Let us examine a set of abortion cases in order to illustrate these problems and to see how a virtue ethicist might approach such cases.

> Case (A): A very frail pregnant woman may die as a result of carrying her pregnancy to term, and her baby might die as well.
>
> Case (B): A healthy but vain pregnant woman will likely need a cesarean section if she carries her pregnancy to term.

A general rule requiring the prohibition of abortion would treat both cases identically by requiring both women to carry their pregnancies to term, regardless of the consequences. A rule such as the one mentioned above, that is, "Upon providing all relevant information, let the patient decide," will be unhelpful. A virtue ethicist would explain that with respect to the doctor's role, in order to provide all relevant information, he must possess the virtue of honesty, he must have enough clinical experience to know what counts as "relevant," and he must have enough moral experience of character-types to know how information may affect the women in the above cases. With respect to the women, a virtue ethicist would eschew a simplistic approach that only takes into account consequences or personal preferences. Each woman should act virtuously (e.g., courageously) in her circumstances, which may dictate different actions to be taken in each case—perhaps to abort in Case (A) and not to abort in Case (B).[5] Rosalind Hursthouse (1991, p. 230) makes the crucial point that "[a]cting rightly *is* difficult, and *does* call for much moral wisdom. . . . Virtue ethics builds this in straight off precisely by couching its rules in terms whose applications may indeed call for the most . . . sensitive judgment." Consequently, a virtue-ethical approach may lead to differential judgments rather than uniform policies.

2. A BRIEF ACCOUNT OF ARISTOTLE'S MORAL THEORY

Aristotle's account of the *phronimos* cannot be understood without some idea of the role it plays in his ethical theory. Accordingly, I first provide an outline of his general theory. His approach to discovering the good for all natural beings is teleological. He determines the highest purpose or end (*telos*) of some being by studying what beings of that kind strive toward and what helps them flourish as the kind of beings they are. Aristotle looks to the physical parts of a being, for example, a foot, and sees that it has a function in relation to the whole organism, namely, to aid in walking. Aristotle implicitly replies "No" to his own rhetorical questions "[B]ut has a human being no function? Is he by nature idle, without any function?" and "Yes" to "Or, just as eye, hand, foot, and, in general, every [bodily] part apparently has its function, may we likewise ascribe to a human being some function apart from all of these?" (*NE* I 7, 1097b31-34). David Bostock (2000, p. 16) explains that "[a]s Aristotle conceives it, the final stage comes when we ask about the body as a whole, with all its various capacities, what *it* is for, and the answer is that it is for a certain kind of life. That is the ultimate goal." Nature (*phusis*) and natural function (*ergon*) are thus essential to Aristotle's understanding of the good for any being.[6] He employs this methodology to determine the human good, and develops a normative ethical theory known as 'virtue ethics' that will help humans attain this good.

Aristotle most generally considers *eudaimonia*[7] to be the highest human good, since it is complete, self-sufficient, and most choiceworthy—all of the features needed for something to be considered as valuable in itself (*NE* I 7, 1097a15-b24). This account of the good is much too broad, though, to be of any use. We have to turn to Aristotle's 'function argument' for more guidance in the realm of ethics.

One must be acquainted, though, with Aristotle's description of the human soul in order to understand his function argument. He thinks that human beings, like all animate beings, have a certain type of soul that defines its nature. He claims that there are three[8] parts of the human soul: (1) a rational part, (2) a nonrational part that responds to reason (the appetitive part), and (3) a nonrational part that does not respond to reason (the vegetative part, which has to do with nutrition and growth) (*NE* I 13, 1102a29-b26).

What *eudaimonia* specifically amounts to will depend on "the special function of a human being" (*NE* I 7, 1098a1). In order to figure out what the human function is, Aristotle compares humans to other sorts of animals. He first rejects the possibility that the human function is nutrition and growth, since this is a function that is shared with plants (which only have vegetative souls). He next rejects sense perception as a candidate, since this is a function that is shared with nonhuman animals (which only have vegetative and appetitive parts to their souls). He finally settles on the human function as "activity of the soul in accord with reason or requiring reason" (*NE* I 7, 1097b25-1098a8).

The next step in the function argument is to figure out how to move from the human function to the human good. What exactly does it mean for humans to strive for *eudaimonia* by actively using the reasoning part of their soul? Aristotle explains this in the following way using the example of a harpist: A harpist's function is to play the harp, and a good harpist's function is to play the harp well according to the virtues[9] of harp playing. In like manner, the human function is to actively use the reasoning part of the soul, and a good human being's function is to do this well, that is, according to the virtues of using the part of the soul that involves reason. Hence, "the human good proves to be activity of the soul in accord with virtue" (*NE* I 7, 1098a8-17).

Aristotle takes the conclusion of his function argument to yield "a sketch of the good" (*NE* I 7, 1098a22). In order to fill out this sketch and see what it means concretely for one to engage in activity of one's soul in accord with virtue, he turns to an examination of virtue. He discusses this in relation to the human soul. Recall that there are three parts of the soul: (1) a rational part, (2) an appetitive part, and (3) a vegetative part. Since the third part does not respond to reason, it does not exemplify virtue. There are, then, two different kinds of virtue that correspond to this structure of the human soul. The first kind includes the virtues of thought, which are virtues of only the rational part of the soul—for example, scientific or mathematical knowledge. The second kind includes the virtues of character, which are manifested by the proper relationship between the rational and appetitive parts of the soul, that is, with the rational part ruling and the appetitive part properly responsive to reason—for example, temperance or courage. 'Moral virtue' is a common translation for this second kind of virtue (*NE* I 13, 1102a5-1103a10).

Moral virtue, as defined by Aristotle, has four key features. The first feature is that it is a mean, and a mean relative to an individual. The mean, however, is not a mathematical mean, but that which is neither excessive nor deficient. For example, courage is a mean between rashness (which is excessive) and cowardice (which is deficient). This mean, though, is relative to an individual. Courage for Hercules would most likely be rashness for the average man on the street, since the average man possesses neither Hercules's great strength nor his battle experience. What is courageous for the average man might be somewhat cowardly for Hercules. Thus, which action counts as a courageous one will depend on the individual performing the action (*NE* II 6, 1106a27-b28).

Though Aristotle's prime case of courage is that of bravery on the battlefield, we can also apply the virtue of courage in some detail to our set of abortion cases. It was suggested above that the pregnant woman in Case (A) could be courageous in having an abortion, while the pregnant woman in Case (B) could be courageous in not having an abortion. We are in a better position now to see why this difference in judgment can occur. For the frail woman, bearing a child would be the excess of rashness, since doing so would likely cause her death (and possibly the death of her baby as well). For the healthy but vain woman, having an abortion would be the deficiency of cowardliness, since doing so is to fear how a scar appears. The mean, then, in each of these cases dictates a different action: the frail woman should have an abortion and the healthy but vain woman should not have an abortion.[10]

The second key feature of moral virtue is that it involves decision. An individual needs to make a decision, upon sufficient deliberation, as to which action should be performed in a given circumstance. This will help ensure not only that the right action will be performed, but also that the individual performing it will be doing so virtuously. For virtue ethics, it is the combination of right action and the right state of character that makes one *be* virtuous rather than merely perform a virtuous action: "[A]ctions are called just or temperate when they are the sort that a just or temperate person would do. But the just and temperate person is not the one who [merely] does these actions, but the one who does them in the right way" (*NE* II 4, 1105b6-9); and "[T]he agent must be in the right state when he does [actions that are in accord with virtue]" (*NE* II 4, 1105a31-32). To be in the right state, a person must meet the following three conditions: "First, he must know [that he is doing virtuous actions]; second, he must decide on them, and decide on them for themselves; and, third, he must also do them from a firm and unchanging state" (*NE* II 4, 1105a32-35).

The third key feature is that an individual must use his reason to figure out the proper moral principles and how to apply them in various situations.[11] Aristotle states: "Virtue, then, is a state that decides, consisting in a mean, the mean relative to us, which is defined by reference to reason, that is to say, to the reason by reference to which the [*phronimos*] would define it" (NE II 7, 1107a1-3). (This feature will be explored more fully in the following sections.)

The fourth key feature has to do with how moral virtue is acquired. It is acquired through the activity of habituation. One cannot sit back passively to acquire virtue, even if one intellectually grasps what it generically means to be virtuous. Having knowledge about moral virtue is not sufficient for being virtuous; one must also act on the knowledge (see, e.g., *NE* I 8, 1098b32-1099a3; VII 10, 1152a8-9; and X 9, 1179b1-4). Habituation is necessary for creating in us the aforementioned "firm and unchanging state" of character, because "[a] state of character arises from [the repetition of] similar activities. . . . It is not unimportant, then, to acquire one sort of habit or another right from our youth" (NE II 2, 1103b21-25). Since habituation takes time, Aristotle claims that the young (for him, those below age twenty-five) are neither ready to learn about morality or politics nor able to be virtuous. This is because "a youth . . . lacks experience of the actions in life, [and] since he tends to follow his feelings, his study will be futile and useless" (*NE* I 3, 1095a2-5). The lengthy process of habituation is what helps someone to get the appetitive part of his soul in line with the rational part, so that he can then perform the appropriate mean and be in a virtuous state. This state is crucial to attain because it is "the state that makes a human being good and makes him perform his function well" (NE II 6, 1106a23-24).

This excursion into Aristotle's general moral theory has been necessary for three reasons. First, we need to be familiar with the subject matter of living well that concerns Aristotle's moral expert. Second, we can see that since habituation plays such an important role in acquiring the virtues, we can guess that the *phronimos* may have something to do with the activity of inculcating proper habituation. *Someone* has to be guiding the habituation of the youth, and a moral expert would be a good candidate for this role. Third, a crucial element of moral virtue—namely, "correct reason" (*NE* II 2, 1103b32)—is supposed to be defined "by reference to which the [*phronimos*] would define it." It is now time, then, to turn to Aristotle's account of the *phronimos*.

3. THE *PHRONIMOS*

Aristotle makes it clear that living the life of moral virtue is an essential part of *eudaimonia*. Thus, acquiring moral knowledge and living in accordance with that knowledge are all-important requisites for living the good life. It ends up, though, that the person who can do these things is indeed a rare person. It is the infrequently existing *phronimos* who is the moral expert to whom people can turn for sagacious moral advice. He is the person who goes beyond the outline of the good (that is articulated in the previous section) and provides us with the specific content that is needed for action.

The *phronimos*—also referred to variously as an excellent or cultivated man (*spoudaios*)—is accorded a significant and unique position by Aristotle: "[T]he excellent person is far superior because he sees what is true in each case, being himself a sort of standard and measure" (*NE* III 4, 1113a33-34); "[V]irtue and the excellent person would seem to be the standard in each case" (*NE* IX 4, 1166a13-14); "[T]he cultivated and civilized person, as a sort of law to himself, will take this [discriminating] attitude" (*NE* IV 8, 1128a33); and "[W]hat is really so is what appears so to the excellent person . . . [and] the good person insofar as he is good, is the measure of each thing" (*NE* X 5, 1176a16-18). It is important to stress that a given action is good to do not simply because the *phronimos* says that it should be done, but because it is objectively good to perform that action. The *phronimos* thus tracks moral truth and is a helpful guide to right conduct for others to follow.

One of the reasons why Aristotle believes that there will be very few moral experts is because it is hard to know what is right to do: "[I]t is also hard work to be excellent. For in each case it is hard work to find the intermediate. . . . Hence doing these things well is rare, praiseworthy, and fine" (*NE* II 9, 1109a25-30). Excess and deficiency come in many forms, but there is only one way for someone to hit the intermediate target in a given circumstance (*NE* II 6, 1106b30-32). Aristotle draws an analogy between the moral expert and a doctor in order to indicate how hard it is to acquire moral knowledge and virtue:

> Knowing how actions must be done, and how distributions must be made, if they are to be just, takes more work than it takes to know about healthy things. And even in the case of healthy things, knowing about honey, wine, hellebore, burning, and cutting is easy, but knowing how these must be distributed to produce health, and to whom and when, takes all the work that it takes to be a doctor (*NE* V 9, 1137a13-18).

There seems to be a hierarchy of both difficulty and types of knowledge or expertise. The knowledge involved in particular skills, such as burning or cutting, is a lower-level sort of knowledge. The knowledge involved in more general professions, such as being a doctor, involves not only the previous sorts of skills, but also the general knowledge of how to use the skills in particular circumstances in order to produce health in patients. Such knowledge takes a great deal of training and experience. The knowledge involved in moral matters is still more general and difficult to acquire since it is about the human good (a subject matter that is imprecise), and it is even more important than scientific medical knowledge (since goods of the soul are more important than goods of the body[12]).

Another reason why Aristotle thinks that there will be few moral experts is because most people allow their appetites (passions or feelings) to overcome their reason. Offering arguments concerning the good to most people will be useless, because such arguments "seem unable to turn the many toward being fine and good. For the many . . . avoid what is base because of the penalties, not because it is disgraceful . . . [T]hey live by their feelings" (*NE* X 9, 1179b10-14). Arguments concerning the good can only be fruitfully offered to those who have been habituated properly: "[T]he soul of the student needs to have been prepared by habits for enjoying and hating finely, like ground that is to nourish the seed" (*NE* X 9, 1179b25-27). Moral habituation can easily go awry, though, and thus ruin one's chances of ever being able to be virtuous.

Since moral knowledge is necessary for living the good life and very few people have the ability to acquire such knowledge, a need for reliance on moral experts arises. Most people will have to despair of ever being virtuous, if they are even among the relatively few who care about moral virtue; they must be satisfied with acting rightly by emulating the actions that the *phronimos* performs: "[I]n everything we clearly must imitate the better person" (*NE* IX 11, 1171b13). Having to emulate the *phronimos*, though a less-than-optimal alternative, is nothing to be ashamed of. Aristotle beckons us to heed the words of Hesiod on this matter: "He who grasps everything himself is best of all; he is noble also who listens to one who has spoken well; but he who neither grasps it himself nor takes to heart what he hears from another is a useless man" (*NE* I 4, 1095b10-13). Besides, trusting experts is not without precedent. Aristotle questions whether one should obey one's father in everything, or concede trust to a doctor in matters of health and vote for someone with military expertise to be a general (*NE* IX 2, 1164b23-25). His query here is a rhetorical one, since it makes sense to consult a doctor about medical matters and place those with military experience in the appropriate offices. Likewise, there are a few moral experts who can help themselves and others on the road to *eudaimonia*, and so we should consult them in moral matters.

As if the *phronimos* were not already accorded a special-enough role in society as the sage to whom people can go for moral advice, Aristotle expands the role of moral expert to being a legislative, or political, expert in the best possible city-state (*polis*). Though many people think that "[*phronêsis*] concerned with the individual himself seems most of all to be counted as [*phronêsis*]" (*NE* VI 8, 1141b30-31), Aristotle thinks that one can have *phronêsis* about the city-state.[13] This is because he believes that it is the function of all those who engage in political science (e.g., legislators) to make their citizens, to the extent possible, "good people who do fine actions" (*NE* I 9, 1199b31-32). Aristotle further claims that "while it is satisfactory to acquire and preserve the good even for an individual, it is finer and more divine to acquire and preserve it for a people and for cities" (*NE* I 2, 1094b9-11).[14] Since the function of the city-state is to make men good, and habituation plays such a crucial role in people being able to be good (or at least to act rightly), the moral expert is also the political expert who devises correct laws, especially ones that guide habituation. For "[i]t is difficult . . . for someone to be trained correctly for virtue from his youth if he has not been brought up under correct laws . . . [L]aws must prescribe their upbringing and practices" (*NE* X 9, 1179b33-36). Aristotle describes law as "reason that proceeds from a sort of [*phronêsis*] and understanding" (*NE* X 9, 1180a22). It thus makes sense, based on Aristotle's views about morality and politics, for the *phronimos*—who has this sort of reason—to serve as the moral expert at the levels of both the individual and the political society.

Now that we know who the *phronimos* is and why he is so important, the next thing to investigate is how it is that the *phronimos* knows what he knows and precisely what it is that he knows. This is where we begin looking through a glass darkly.

4. *PHRONÊSIS*

It was noted above that the *phronimos* is the person who is able to provide us with the specific content that is needed to determine good human action. This is useful for the *phronimos* and for those who can at best emulate the *phronimos*, but what about someone who has the capacity for becoming a *phronimos* himself? How is it that he can become a moral expert? Telling him simply to act as the *phronimos* would act—that is, in the state in which the *phronimos* acts—is not sufficient. Aristotle again draws an analogy between moral virtue and medicine in order to make his point: "But knowing only this, we would be none the wiser about, for instance, the medicines to be applied to the body, if we were told we must apply the ones that medical science prescribes and in the way that the medical scientist applies them" (*NE* VI 1, 1138b30-33). What must finally be done is to define the all-important "correct reason," that is, to understand *phronêsis*.

In order to see what *phronêsis* is, Aristotle first discusses a more refined, four-part division of the human soul. He divides the rational part of the soul itself into two subparts: (a) a part called the scientific part, which studies eternal things, and (b) a part called the rationally calculating part, which studies and deliberates about those things that can be otherwise (*NE* VI 1, 1139a5-15). (Recall that the other two parts of the soul are the appetitive part and the vegetative part.) This subdivision of the rational part of the soul is an important move in distinguishing Aristotle's account of moral expertise. He holds that Thales and other sages had wisdom (*sophia*) but not *phronêsis* (*NE* VI 7, 1141b4-9). This move is what enables him to reject Plato's idea of the moral expert: the philosopher-king. We use reason to deliberate about things that can be otherwise, human action (and thus moral action) can be otherwise, and there is a truth about what is good for humans, so *phronêsis* "is a state grasping the truth, involving reason, concerned with action about things that are good or bad for a human being" (*NE* VI 5, 1140b5-7). As such, *phronêsis* is the virtue of the rationally calculating part of the human soul.

Phronêsis, then, is one of the virtues of the rational part of the soul (and so is a virtue of thought), and it is the one by which one can fathom what morally ought to be done. How does this "eye of the soul" (*NE* VI 12, 1144a30) operate, though, and what exactly does it see? Just what kind of thing is it—a faculty, an ability, a sense? Responding to these questions is far from easy and gives rise to more questions than answers.

4.1. What Is Phronêsis*?*

The correct interpretation of virtually every element of Aristotle's account of *phronêsis* is a matter of deep contention. Most generally, the virtue of *phronêsis* is described as "the best state of the rationally calculating part" (*NE* VI 1, 1139a17-

18), that is, as good deliberation about what ought to be done. What is involved in this state of good deliberation is unclear. The two major interpretations hold that either (1) *phronêsis* involves perceiving the right end and the means to that end, or (2) *phronêsis* involves perceiving the correct means to some end. (A third possible interpretation will be mentioned below.)

There is some textual evidence for the first interpretation. At one point, Aristotle explains that in order to have *phronêsis*, it is not sufficient simply to do as (what he calls) 'the clever (*deinotes*) person' does—namely, "use rational calculation to reach what he proposes to see" (*NE* VI 9, 1142b19). Simply to take any given end and figure out what means will best attain that end is merely to be clever. Some ends are bad. Since *phronêsis* is about what one ought to do, it cannot be equivalent to cleverness. Rather, "[i]f, then, having deliberated well is proper to a [*phronimos*], good deliberation will be the type of correctness that accords with what is expedient for promoting the end about which [*phronêsis*] is true supposition" (*NE* VI 9, 1142b32-35). According to Terence Irwin's translation and interpretation of this passage, *phronêsis* involves both correctness about the end and correctness about the means to the end.[15]

Aristotle complicates matters, though, by talking about *phronêsis* in a different way elsewhere. He states: "[W]e fulfill our function insofar as we have [*phronêsis*] and virtue of character; for virtue makes the goal correct, and [*phronêsis*] makes the things promoting the goal [correct]" (*NE* VI 12, 1144a8-10); and "[virtue] makes us achieve the end, whereas [*phronêsis*] makes us achieve the things that promote the end" (*NE* VI 13, 1145a6-7). In both of these claims, Aristotle has removed the correct end from the jurisdiction of *phronêsis* and given it over to virtue. What we are left with is the option that *phronêsis* has to do with correctly deliberating about the means to an end. The skill that enables one to "achieve the things that promote the end" is cleverness, though, so we are left on this second interpretation with a puzzle about how to keep *phronêsis* from collapsing into cleverness.

It should be noted that the second interpretation involves no less difficulty than the first interpretation, a situation which actually suggests a third possible interpretation of what *phronêsis* is. If it is virtue of character that allows us to see the good end, and virtue and *phronêsis* are not the same thing, then what is *phronêsis* after all? The first interpretation suggests that *phronêsis* gives us both the means and the end. If, though, virtue gives the correct end and *phronêsis* is not simply cleverness (despite the complications raised by the second interpretation), then perhaps '*phronimos*' is just the name given to someone who is both virtuous and clever and *phronêsis* is not some distinct state of its own after all.

4.2. What Enables Us to Develop Phronêsis*?*

Whatever *phronêsis* is, there are two key things that enable one to acquire it. The first is experience, and the second is the possession of a good moral character.

The role of experience was briefly alluded to above when explaining why it is that youths are not good students of morality and politics. The young[16] can memorize universal moral claims and parrot them back, but they will not possess moral knowledge or know how to use it. As such, they will not *really* have moral knowledge, since moral knowledge is supposed to be practical and not merely theoretical: someone does not have *phronêsis* "simply by knowing; he must also act on his knowledge" (*NE* VII 10, 1152a8-9). In order to act on one's moral knowledge, one must at least have significant life experience: "[*Phronêsis*] is concerned with particulars as well as universals, and particulars become known from experience, but a young person lacks experience, since some length of time is needed to produce it" (*NE* VI 8, 1142a13-16); and "[W]e must attend to the undemonstrated remarks and beliefs of . . . [the *phronimoi*[17]] . . . For these people see correctly because experience has given them their eye" (*NE* VI 11, 1143b12-14). Once again, Aristotle provides us with an analogy between ethics and medicine in order to persuade us of why real-life experience with one's subject matter should facilitate the successful practice of one's subject matter. He reminds us that doctors do not appear to be medical experts simply by reading books; they must engage in practices that give them the necessary experience over time that qualify them as medical experts (see *NE* X 9, 1181a20-b3). The implication is that the same reasoning holds for those who develop moral expertise.[18]

The role that character plays in acquiring *phronêsis* seems to be even more important than the role that experience plays. One can garner all sorts of experience, but if one also has a bad character, one can never become a *phronimos*. Aristotle claims that the "[best good] is apparent only to the good person; for vice perverts us and produces false views about the principles of action. Evidently, then, we cannot [have *phronêsis*] without being good" (*NE* VI 12, 1144a35-b1), and "we cannot . . . [have *phronêsis*] without virtue of character" (*NE* VI 13, 1144b32). These observations that he makes about character vis-à-vis *phronêsis* lend some credence to the first interpretation (or perhaps to the third) of what *phronêsis* is. If *phronêsis* were no more than cleverness, then having the correct end would not be necessary for having *phronêsis*. But it is, so the second interpretation does not fit well with the rest of what Aristotle says about *phronêsis*.

4.3. What Does Phronêsis *Enable Us to Perceive?*

At this point, it is clear that *phronêsis* is supposed to enable one (somehow) to acquire moral knowledge that one acts on. But what sort of thing is this knowledge about? The responses to this question have been varied, and they range from universal principles that one either reasons toward (Wallace, 1989, p. 23) or directly intuits (Ross, 1930, pp. 41-42) to particular actions that are either surmised casuistically (Brody, 1988, pp. 39-40, 96; Broadie, 1991, pp. 179-266, esp. pp. 249-50) or directly perceived by some moral sense (McDowell, 1979, pp. 124-27, 134-35, 140) to some combination of universal principles and particulars (Reeve, 1995, pp. 67-98; Irwin, 1999, pp. 249, 252; Bostock, 2000, pp. 83-88).

In several places, Aristotle states that *phronêsis* has to do with both universals and particulars. For example, "Nor is [*phronêsis*] about universals only. It must also acquire knowledge of particulars, since it is concerned with action and action is about particulars" (*NE* VI 7, 1141b15-17); and "[*Phronêsis*] is concerned with particulars as well as universals, and particulars become known from experience" (*NE* VI 8, 1142a13-14). The relationship between universals and particulars is not clarified in these passages, however. The picture dims even more when we consider the previous discussion of whether (and how) *phronêsis* involves deliberation about means and ends or just means.

If the first or third interpretation about what *phronêsis* is is correct, then the view that the *phronimos* apprehends some combination of universals and particulars would be correct. The universals would be the correct ends that are perceived, and the particulars would be the specific actions that are prescribed upon deliberating about how best to attain the correct end in some circumstance. This sort of reading could support a principlist interpretation of *phronêsis* (of either the intuitionist or nonintuitionist variety). *Phronêsis* would serve the dual role of allowing us to see the end (through intuition or reason) and helping us to apply the principle or rule properly.

If the second interpretation about what *phronêsis* is is correct, then virtue is what gives us the end and *phronêsis* is thus only about deliberating well about the means to that end. What we would perceive with *phronêsis*, then, is not principles themselves, but rather how best to apply the virtue-supplied principles in a given circumstance. This sort of reading could support a type of principlist interpretation of *phronêsis*. We are given principles by something other than *phronêsis*, and then *phronêsis* allows us to apply them properly.

An additional complication crops up when we consider the role that experience is supposed to play in the acquisition of *phronêsis*. According to Aristotle, we can have principles on hand (and thus have the universal), but this is not yet enough for having *phronêsis*. It is only through experience that one becomes acquainted with particulars. What this acquaintance amounts to is not fully determinate. On one reading, it could be that one simply sees the right particular action with one's "eye of the soul"—this could be used to support some kind of intuitionist/moral sense reading of *phronêsis*. On another reading, the role allotted to experience could be understood as experience of lots of similar situations over time. This sort of experience could allow the *phronimos* to cull out some conclusions concerning what action ought to be done in a particular type of circumstance—this could be used to support a casuist reading of *phronêsis*.

4.4. Some Problems with Phronêsis

Despite the appeal of the promise of moral expertise, there are a few significant difficulties with Aristotle's account of the *phronimos* and *phronêsis* that should be noted briefly. Some of these problems would still exist even if the textual ambiguities noted above were resolved.[19]

1. Circularity. One difficulty has to do with a claim that Aristotle repeats several times throughout the *NE*: "[W]e cannot be fully good without [*phronêsis*], or [have *phronêsis*] without virtue of character" (VII 13, 1144b32-33). This has the ring of circularity to it. Having *phronêsis* is what is supposed to enable one to figure out what virtue is so as to lead a good life. However, it sounds as though one must be virtuous in order to acquire *phronêsis*. One cannot *be* virtuous, though, unless one is in the same state as the *phronimos*, which means that one already has *phronêsis*.

2. Identifying the phronimos. It is perplexing to think that the many who do not have *phronêsis* should know how to identify the *phronimos*. How can those without moral expertise accurately identify the *phronimos*—who can only be picked out by his actions and advice—when they themselves do not know which actions and advice are morally correct? Aristotle seems not to provide a criterion by which a non-*phronimos* can judge who the *phronimos* is. Only fellow *phronimoi* can accurately identify another *phronimos*, a situation which gives no one else any reason to trust the claims of the self-proclaimed *phronimos*. Alvin Goldman (1999, p. 268) nicely states the problem in the following way: "He might ask *other* authorities, but this just pushes the problem back one step further: how can he identify *them* as authorities on the authorities?"

Goldman (1999, pp. 267-71) tries to resolve this very problem that he poses for accounts like Aristotle's. He focuses primarily on how to resolve this problem in science, but extrapolates it more broadly to cover some other domains. His key claim is that while a nonexpert cannot directly *recognize* an expert, a nonexpert can directly *determine* by means of inference over time that someone is an expert. For example, I, a car novice, do not know whether Mr. Fix-It is the car expert that he claims to be. One way for me to know is for Mr. Fix-It to claim that replacing my car's suspension will cause the clanking noise to cease, I observe my car's suspension being replaced, and I observe that the clanking noise ceases.[20]

Granted, it is easier for one to recognize the kind of scientific expert that Goldman discusses than it is to recognize Aristotle's moral expert. However, Aristotle actually does employ a tactic (somewhat similar to Goldman's) for justifying the moral expert's moral principles to those who are not moral experts: a dialectical method that starts from common beliefs (*endoxa*). Aristotle summarizes his dialectical method in the *NE*:

> [W]e must set out the appearances, and first of all go through the puzzles. In this way we must prove the common beliefs about these ways of being affected—ideally, all the common beliefs, but if not all, most of them, and the most important. For if the objections are solved, and the common beliefs are left, it will be an adequate proof (VII 1, 1145b4-8).

C.D.C. Reeve (1995, p. 35) draws our attention to what Aristotle counts as common beliefs: "*Endoxa* are those opinions accepted by everyone or by the majority or by the wise—either by all of them or by most or by the most notable and reputable" (*Topics* 100b21-22). Reeve (1995, p. 40) states that Aristotle's dialectical method "takes principles already justified by experience and made true by reality and 'clarifies' them. And this is as true in ethics, where the object of inquiry is the good, as it is in science." Someone who is not a moral expert, then, can be offered *some* kind of argument for trusting an alleged moral expert, even if he does not have the same understanding that the moral expert has.

3. The regress of habituation. There is also something unexplained about the institution of moral education through habituation. Proper habituation presupposes that those who are in charge of habituation are moral experts. However, these moral experts were themselves habituated to their moral condition. There is either going to be an infinite regress of habituation, or there will be a first person who acquired *phronêsis* without first being habituated. The former option is problematic because we can never be sure that those who are doing the habituating have gotten it right—the cycle of habituation could simply be relative to a particular society, which flies in the face of Aristotle's claim that there are objective moral standards.[21] The latter option is problematic because Aristotle leaves no room in his theory for acquiring *phronêsis* in the absence of character habituation.

5. CONCLUSION

While it is not entirely clear what *phronêsis* is, how we acquire it, or what it enables us to perceive, it is clear that Aristotle's *phronimos* was accorded a special and powerful place in morality and politics. Given the prominent and influential role granted to such a moral expert, and given that some individuals—particularly in the field of bioethics—have looked to adapt Aristotle's views to the circumstances of our own time, we would do well to be acquainted with the virtues and vices of Aristotle's account of the moral expert.

NOTES

* I would like to thank Khalil S. Khan and Fred D. Miller, Jr., for providing valuable feedback on an earlier draft of this essay, and Maureen C. Kelley and Lisa Rasmussen for providing me with the opportunity to work on this project.

1. The *phronimos* is the practically wise man. The virtue that the *phronimos* possesses is *phronêsis*, which is practical wisdom. Irwin translates *ho phronimos* as 'the prudent man' and *phronêsis* as 'prudence'. I shall retain the ancient Greek terms, and understand their translations to be as I describe here. 'Prudence' has too many contemporary connotations—such as shrewdness or being economical—to convey accurately what *phronêsis* means.

2. *Nicomachean Ethics* will hereafter be referred to as *NE*. All translations of the *NE* are Irwin's unless otherwise noted.

3. *Newsweek*, May 6, 2002. The question was posed as a possible way of going about how to decide what to do with Catholic priests who are accused of sexual abuse of children.

4. Casuists (Jonsen and Toulmin, 1988, pp. 19, 304-43; Brody, 1988, pp. 77-79, 94-96, 100-236; Toulmin, 1981, pp. 37-38; Toulmin, 1994, pp. 311, 314-15, 318), principlists (Beauchamp and Childress, 1994, pp. 62-69, 468-69, 490-93), and intuitionists—either indirectly through Ross (1930) (Gillett, 1994) or directly (Brody, 1988, pp. 12-14, 35-42)—alike who work in the field of bioethics lay claim to being influenced by Aristotle. There are some virtue ethicists (Hursthouse, 1991, 1995; Drane, 1994) generally who work on issues in bioethics. Additionally, there are Catholic thinkers (Wallace, 1989) who work in bioethics and have been either directly influenced by Aristotle or indirectly influenced by him through St. Thomas Aquinas.

Casuistry rejects formulating moral theories/principles and then applying them uniformly to cases. Toulmin (1994, pp. 310, 318) explains that it is "the direct analysis of cases" and has us "stay close to the particular insights we accumulate in the course of our pedestrian, concrete practical experience." We may be able to generalize to a certain extent about types of cases that have similar features, but additional experience and changes in circumstances could affect the moral assessment of what to do.

Churchill (1994, p. 321) states that *principlism* generally is "the application of antecedently derived principles to problem cases." Childress (1994, p. 73) offers a more nuanced understanding of principlism: It refers to "various principles-oriented approaches" where "general moral norms [are viewed] as central in biomedical ethics" and applied to cases in order to specify the appropriate action. These principles or rules can be known either through some reasoning process or through intuition (see below).

Brody (1988, p. 12) explains how *intuitionism* comes in two varieties: (1) "fundamental moral intuitions are indubitable and evident intuitions of the truth or falsity of purported moral rules," which are then applied to cases, or (2) "fundamental moral intuitions are judgments about the rightness or wrongness of particular actions." Sometimes, intuitionism is described as a "moral sense" theory, where we come to know what is right much in the way that we use our eyes to see what objects exist. McDowell (1979, repr. 1999, pp. 122, 140) states that the "sensitivity is, we might say, a sort of perceptual faculty" that is activated properly "by being a certain sort of person: one who sees things in a certain distinctive way."

5. I shall return in Section II below to a discussion of how the virtue of courage can be manifested differently by the women in this set of cases.

6. Some neo-Aristotelian natural philosophers and Catholic moral theologians rely heavily on this natural teleology in developing "natural norms for medical ethics" (Wallace, 1989, p. 47). Female circumcision is an example of a medical procedure that they would condemn. The clitoris is a natural part of female anatomy, and serves a function in relation to the overall organism, namely, to make sexual activity more pleasurable and the female thus more likely to reproduce and perpetuate the species. Interfering with this function would go against nature and thus be wrong.

7. This is usually translated as 'happiness' or 'flourishing', but a fuller understanding of what the term conveys might be something like 'the good fortune of living well'.

8. Elsewhere, Aristotle divides the human soul into four parts, because he subdivides the rational part. This division will be discussed in Section IV below in relation to *phronêsis*.

9. The Greek word for 'virtue', *aretê*, also connotes 'excellence'. Also, Aristotle's term *spoudaios* means 'excellent' as well as 'virtuous'.

10. This set of cases is oversimplified as presented. There certainly could be (and often are) other particularities of abortion cases that affect the final judgment about what one should do.

11. This general claim is not precise and it is quite contentious, as will become clear below in the discussion of *phronêsis*.

12. Aristotle (*NE* I 8, 1098b14-15) states: "We say that the goods of the soul are goods most fully, and more than the others [i.e., external goods and goods of the body]."

13. This should not be entirely surprising, because the ancient Greeks often believed that the moral and political realms should overlap to a much greater degree than political liberals believe they should today.

14. This is because of Aristotle's elaborately developed account of the political nature of humans in his *Politics* (see, especially, *Politics* I 1-2, III 9, and VII-VIII, trans. Lord).

15. Irwin (1999, p. 249) notes that this passage could possibly be interpreted differently: "It is grammatically possible, though implausible, to take 'what is expedient . . .' rather than 'the end' to be the antecedent of 'about which' (so that Aristotle does not affirm that [*phronêsis*] is correct supposition about the end)." Irwin (1999, p. 249) claims that his is the more plausible interpretation because he thinks that the phrase 'promoting the end' has a "broad scope": "As a result of deliberating about what promotes happiness, we discover its constituents, and so we have a more precise conception of happiness. . . . Deliberation both precedes this true conception of the end and follows it (since a fairly precise conception of happiness is the basis for further deliberation about what to do)."

16. One can be young in two ways, according to Aristotle: "young in years or immature in character" (*NE* I 3, 1095a6-7).

17 .This is the plural of *phronimos*.

18. It should be stressed that not all virtue ethicists who integrate virtue ethics into bioethics sharply separate the moral expert from the medical expert (as Aristotle does). Some recommend that "professional ethicists" (usually, those who hold degrees in moral and/or applied philosophy) serve on hospital ethics committees, while others hope to inculcate virtuous habits and ways of thinking into medical professionals themselves.

19. Some of them would not exist, though, depending on how one interprets Aristotle. For example, the regress problem discussed in the text below would not arise in the way discussed if Aristotle is an intuitionist. I thank Fred Miller for bringing this point to my attention.

20. This assumes that what is occurring between the two events is causation rather than correlation or coincidence.

21. This problem causes the problem of identifying the *phronimos* to resurface, despite Aristotle's dialectical method of justifying moral principles. This is because the source of the various *endoxa* that serve as starting points in dialectic is habituation. Of course, one has to start from *somewhere*, but Aristotle's set of *endoxa*—especially since it allows in the *endoxa* of the many—is dependent upon the society in which a dialectical discussion occurs.

REFERENCES

Beauchamp, T. & Childress, J. (1994, 4th ed.). *Principles of Biomedical Ethics.* New York: Oxford University Press.

Bostock, D. (2000). *Aristotle's Ethics.* Oxford: Oxford University Press.

Broadie, S. (1991). *Ethics with Aristotle.* New York: Oxford University Press.

Brody, B. (1988). *Life-and-Death Decision Making.* New York: Oxford University Press.

Childress, J. (1994). 'Principles-Oriented Bioethics.' In: E. DuBose, R. Hamel, and L. O'Connell (Eds.), *A Matter of Principles? Ferment in U.S. Bioethics* (pp. 72-98). Valley Forge, PA: Trinity Press International.

Churchill, L. (1994). 'Rejecting Principlism, Affirming Principles.' In: E. DuBose, R. Hamel, and L. O'Connell (Eds.), *A Matter of Principles? Ferment in U.S. Bioethics* (pp. 321-31). Valley Forge, PA: Trinity Press International.

Drane, J. (1994). 'Character and the Moral Life: A Virtue Approach to Biomedical Ethics.' In: E. DuBose, R. Hamel, and L. O'Connell (Eds.), *A Matter of Principles? Ferment in U.S. Bioethics* (pp. 284-309). Valley Forge, PA: Trinity Press International.

Foot, P. (2001). *Natural Goodness.* New York: Oxford University Press.

Gillett, G. (1994). 'Killing, Letting Die, and Moral Perception.' *Bioethics, 8,* 312-28.

Goldman, A. (1999). *Knowledge in a Social World.* Oxford: Clarendon Press.

Hursthouse, R. (1991). 'Virtue Theory and Abortion.' *Philosophy and Public Affairs, 20,* 223-46.

Hursthouse, R. (1995). 'Applying Virtue Ethics.' In: R. Hursthouse, G. Lawrence, and W. Quinn (Eds.), *Virtues and Reasons* (pp. 57-75). Oxford: Clarendon Press.

Irwin, T. (Trans.) (1999, 2d ed.). *Aristotle: Nicomachean Ethics.* Indianapolis: Hackett Publishing Company, Inc.

Jonsen, A. & Toulmin, S. (1988). *The Abuse of Casuistry: A History of Moral Reasoning.* Berkeley: University of California Press.

Lord, C. (Trans.) (1984). *Aristotle: The Politics.* Chicago: University of Chicago Press.

McDowell, J. (1979, repr. 1999). 'Virtue and Reason.' In: N. Sherman (Ed.), *Aristotle's Ethics: Critical Essays* (pp. 121-43). Lanham, MD: Rowman & Littlefield Publishers, Inc.

Reeve, C.D.C. (1995). *Practices of Reason: Aristotle's Nicomachean Ethics.* Oxford: Clarendon Press.

Ross, W. D. (1930). *The Right and the Good.* Oxford: Clarendon Press.

Sherman, N. (1991). *The Fabric of Character.* New York: Oxford University Press.

Toulmin, S. (1981). 'The Tyranny of Principles.' *Hastings Center Report, 11,* 31-39.

Toulmin, S. (1994). 'Casuistry and Clinical Ethics.' In: E. DuBose, R. Hamel, and L. O'Connell (Eds.), *A Matter of Principles? Ferment in U.S. Bioethics* (pp. 310-18). Valley Forge, PA: Trinity Press International.

Wallace, W. (1989). 'Nature and Human Nature as the Norm in Medical Ethics.' In: E. Pellegrino, J. Langan, and J. Harvey (Eds.), *Catholic Perspectives on Medical Morals* (pp. 23-53). The Netherlands: Kluwer Academic Publishers.

CHAPTER 4

CHRISTOPHER TOLLEFSEN

HUME ON TRUE AND FALSE PHILOSOPHY

1. INTRODUCTION

What would David Hume's reaction be to the contemporary phenomenon that I shall call expertise in the moral sciences, that is, expertise in law, morality and politics? Claims to such expertise are common enough in our day, with Constitutional scholars providing expert testimony on a variety of legal issues, bioethics experts serving as consultants to hospitals and biotech companies, and expert political theorists weighing in on all manner of social and political controversies. The answer to this question will certainly vary according to the interpretation of Hume. According to an early to mid-twentieth century picture, Hume was a forerunner of the logical positivists. On this picture of Hume, there would presumably be little room for such expertise. In the area of moral claims, for example, there would be no room at all, for moral claims are, as Ayer argues in defense of the positivist position, meaningless (Ayer, 1952). Law and politics would likewise suffer from reductive analysis, leaving little possibility for expertise.

More recent approaches to Hume's moral psychology and ethics would perhaps leave more room for expertise. "Reason is and ought to be," Hume famously says, "the slave of the passions" (Hume, 1973, p. 415). On a common reading of Hume's moral psychology based on this and other passages, reason has no room for deliberation about ends, but only about the best means to whatever ends are set by the passions: "Reason, being cool and disengaged, is no motive to action, and directs only the impulse received from appetite or inclination, by showing us the means of attaining happiness or avoiding misery" (Hume, 1983, p. 88). On this view, one might not be an expert vis-à-vis one's passions, but clearly, assessment of the rationality of means can be more or less developed. So perhaps a moral expert will be one who has a particularly acute awareness of the best means to his ends.

Such approaches to the question of expertise seem, unfortunately, somewhat limited, by addressing only a rather small aspect of Hume's work – his moral psychology, say, or his expressivism. But there is another picture of Hume's comprehensive view of philosophy that has emerged recently in the writings of Donald Livingston (Livingston, 1998). In this picture, it remains true that the passions are the source of human motivation. But there are radical differences

L. M. Rasmussen (Ed.), Ethics Expertise: History, Contemporary Perspectives, and Applications, 55-71.

between the ways in which philosophy can approach, theoretically and practically, our world and our selves. It is here, in the distinction between false philosophy and true that, I suggest, we should see resonance with the contemporary phenomenon of moral, legal, or political expertise. The resulting answer to the initial question will be, accordingly, richer and more complex.

Hume's account of the nature of philosophy is complex, for it includes an account both of false, and of true philosophy, the latter being much rarer than the former. It is the thesis of this paper that Hume's diagnosis of false philosophy is such that he would consider many instances of contemporary expertise – moral, political, or legal – as manifestations of false philosophy. To make this case, Hume's picture of false philosophy, and his account of the transition from false to true philosophy, must be explained. In depicting this transition, I illustrate Humean false philosophy with three examples, thus showing how the Humean category of false philosophy, and the modern category of "expertise", significantly overlap. I then discuss Hume's conception of true philosophy. The true Humean philosopher might, I suggest, also make a kind of claim of expertise. However, this claim will be considerably different from the claim of the false philosopher-expert. The true philosopher, and hence the true expert, will be seen to be a sort of rhetorician. I conclude the paper with some critical considerations about Hume's understanding of false and true philosophy, and hence about the Humean understanding of practical "expertise."

2. THE SUBVERSION OF REASON

When Hume retired to France in his early twenties to write the *Treatise of Human Nature*, it was, apparently, with a single guiding idea. Against the rationalists of his day, Hume was convinced "that all our reasonings concerning causes and effects are deriv'd from nothing but custom; and that belief is more properly an act of the sensitive, than of the cogitative part of our natures" (Hume, 1973, p. 183). This dichotomy, Hume believed, between reason and passion, and between a radically autonomous rationalistic philosophy and "custom" or "common life" was not the province of theoretical reason alone. "There has been a controversy started of late…concerning the general foundation of MORALS; whether they be derived from REASON, or from SENTIMENT" (Hume, 1983, p. 13). While Hume does not eliminate reason's role in morals, it must be seen as grounded in sentiments, and shaped by custom and common life.

The effect of this insight, in theoretical and practical reasoning, was to be, for Hume, liberating. Philosophy carried out under the misconception that abstract reasoning was the source of belief or action was not simply wrongheaded, but potentially dangerous and divisive. For Hume, this was especially the case when false philosophy came to be allied with religion – a circumstance Hume thought widely obtained. What Hume came to call "false philosophy," by attempting to transcend custom and sentiment, an impossible task, could only end up seizing covertly upon some principle of common life, and transforming it into an overarching principle of reason intended to rule over common life. This false philosophy was for Hume a kind of pathology, a spiritual disorder. True philosophy, by contrast, for Hume, was "nothing but the reflections of common life, methodized and corrected" (Hume, 1977, p. 112). When the true philosopher returned to common life, he would leave behind a scene "full of obscurities and perplexities…an ocean of doubt, uncertainty, and contradiction" (Hume, 1977, p. 69).

Book One of Hume's *Treatise*, and especially Part IV of that work, reveals, as Livingston has argued, how a true philosophy emerges from the self-destructive dialectic of false philosophy. False philosophy, if honest, eventually generates scepticism; scepticism, if honest, mandates a return to common life.

Hume traces the sceptical path of philosophy in several sections of Part IV of Book One. In each case, we see the pretensions of a false philosophy, detached from sentiment and custom, attempting to find an autonomous standpoint. In each case, however, pure reason undermines itself; the rationalist philosopher ends up subverting his own attempt to transcend sentiment, imagination and custom, while covertly relying on these very phenomena. Thus, in Section I, "Of scepticism with regard to reason," Hume considers the possibility that belief is founded upon reason. Yet, considered as a cause of belief, reason is plainly fallible, capable of going wrong in various ways. "By this means, all knowledge degenerates into probability; and this probability is greater or less, according to our experience or the veracity or deceitfulness of our understanding, and according to the simplicity or intricacy of the question" (Hume, 1973, p. 180).

The natural response to this fallibility is to double check, and to break down complex reasoning into simpler components, to verify accuracy. But in this way the number of fallible judgments actually increases, all of which judgments themselves need to be checked. Hume writes,

> But this decision, tho' it shou'd be favorable to our preceeding judgment, being founded only on probability, must weaken still further our evidence, and must itself be weaken'd by a fourth doubt of the same kind, and so on *in infinitum*; till at last there remain nothing of the original probability, however great we may suppose it to have been, and however small the diminution by every new uncertainty (Hume, 1973, p. 182).

In this way, Hume thinks, reason subverts itself. Were all our belief founded on reason, we would thus be incapable of belief. With regard to reason, scepticism rules.

Yet clearly we are not reduced to sceptical doubt every time we give our consideration to some matter or other; the conclusion Hume draws is that belief is not, after all, founded on reason, but on custom, the shaping of our natural sentiments in common life. Belief is natural to us, in a way that the pressing of pure reason to its logical conclusion is not.

We here see a theme common in Hume: in direct proportion as reason becomes more "dogmatical", it inevitably ends in scepticism:

> The sceptical and dogmatical reasons are of the same kind, tho' contrary in their operation and tendency; so that where the latter is strong, it has an enemy of equal force in the former to encounter…'Tis happy, therefore, that nature breaks the force of all sceptical arguments in time (Hume, 1973, p. 187).

This dialectic, by which philosophy seeks to transcend "the vulgar" mode of thought is repeated several times in Book IV. In Section Two, Hume traces the development of a philosophical, but unsupportable, view of the objects of our senses. The philosophical problem arises through reflection on a contradiction that common sense manages to conceal. Hume, with the rest of the modern philosophers, believes that our primary access as sentient creatures is to "perceptions." But perceptions are clearly dependent upon us, and existent for only as long as we are having them. Whence then, Hume asks, comes the vulgar's idea of a continued and external existence?

The common sense idea that our perceptions have independent and continued existence is, thinks Hume, quite at odds with the deliverances of either sense or reason; all perceptions are internal as far as we can tell, and as we have no experience of our perceptions except when we are perceiving them, so we have no access to their independence either. Our belief in the continued independent existence of our perceptions is more the result of imagination, seeking to generate a coherence and consistency in what would otherwise be a welter of perceptions.

So far from being the work of reason, this belief in the continued existence of perceptions actually involves a contradiction, according to Hume, for we are forced to attribute identity to patch over the manifest diversity of our perceptions through time. But identity on Hume's account is strict: any difference in properties between two possibly identical objects suffices to rebut the claim of identity. Thus the contradiction: we attribute, in an act of the imagination, identity to two perceptions – apple X at t and apple Y at $t + n$, precisely because we are faced with diversity.

Nonetheless, this act of imagination is very natural to us; indeed, the vulgar are not troubled at all by what Hume calls a "manifest contradiction". It is only with the onset of philosophical reflection that the falsity of the common sense position becomes obvious. Simple experiments serve initially: we press our fingers to our eyes and "immediately perceive all the objects to become double" (Hume, 1973, p. 210). Obviously, two objects do not come into being, so it is likewise, Hume thinks, obvious that "the doctrine of the independent existence of our sensible perceptions is contrary to the plainest experience" (Hume, 1973, p. 210).

The philosophical response – found throughout much of modern philosophy – is to now distinguish between perceptions and objects, a distinction Hume thinks alien to the vulgar. Philosophers now accept that the existence of their perceptions is dependent upon their own organs, but attribute continued and independent existence to what they call "objects." Where before there was one reality, there are now two.

Again, however, the philosophical position cannot stand; Hume calls it the "monstrous offspring of two principles, which are contrary to each other" (Hume, 1973, p. 215), namely, the desire of the imagination for continued existence of perceptions, and the verdict of reason that perceptions have no such independent existence. The result is again scepticism, in this instance, in the form of a crisis of confidence in the senses: "I feel myself *at present*…inclined to repose no faith at all in my senses, or rather, imagination, than to place in it such an implicit confidence" (Hume, 1973, p. 217). Hume hints at the end of this passage what an appropriate response to this scepticism is ("whatever may be the reader's opinion at this present moment,…an hour hence he will be persuaded that there is both an external and an internal world" (Hume, 1973, p. 218)), but returns to it more fully later in Part IV. At the same time, this passage gives some indication of the anxiety produced by false philosophy, an anxiety that eventually led Hume himself to something like a nervous breakdown.

A similar dialectical subversion of philosophy's attempt to transcend vulgar opinion is presented in Sections III and IV, on ancient and modern philosophy, respectively. In the former section, the doctrine of matter, substance and accident is attacked; in the latter, the distinction between secondary and primary qualities is shown to generate the view that primary qualities are just as unreal as secondary, since they are known only through the supposedly illusory secondary qualities. Section III is notable for containing the following explicit picture of the ascent, as it were, of philosophy:

> ...we may observe a gradation of three opinions, that rise from each other, according as the persons, who form them, acquire new degrees of reason and knowledge. These opinions are those of the vulgar, that of false philosophy, and that of the true; where we shall find upon inquiry that the true philosophy approaches nearer to the sentiments of the vulgar, than to those of mistaken knowledge (Hume, 1973, pp. 222-223).

3. THE NATURE OF FALSE PHILOSOPHY

We need here to ask three questions. First, what is it that is characteristic of the operations of false philosophy? Second, and third, how does Hume think that scepticism allows for a restoration of sanity, and what is involved in that restoration?

False philosophy is generated by philosophical reflection on assumptions of common life, those assumptions which are shared widely by the vulgar, and which contribute to their common form of life. But while Hume thinks these assumptions inescapable, and also in a certain way innocent, it is also the case that that the contradictions which give rise to philosophical reflection really exist in common life. Although Hume's own reasoning, in regards to the continued independent existence of our "perceptions", might be suspect and overly indebted to the Cartesian picture of sensation, Hume seems genuinely to think that common sense conceals a latent contradiction.

Moreover, the philosophical impulse is in some way or other natural; at certain times – when "tir'd with amusement," Hume writes, "I feel my mind all collected within itself, and am naturally *inclin'd* to carry my view into all those subjects, about which I have met with so many disputations" (Hume, 1973, p. 271). This natural inclination, as Hume describes it, is for the discovery of principles, whether in the origins of government, of morals, and so on.

Where philosophy seems to go wrong is in its positive assumption that it can solve these latent conflicts by introducing some or other fixed principle, which is then treated as an absolute starting point, an Archimedean point. Philosophy claims for its Archimedean points the authority of pure reason; the opinions of the vulgar are to be transcended, and nothing accepted from them. Philosophy thus seeks to exert a form of dominion over common life. But Hume's arguments in the *Treatise* and elsewhere proceed by showing that philosophy's principled solution to the difficulties of common life is no more securely grounded in reason than is common life, and considerably less securely grounded in other ways. The Double Existence solution to the problem of the continued and independent existence of our perceptions is guilty on several counts. It has no support in reason, for we have no experience of objects independent of our perceptions; and it is unstable, as it tries to force us out of our ordinary mode of belief. Even the philosopher cannot maintain this position outside of the reflective mode. It is natural that, seeing no solution in pure reason to the problems unearthed by reflection, an honest philosopher should be driven to a sceptical moment, in which he despairs of any solution.

In a less honest moment, however, false philosophy conceals from itself its radical insufficiency. Hume is especially sensitive to dogmatism in philosophy, and seems to think that this is a form of false consciousness. Yet false philosophy cannot even convince its own practitioners unless it relies in some way upon common life. Thus, philosophers conceal from themselves the instability of their views on the independence of perceptions only because their doctrines appeal to the imagination – the faculty which has allowed the vulgar to paper over the disruption of sense they experience every time they close their eyes or turn their heads. Indeed, false philosophy's principles seem largely to emerge by theft of the principles of common life, and transformation of those principles from guides to life into abstract, autonomous and authoritative rules. False philosophy is a totalizing and almost totalitarian phenomenon; seizing upon some aspect of common life, it grants it supreme status, and refuses to recognize any other principle: "When a philosopher has once laid hold of a favorite principle, which perhaps extends to many natural effects, he extends the same principle over the whole creation, and reduces to it every phenomenon, though by the most violent and absurd reasoning" (Hume, 1985, p. 159).

It will be helpful, for reflecting on Hume's likely response to the phenomenon of expertise in the moral sciences, to identify some instances of false philosophy at work in morals and politics. Of the following, the first example is taken from Hume, while the second and third are contemporary. It should be clear, however, that Hume would see in these examples the abstractness and false claims to authority characteristic of false philosophy.

In his own day, Hume was concerned that the new contract theory of government displayed an unhealthy adherence to principle over custom, and a false consciousness. The latter suspicion is displayed in the introduction to Hume's essay "Of the Original Contract:"

> As no party, in the present age, can well support itself, without a philosophical or speculative system of principles, annexed to its political or practical one; we accordingly find, that each of the factions, into which this nation is divided, has reared up a fabric of the former kind, in order to protect and cover that scheme of actions, which it pursues (Hume, 1985, p. 465).

Hume finds two such principles playing a dominant role in English political thought: that of the divine origin of government, and that of the original contract. Hume allows that both views are not without merit: he does not deny the origin in the divine of everything, nor does he deny that men originally came together in society from mutual consent, given the necessity of living together in some sort of rudimentary society. But:

> ...philosophers, who have embraced a party...are not contented with these concessions. They assert, not only that government in its earliest infancy arose from consent, or rather the voluntary acquiescence of the people; but also, that, even at present, when it has attained full maturity, it rests on no other foundation. They affirm that all men are still born equal, and owe allegiance to no prince or government, unless bound by the obligation and sanction of a promise (Hume, 1985, p. 469).

Hume argues at some length that this view is at odds with evident facts. Throughout history, Hume finds princes demanding obedience and subjects giving it without either party believing that consent is necessary. Moreover, and again throughout history, the pattern of government established through usurpation and conquest has played itself out. In such cases, early submission to authority is a result of "fear and necessity," while later it is from subjects becoming accustomed to their rule. Again Hume finds little evidence of consent, an absence mirrored in philosophy in which, except in Plato's *Crito*, the contract theory is not found until modern times. Hume concludes: "If scarce any man, till very lately, ever imagined that government was founded on compact, it is certain that it cannot, in general have any such foundation" (Hume, 1985, p. 487).

Not only is the doctrine unsupported, but it has potentially pernicious consequences, for it renders what Hume considers clearly legitimate governments, in particular absolute monarchies, illegitimate. Hume's own account of the legitimacy of authority and obedience is, by contrast, less speculative: it is the very existence of society which requires obedience, and the danger of societal collapse that gives particular regimes a presumption of authority:

> The general obligation, which binds us to government, is the interest and necessities of society; and this obligation is very strong. The determination of it to this or that particular prince or form of government is frequently more uncertain and dubious. Present possession has considerable authority in these cases, and greater than in private property; because of the disorders which attend all revolutions and changes of government (Hume, 1985, p. 486).

Three points are worth making here. The first is to draw attention to Hume's description, immediately after this passage, of the relationship between the authority of custom and the common sentiments, on the one hand, and the form of speculative moral philosophy under discussion. With regard to the deliverances of "general opinion," Hume writes:

> ...in all questions with regard to morals, as well as criticism, there is really no other standard, by which any controversy can ever be decided. And nothing is clearer proof, that a theory of this kind is erroneous, than to find, that it leads to paradoxes, repugnant to the common sentiments of mankind, and to the practices and opinions of all nations and all ages (Hume, 1985, p. 486).

We see here Hume's opposition to the removal of philosophy from the authority of the opinions and practices of the vulgar, a removal that he has already accused, in the paragraph from the beginning of the essay quoted above, of being a "cover" for a particular faction's "scheme of action."

The second point to make is that although common life and custom has this authority, this does not, again, mean that this authoritative practice is without ambiguity or unclarity—"uncertain, ambiguous, and arbitrary," Hume calls these rules (Hume, 1985, p. 482). It is thus natural, in a way, for philosophical reflection to find this state of affairs distressing, and seek some more fixed point for the arbitration of political or moral questions.

This raises the third point, which is that Hume is not calling for a complete quietism about these issues. The uncertainty, ambiguity and arbitrariness of the rules regarding private property, or the rightful succession, are genuine problems, recognized by the true philosopher, as well as by the false. While more modest, in refusing to overthrow the principles of common life, the true philosopher still has the task of "methodizing and correcting." I return to this below.

First, however, I wish to introduce two further possible examples of what Hume might consider false philosophy, taken from the contemporary moral and political scene. The first is the primacy of utilitarianism among those who act as ethical consultants and theorists; the second is the strong tendency of the courts, seen in the Ninth U.S. Circuit Court of Appeals pledge ruling, and the Supreme Court's dissenting opinion of the constitutionality of vouchers, to erect a rigid "wall of separation" between Church and state.

Contemporary utilitarian and consequentialist thinking exercises a dominant rule, at present, over much academic moral theory, as well as over much applied ethics. The influence in applied ethics in turn spills over into the practice of ethical consultation, whether in hospitals, research institutions, or governmental panels. Its reach is long (see Oderberg and Laing, 1997, for discussion of the dominance of consequentialism in this context.)

Utilitarians are thus first among those who, in practical life, assert a claim of moral expertise. This expertise is both theoretical and technical; utilitarians are "expert" both inasmuch as they have the true moral theory, and inasmuch as they can divine the consequences and the balance of good over evil that those consequences will have in a proposed course of action.

This state of affairs seems to me to meet all the marks of Hume's conception of false philosophy. There is no doubt that consequentialist thinking plays a role in common sense and common life. This role has of course increased since the advent of ultitarianism as a dominant mode of moral theory, but even apart from this, consideration of the consequences is a typical concern among human agents.

Clearly, however, it is not the only concern. Consequentialist theorists themselves recognize that a variety of non-consequentialist concerns permeate common morality, some of which seem obviously in conflict with strict consequentialism. In accordance with the direction of a Humean false philosophy, however, consequentialists typically are not bothered by this gap. Rather, consequentialists such as Peter Singer and Shelly Kagan urge that it is high time we give up our ordinary or common sense morality, with, for example, its notion of the sanctity of human life, and embrace the radically impartial and highly demanding consequentialist ethic (see Singer, 1993 and Kagan, 1989). Consequentialism thus detaches from common life one part of its morality, assigns to it the status of supreme rational principle, and demands that common life conform to this principle or be condemned as irrational.

It would take us too far afield here to reiterate the variety of ways in which consequentialism has been found theoretically wanting. The problems of the incommensurability of goods (Grisez, 1978; Finnis, 1980; Raz, 1986), of failure to treat persons as individuals (Rawls, 1972), of the incompatibility of utilitarianism with freedom (Finnis, 1991), and others remain difficulties which utilitarianism has not, in my view, successfully answered.

Moreover, there is an abundantly clear practical problem with consequentialism, and with all claims of expertise based upon it, namely that of ignorance of consequences (Donagan, 1974). It thus seems reasonable to infer that Hume would have considered consequentialism, and any claim of moral expertise erected entirely upon a consequentialist foundation, to be a species of false philosophy.

Something like Hume's conception of false philosophy can perhaps also be seen at work in a recent event, the Ninth U.S. Circuit Court of Appeal's ruling that it is unconstitutional for the pledge of allegiance to be said in schools in the morning (Ninth U.S. Circuit, 2002). This followed shortly after a Supreme Court ruling that school vouchers which allowed poor students to attend private, including parochial schools was not unconstitutional. But the dissenting judges, in a 5-4 decision, argued strenuously for much the same position as was taken in the 9th Circuit case, namely, the strict separation of church and state.

As commentators have pointed out, the dissenting judges in this ruling held that this strict principle trumped all other considerations. Justice Souter wrote, "If there were an excuse for giving short shrift to the Establishment Clause, it would probably apply here. But there is no excuse. Constitutional limitations are placed on government to preserve constitutional values in hard cases, like these" (quoted in Berkowitz, 2002, p. 2).

Moreover, Peter Berkowitz has effectively argued that despite the Court's attempt to root its interpretation of the Constitution as mandating a strict separation, this interpretation is a relatively recent one, out of kilter both with tradition and traditional understanding of the Constitution (Berkowitz, 2002).

The pledge case likewise involves a wholesale rejection of common life – few court decisions in recent years have led to such a wide public outcry. But in the dissent in the vouchers case, and in the pledge case, what emerges is the elevation of a single abstract principle – in this case, the strict separation between church and state, into an absolute Archimedean point of Constitutional interpretation, given absolute authority over tradition and custom. Moreover, it seems possible that the strict separation interpretation is, taken to its logical limit, self-subverting. For it is of a piece with an understanding of liberalism which makes tolerance the supreme virtue, but allows no tolerance for religion. It is impossible here to pursue this line of thought, however. And finally, we see that, as with the utilitarian-as-expert model above, there is some reason to think that the legal expert who judges according to the rigid establishment of an abstract principle would for Hume be another instance of the false philosopher at work. (Lest it be thought that Hume would contest this analysis, given his apparent hostility to religion, two points may be made. First, the analysis is, in a way, content-neutral; it is the way in which the courts have made use of the separation principle that would be objectionable from a Humean point of view. And second, Hume actually would not be sympathetic to the content; a recurring theme of the *History of England* is Hume's opposition to the independence of the clergy from the state. See also Hume, 1995, pp. 512-529.)

4. FROM FALSE TO TRUE PHILOSOPHY

It is necessary at this point to return to Hume. Hume offers an image of false philosophy, but he also offers an account of the natural telos of false but honest philosophy: scepticism followed by a return to common life. We have seen that many claims of moral, or political, or legal "expertise" could very well be seen by Hume as instances of false philosophy. But if there is an image of true philosophy, and of the philosopher returned to custom and common life, then it might be possible to find a more benign form of expert, albeit one very different from the utilitarian or strict separationist.

How is it that the honest false philosopher becomes a sceptic? The answer is implicit in what has been said already: philosophy that begins by rejection of common life and elevation of abstract principle is inevitably, on Hume's account, a failure. When we are "necessitated by reasoning to contradict or depart from the primary instincts of nature, and to embrace a new system…philosophy finds herself extremely embarrassed" (Hume, 1977, p. 105). But this failure is often concealed by philosophers from themselves, just as they conceal from themselves the fact, as Hume sees it, that their own principles are nothing but "cover" for some bit of common life that they have smuggled into their system. The continuation of false philosophy can only come about through self-deception and dishonesty.

Hume's exercises in Part IV of the *Treatise* might thus be seen as attempts to shake the false philosopher out of his self-deceptive complacency; systems based upon abstract principle fail when reason is pressed. The result, faced honestly in each case, is a moment of scepticism: scepticism regarding reason, scepticism regarding the senses, and so on. Or again, in the conclusion, "We have, therefore, no choice left but betwixt a false reason and none at all" (Hume, 1973, p. 268).

So scepticism is the natural outcome of the subversion of reason. The big question is, what is the natural outcome of scepticism? By natural here is not simply meant "expected, in the natural course of things," but also, for Hume, "normatively, to be expected." Hume has frequently been interpreted as holding to a dangerous scepticism that is nothing but corrosive, an interpretation which came to haunt him when he applied for the chair of Ethics and Pneumatic Philosophy at the University of Edinburgh. Hume devotes a section of his "Letter from a Gentleman to His Friend in Edinburgh," not to rebutting the charge of scepticism, but to rebutting the charge that scepticism has noxious consequences. The aim of scpeticism is not universal doubt, he writes, but "Modesty and Humility, with regard to our natural Faculties, [that] is the Result of Scepticism" (Hume, 1977, p. 116). As Hume points, out, in the *Treatise*, when faced with his own attack on reason, his moment of scepticism is referred to as a "Philosophical melancholy and Delusion" (Hume, 1977, p. 117). It is a self-reflexive moment: "nothing can be more sceptical, or more full of doubt and hesitation, than this scepticism itself" (Hume, 1977, p. 109).

The natural outcome of Hume's scepticism, in both senses of natural above, is this: "'Tis evident, that so extravagant a Doubt as that which Scepticism may seem to recommend, by destroying every Thing, really affects nothing, and was never intended to be understood seriously, but was meant as a mere Philosophical Amusement, or Trial of Wit and Subtilty" (Hume, 1977, p. 117).

Because of the context in which wrote this Letter, namely, as a defense against the charge of atheism, it is not surprising that the province of custom and common life to which Hume returns after the sceptical moment is one permeated by religious belief. For example:

> In Reality, whence come the various tribes of hereticks...but from too great a Confidence in mere human Reason, which they regard as the Standard of every Thing, and which they will not submit to the superior Light of Revelation? And can one do a more essential Service to Piety, than by showing that this Boasted Reason of theirs, so far from accounting for the great Mysteries of the Trinity and Incarnation, is not able fully to satisfy itself with regard to its own Operations, and must in some Measure fall into a kind of implicite Faith, even in the most obvious and familiar principles (Hume, 1977, p. 118).

But this "implicite Faith" in obvious and familiar principles is not strictly a religious faith. This is consistently that to which Hume returns after his destructive reasonings; there is no abstract principle by which all questions of succession, or property, may be settled, nor by which the existence of external objects or even of a creative divinity can be proved. But these latter two beliefs are natural to us and cannot be eroded by sceptical arguments, and in common life we have ways and means of negotiating among the various principles that determine rightful succession and just transfer of property. It is an act of an overwhelming pride to throw over these aspects of common life, and the sceptical moment serves to persuade the honest philosopher of this truth.

What then, of the true task of philosophy? I believe that Hume presents us with a complex picture, which includes a negative and a positive component.

The first, negative, side is seen throughout Hume's writings, but clearly in the *Treatise*. Hume believes there is a family resemblance between philosophy and religion, or "superstition". Both are a result of the natural tendency of the mind to move beyond what is legitimated by custom and experience. Superstition differs in two respects from philosophy, however. First, it is more all-encompassing, seeking not just ultimate principles, but opening up an entire "world of its own" (Hume, 1973, p. 271). Further, superstition seizes more violently upon the mind and is thus "often able to disturb us in the conduct of our lives and actions," while philosophy, even of the worst sort, is seldom able to "interrupt the course of our natural propensities" (Hume, 1973, p. 272). "Generally speaking, the errors in religion are dangerous; those in philosophy only ridiculous" (Hume, 1973, p. 272).

There is some evidence that Hume later came to think false philosophy more than merely ridiculous; certainly the alliance of false philosophy with religion in the religious wars in England was seen as a toxic combination, and Livingston argues that towards the end of his life, Hume was increasingly convinced that false philosophy posed a serious threat to society (Livingston, 1998, ch. 9). But one task of true philosophy, directed both at false philosophy and at superstition, is the constant abatement of pride that drives the construction of systems and theories. In the essay "On Superstition and Enthusiasm" Hume writes, "As superstition is a considerable ingredient in almost all religions, even the most fanatical…[there is] nothing but philosophy able entirely to conquer these unaccountable terrors" (Hume, 1985, p. 75).

In the *Enquiry Concerning Human Understanding* Hume describes a moderated, or academical scepticism which mitigates the violence of "dogmatical reasoners":

> But could such dogmatical reasoners become sensible of the strange infirmities of human understanding, even in its most perfect state, and when most accurate and cautious in its determinations; such a reflection would naturally inspire them with more modesty and reserve…The illiterate may reflect on the disposition of the learned…And if the learned be inclined…from their natural temper, to haughtiness and obstinacy, a small tincture of PYRRHONISM might abate their pride, by showing them that they are but inconsiderable, if compared with the universal perplexity and confusion, which is inherent in human nature (Hume, 1977, p. 111).

A "second species" of moderate scepticism, which here seems equivalent to true philosophy, has a more positive task, though a still limited scope, for it "confines itself to common life, and to such subjects as fall under daily practice and experience" (Hume, 1977, p. 112). Hume continues: "Those who have a propensity to philosophy, will still continue their researches, because they reflect, that, besides the immediate pleasure, attending such an occupation, philosophical decisions are nothing but the reflections of common life, methodized and corrected" (Hume, 1977, p. 112).

This view is, incidentally, Hume claims, supported by an enquiry into the principles of the human mind – the source of belief in experience and imagination, of morals in custom and sentiment – in short, it is supported by the sort of naturalistic investigation Hume conducts in the *Treatise* and the two *Enquiries*. For this investigation is deeply anthropocentric, leading us away from first principles in the things themselves—substance, independent existence, eternal and immutable relations of right—and towards the source of our beliefs, theoretical and practical, in the workings of our own sentiment and custom.

It should not, perhaps, be surprising, if there are limits to what can be said positively about this role of the philosopher as a methodizer and corrector of common life, if only because the nature of philosophy on this account is so explicitly less systematic and more modest than on many other accounts.

I want to suggest here that this second species of true philosophy, that which is concerned with common life, has a number of components. In what follows, I briefly describe several, though not exhaustively, showing the relationship of each to common life. I then describe the manner of the true philosopher in his approach to these concerns. I conclude with some critical reflections on Humean true and false philosophy.

A first form of the second species of true philosophy creates a sort of bridge from philosophy as the abatement of pride to philosophy as the methodizing and correcting of common life. In this role, described at length in Hume's essay "The Sceptic," the practice of philosophy serves, both directly and indirectly, to refine and moderate the passions generally, a moderating which is conducive to happiness.

Hume's essay is full of warnings as to the limits of philosophy's power in moderating the passions; and also, warnings that where philosophy is effective, it is often destructive of virtuous passions as well as ill-considered passions. But what I wish to draw attention to here is Hume's claim, placed in the mouth of the Sceptic, that philosophy will effectively "tranquilize and soften all the passions" (Hume, 1985, p. 177) by bringing forth considerations "drawn from common life" (Hume, 1985, p. 176).

"Philosophy" says Hume, "greedily seizes these [considerations], studies them, weighs them, commits them to the memory, and familiarizes them to the mind: And their influence on tempers, which are thoughtful, gentle, and moderate, may be considerable" (Hume, 1985, p. 177). Among these considerations of common life, to which philosophy draws our attention, are "When we reflect on the shortness and uncertainty of life, how despicable seem all our pursuits of happiness?" And "Is it not certain, that every condition has concealed ills? Then why envy any body?" These reflections play a role in a Humean "consolations of philosophy".

The second form of the second species of true philosophy is a constructive investigation into the principles at work *in common life*, as well as in habit, custom, and experience, of our understanding of the world, and our practical approach to the world. This form of investigation may be seen in Hume's positive account of the nature of justice.

The authority of common life and custom is seen in Hume's account of justice as an artificial virtue. To say that it is artificial is not to say, as Hume is at pains to explain in the "Letter to a Gentleman" that there is no, or only an arbitrary, difference between just and unjust, right or wrong. Like all moral distinctions for Hume, that between justice and injustice is founded upon natural sentiment. Unlike, however, compassion or generosity, justice requires that certain natural sentiments both shape and be shaped by society; thus he refers to the virtue of justice as both an artificial virtue, and as founded upon indirect passions.

Hume compares this artificiality to that of language, from which we can see the authority given to common life – language is not spoken by attributing authority to abstract principle, but by attributing it to the custom and tradition of the social group of which we are a part.

Hume's account of justice is complex, and cannot be given an adequate rendering here. The account begins with a natural sentiment – self-interest – and a natural condition, the instability of external goods. Tutored by living in familial society, men realize that their passion for acquiring and keeping external goods is best served by its own restraint in conjunction with the restraint of others. So rules are developed with a view to fixing, stabilizing external goods, and restraining those who would take from others. The enforcing of this system of restraints is, Hume claims, the primary function of government.

There is thus a first step of artificiality to justice: the rules of justice are neither previous matters of fact, nor eternal relations, but are factitious, generated by society for the sake of society's own preservation. Justice is not yet, for that reason, a moral virtue. A virtue, for Hume, is a disposition which gives an impartial, or detached spectator pleasure. It is this detached pleasure which counts as moral approbation, as opposed to the pleasure given by that which is to our personal advantage. Of course, our own justice is, generally, to our personal advantage, but not always; sometimes our own injustice benefits. But we are always aware of the harm injustice does to us, and this, Hume argues, is the foundation for our sympathetic "unease" at the harm injustice does to others, and to society. In turn, this frees the mind up to that detached view, whereby we find justice pleasing:

> We partake of their uneasiness by sympathy; and as everything, which gives uneasiness in human actions, upon the general survey, is call'd Vice, and whatever produces satisfaction, in the same manner, is denominated Virtue; this is the reason why the sense of moral good and evil follows upon justice and injustice…Thus self-interest is the original motive to the establishment of justice: but a sympathy with public interest is the source of moral approbation, which attends that virtue (Hume, 1973, pp. 499-500).

Thus the investigation into the nature and origins of justice forms a part of the true philosophy, seeking as it does for those origins in a human sentiment shaped by social life, rather than in abstract principle or eternal relations. This investigation is, of course, methodologically continuous with all Hume's investigations in the *Treatise* and the two *Enquiries*.

A third form of positive true philosophy may be described as the philosopher's taking his place in the world of affairs, entering into public conversation on matters of societal import. This task is importantly subsequent to the previous investigation, for without understanding of, e.g., the artificial nature of justice, the tendency to false philosophy will recur, and the imposition of abstract principle will reemerge as a temptation.

In "Of Essay Writing," Hume divides the "elegant" part of mankind into the Learned, and the Conversible, and laments that "The Separation of the Learned from the conversible World seems to have been the great Defect of the last Age" (Hume, 1985, p. 534). Hume writes of philosophy, under these conditions of divorce from the world of conversation, that it

> …went to Wrack by this moping recluse Method of Study, and became as chimerical in her Conclusions as she was unintelligible in her Stile and Manner of Delivery. And indeed, what cou'd be expected from Men who never consulted Experience in any of their Reasonings, or who never search'd for that Experience, where alone it is to be found, in common Life and Conversation (Hume, 1985, p. 534-5).

Hume describes his own goal in the Essay as that of being "a Kind of Resident or Ambassador from the Dominions of learning to those of Conversation" (Hume, 1985. p. 535). It should be clear by now that Hume does not think of this difference between the learned and the conversible as simply a difference of expression; rather, it reflects the methodological difference between a philosopher who elevates abstract principle and a philosopher whose considerations are grounded in experience and common life. Hume's ambassadorial work may be seen to good effect in a series of essays, "Of the Original Contract," "Of Passive Obedience," "Of the Coalition of Parties," and "Of the Protestant Succession."

In addition to attempting an answer at a number of vexed substantive questions in these essays, Hume also portrays the appropriate conduct of a philosophical investigation in describing how a true philosopher will approach the question of "the protestant succession," that is, the question of whether the succession of English rule should be fixed in the Hanover or Stuart line:

> It belongs, therefore, to a philosopher alone, who is of neither party, to put all the circumstances in the scale, and assign to each of them its proper poise and influence. Such a one will readily, at first, acknowledge that all political questions are infinitely complicated, and that there scarcely ever occurs, in any deliberation, a choice, which is either purely good, or purely ill. Consequences, mixed and varied, may be foreseen to flow from every measure: And many consequences, unforeseen, do always, in fact, result from every one. Hesitation, and reserve, and suspense, are therefore, the only sentiments he brings to this essay or trial. Or if he indulges any passion, it is that of derision against the ignorant multitude, who are always clamorous and dogmatical (Hume, 1985. p. 507).

We should note at the end that Hume's antipathy is towards the dogmatical attitude of the vulgar, which in this context is expressly formed by religious prejudice and superstition; it does not call into question his sense of the authority of common life in other respects.

What emerges from all this, I suggest, is a portrait of the true philosopher as a kind of rhetorician. The true philosopher must be eloquent and persuasive, sensitive to the *topoi* and *doxa* of common life. Hume's own affinity for Cicero places him in a tradition of high rhetoric in which eloquence and wisdom are of a piece with one another. Livingston writes in this connection,

> ...eloquence is a virtue of the true philosopher. The world of common life is a social world constituted by the communication of sentiments through sympathy. Eloquence is the art of discriminating among sentiments and communicating them. Since the true philosopher is a critical participant in custom, he must develop the art of eloquence (Livingston, 1998, p. 87).

Now the contemporary moral, political, or legal expert, as described above, seeks also to be an active and critical participant in the social world. But, I have argued that many recognizable forms of such expertise are more like Hume's false philosopher than the true, in their elevation of a single principle over common life, and authoritarian application of it to common life. It would seem that there can be no expectation that the "learned" would escape the snares of false philosophy, on Hume's account, save also by becoming conversible. "Experts", were Hume to evaluate them, would need to be true, not false philosophers, to warrant their role in the social world.

To illustrate this Humean view, one might see the field of bioethics, as it has developed over the last thirty-odd years, as a constant scene of the sort of struggle between true and false philosophy portrayed by Hume. As we have seen, utilitarian, or liberal-theoretic, approaches to bioethics have conformed in various ways to the pattern of false philosophy, by elevating a single abstract principle and imposing it from above. Moreover, the social role played by experts on presidential councils and commissions, or on more humble hospital committees, has often seemed to be removed from the opinions and ways of life of those they were supposed to serve, and inclined to a thinly disguised dominance over those lacking "expertise." And in its resolute insistence on problem *solving*, bioethics has often foresworn attention to the emotions and virtues, as well as any consolatory role philosophy might play.

And yet there have been considerable moves in the opposite direction: bioethics has in recent years seen a move to new attention to clinical practice, and to the mores and commitments of those who give, and those who are in need of, medical care, and a move away from abstract principles. Many bioethicists have recognized the need for more public conversation, rather than backroom decision making. And ethics generally, and bioethics in consequence, has seen renewed attention to the emotions and virtues, after a fashion that is arguably Humean. In thinkers such as Edmund Pellegrino, Leon Kass, or Wiliam May, one may perhaps see a new, more humanistic and "conversible" form of bioethics taking place, with a consolatory emphasis on the nature and limits of the human condition. Such thinkers seem indeed to strive to be ambassadors from the dominions of learning to those of bioethical conversation.

5. CRITICAL REFLECTIONS

There is more to be said concerning the Humean distinction between true and false philosophy than can be attempted here. And I have only provided the briefest of sketches above of *examples* of expertise, without attempting to diagnose the entire field. So I cannot claim to have shown that all those who would claim or be attributed expertise in the practical sciences would be best characterized as Humean false philosophers. Nor can I claim to have given anything like a full description of what a true philosopher/expert in the modern world of bioethics, or law, or politics, would look like. Nonetheless, with much still unsaid, I shall conclude with three critical reflections on Hume's distinction, and its possible application in the contemporary world. My considerations proceed in ascending order of severity.

First, we might be inclined to question the plausibility of the connection of eloquence and wisdom. Hume has a place in a tradition of thought about rhetoric, a tradition which spans Cicero as well as a number of Renaissance thinkers, which sees beauty and eloquence of speech as somehow conducive not simply to persuasion, but to truth. But we might well be sceptical of this tradition, given the ability of rhetoricians, like poets, to make, as Plato complains, the weaker argument appear stronger, and the unjust appear just.

Second, Hume's conception of common life, and true philosophy's connection to common life might be called into question by the radical disagreement present in common life today. Here an instructive comparison might be made between Hume, advocating a return to common life at large in consequence of a scepticism regarding reason, and contemporary thinkers such as H. Tristram Engelhardt. Like Hume, Engelhardt engages in sceptical reasoning to abate reason's pride, arguing that the various attempts to ground morality on a neutral foundation, established by pure reason, fail. Again, like Hume, Engelhardt advocates a return to the content-rich forms of community within which moral discourse can rise above merely formal constraints. Unlike Hume, however, Engelhardt recognizes the "common" social unit as considerably smaller, more local, and more religious, than Hume did (Engelhardt, 1996). For Engelhardt sees in larger units the same radical disagreements present in moral philosophy, and hence an inability to settle substantive moral disputes. It might be that there is, now, no longer any genuinely common life in the modern West.

And finally, just as one can question the efficacy of Engelhardtian attacks on reason, one can certainly question whether Hume's sceptical attacks in, e.g., the *Treatise*, are effective. Hume's arguments rely, notoriously, on an implausible atomism about experience (Anscombe, 1981), and a subjectivist view of perception, popular in post-Cartesian philosophy, but not, for that reason, philosophically attractive. Likewise, Hume's attacks on reason as the foundation of morals rely on a number of suspect arguments, and need not be taken as undefeatable.

In these three ways, though not, probably, all three ways at once, one might bring into question the picture of expertise that would emerge from a Humean understanding of true and false philosophy. Nonetheless, it seems that Hume does offer a plausible, if disputable, way of understanding the claims and pretensions of a certain style of public philosophy, whether moral, political, or legal. Hume should certainly remain in the forefront of our awareness and analysis when we consider the claims, and limits, of experts in the moral sciences.

REFERENCES

Anscombe, G.E. (1981). 'Times, Beginnings and Causes.' In: *The Collected Papers of G.E.M. Anscombe, II: Metaphysics and the Philosophy of Mind* (pp. 148-162). Minneapolis: University of Minnesota Press.

Ayer, A.J. (1952). *Language, Truth and Logic*. New York: Dover Publications.

Berkowitz, P. (2002). 'Liberals Versus Religion.' *The Weekly Standard* 7(42). Available at *http://www.weeklystandard.com/Content/Public/Articles/000/000/001/431slfjp.asp*.

Donagan, A. (1977). *The Theory of Morality*. Chicago: University of Chicago Press.

Engelhardt, H.T. (1996). *The Foundations of Bioethics*. Oxford: Oxford University Press.

Finnis, J. (1980). *Natural Law and Natural Rights*. Oxford: The Clarendon Press.

Finnis, J. (1991). *Moral Absolutes: Tradition, Revision, and Truth*. Washington, D.C.: The Catholic University of America Press.

Grisez, G. (1978). 'Against Consequentialism.' *American Journal of Jurisprudence* 23, 21-72.

Hume, D. (1973). *A Treatise of Human Nature*. L.A. Selby-Bigge (Ed.). Oxford: The Clarendon Press.

Hume, D. (1977). *An Equiry Concerning Human Understanding* with *A Letter from a Gentleman to his Friend in Edinburgh*. Eric Steinberg (Ed.). Indianapolis: Hackett Press.

Hume, D. (1983). *An Enquiry Concerning the Principles of Morals*. J.B. Schneewind (Ed.). Indianapolis: Hackett Press.

Hume, D. (1983). *The History of England*, Vols. I-VI. William Todd (Ed.). Indianapolis: Liberty Fund.

Hume, D. (1985). *Essays: Moral, Political and Literary*. Eugene F. Miller (Ed.). Indianapolis: Liberty Classics.

Kagan, S. (1989). *The Limits of Morality*. Oxford: Oxford University Press.

Livingston, D. (1998). *Philosophical Melancholy and Delerium*. Chicago: University of Chicago Press.

Ninth U.S. Circuit Court of Appeals (2002). *Newdow vs. U.S. Congress*. Available at caselaw.lp.findlaw.com/data2/circs/9th/0016423p.pdf.

Oderberg, D. & Laing, J. (Eds.) (1997). *Human Lives: Critical Essays on Consequentialist Bioethics*. Houndmills: Macmillan Press.

Rawls, J. (1972). *A Theory of Justice*. Cambridge: Harvard University Press.

Raz, J. (1986). *The Morality of Freedom*. Oxford: Oxford University Press.

Singer, P. (1993). *Practical Ethics*, 2nd ed. Cambridge: Cambridge University Press.

CHAPTER 5

DALE E. MILLER

MORAL EXPERTISE: A MILLIAN PERSPECTIVE*

1. INTRODUCTION

Although the utilitarian moral theorist John Stuart Mill (1806–73) does not refer explicitly to the concept of moral expertise, it nevertheless plays an important role in his thought. In order to explain what this role is, it will be useful to begin by considering in more general terms what utilitarians have to say about moral expertise. With this as a foundation it will then be possible to introduce some of the distinctive features of Mill's utilitarianism and to construct a Millian account of moral expertise.

2. A BRIEF PRIMER ON UTILITARIANISM

It may be helpful to start with an overview of utilitarianism. Utilitarianism can be thought of as a family of moral theories. These theories all link the moral standing of actions (i.e., their rightness or wrongness) with the promotion of utility. This characterization is extremely vague, but as soon as one starts to make the definition any more precise one begins to run into controversies that it would be best to avoid here as far as possible. A few small clarifications are necessary, however, although even they may be a little contentious. Utilitarians typically use 'utility' as a synonym for 'well-being.' Well-being can be conceived of in a variety of ways: hedonistically, in terms of pleasure and the absence of pain; in terms of the satisfaction of desire; or in terms of the possession of whatever it is objectively good to have. Promoting utility is usually understood to mean maximizing either the aggregate or the average level of it.

Act utilitarianism, the best-known utilitarian theory, says that an action is morally right just if there is no other action that the agent could perform instead that would yield more utility. On this view, whether an action is right depends not only on its consequences but also on all of the consequences of all of the actions that the agent could perform instead. Act utilitarianism is often criticized on the grounds that mere humans could seldom, if ever, know whether a given action would maximize utility or not. In order to be certain that an action one was contemplating was utility-maximizing, one would apparently have to calculate how much utility would be enjoyed over the entire future course of history if one were to perform it, and then do the same for everything else one might do instead. And one would have to do all of this quickly

L. M. Rasmussen (Ed.), Ethics Expertise: History, Contemporary Perspectives, and Applications, 73-87.

enough that the opportunity for acting would not be lost! In order to appreciate why this is utterly impossible, in most situations, one must bear several points in mind. First, act utilitarianism does not merely say to do an action with a good outcome; it demands perfection. An action is morally wrong unless nothing the agent could have done instead would have had a better outcome. Second, people typically only consciously consider doing the tiniest fraction of the things that it would be possible for them to do at any given time, and actions that they do not consider may often yield far more utility than any that they do. Suppose, for example, that someone is just about to die because he or she is asleep in a burning house, and that a ringing telephone would wake him or her. This scenario must arise several times each day, somewhere in the world. As long as you had access to a telephone yesterday, therefore, there must have been several different moments at which by simply dialing the right number you could have saved a life. Yet the odds against you having even once given actual consideration to dialing just the right number at just the right time are astronomical. Third, the actions of even the most ordinary individuals can have consequences centuries in the future. If a couple decides to have children, for example, then their children may themselves someday have children, and so on for generations. And all sorts of seemingly innocuous actions might turn out to make a difference to whether one becomes a parent, such as putting on a particular outfit to wear to a party, one that will catch the eye of a potential mate among one's fellow partygoers . . . or having just one more drink once one gets there (see Lenman, 2000). Fourth, and finally, it is not just the causal consequences of one's actions that have to be taken into account. If Andy trips, and Betty chooses simply to watch rather than to keep him from falling, then the pain he suffers has to be counted among the outcomes of her choice for the purposes of utilitarian calculations even though she did not cause his pain.

Despite the fact that we can hardly ever know what the theory requires from us, a number of philosophers do subscribe to act utilitarianism; J. J. C. Smart and R. M. Hare, for example, are among its best-known recent defenders (see Smart, 1973, pp. 42-45; Hare, 1981, pp. 25-64; and Shaw, 1999, 145-163). The usual way of coping with the fact that humans can only apply the theory very imperfectly at best is to point out that it only purports to be an account of what makes actions right or wrong. It does *not* purport to tell agents how they should decide what to do, and in particular it does not say that they should always attempt to calculate which of the actions that it is possible for them to perform would maximize utility. Sophisticated act utilitarians say that agents should instead follow some alternative "decision procedure" or other, one that requires them to perform utility calculations only under unusual circumstances (if at all). Usually these alternative decision procedures consist of a set of "summary rules" or "rules of thumb". Whatever rules it is proposed that agents should follow must be simple and few enough in number for ordinary people to learn, and they must be rules that can be properly applied with the amount of information that agents usually have available to them when they must act. The better formulated these rules, the better the behavior that the people who follow them will be when judged in utilitarian terms. No set of rules that beings with human cognitive limitations could learn and apply could possibly direct us to perform the utility-maximizing action in every situation we encounter, however; at best, they could only prevent us from performing actions that are too sub-optimal too often.

According to the act utilitarian, an action that fails to maximize utility is morally wrong, even if the agent properly followed the best rules of thumb that had been devised. This is the point that distinguishes act utilitarianism from a second version of utilitarianism, rule utilitarianism. Rule utilitarianism might be characterized in terms of the following two tenets:

1. Whether an action is right or wrong depends upon whether the authoritative set of moral rules would permit or forbid actions of that general kind, and
2. The authoritative set of moral rules is the one whose general acceptance would yield more utility than the general acceptance of any other set.

The rule utilitarian's authoritative moral rules are "true" moral rules in a sense in which the act utilitarian's summary rules are not; violating these rules is constitutive of acting immorally. The utilitarian element of rule utilitarianism comes in its way of specifying which collection of moral rules is the authoritative one: different sets of moral rules are compared in order to see which one is "best" in utilitarian terms.[1]

3. UTILITARIANISM AND MORAL EXPERTISE

So what, then, do utilitarians have to say about moral expertise? To begin, a fundamental distinction can be drawn between two primary varieties of moral expertise. First, there is what might be called "theoretical" moral expertise. Being a moral expert in this sense means being able to evaluate the arguments for and against rival moral theories and to pick out the one for which the strongest case can be made. There is not any specifically utilitarian account of the attributes required for theoretical moral expertise, although individual utilitarians may have views about this. Second, there is what might be called "practical" moral expertise. Being a moral expert in this sense means being able to "apply" a moral theory, in order to determine in concrete terms what it requires from us. The attributes that a person would need in order to apply one theory successfully may be different from those needed to apply another, so ascriptions of practical moral expertise are relative to a particular theory. Applying some moral theories may be so straightforward that everyone can do it perfectly well, in which case talk about moral expertise is somewhat out of place. When it comes to utilitarianism, however, talk about practical moral expertise may appear out of place for exactly the opposite reason: if the utilitarian approach to morality is the right one, then this may entail that no one could have enough of this expertise to be worth talking about.

This appearance contains the most reality in the case of act utilitarianism, for it is clear from what was said above that there is no way agents with merely human cognitive abilities could avoid performing many actions every day that act utilitarianism says are wrong. Nor could they avoid sometimes performing actions that it says are very wrong, inasmuch as they fall far short of maximizing utility. This certainly seems to suggest that, from the perspective of act utilitarianism, no one *really* deserves to be called a practical moral expert. If some people are slightly better than the rest of us at finding utility-maximizing actions, or at least at avoiding actions whose consequences are very bad in utilitarian terms, then this might make them practical moral experts in a relative sense. In absolute terms, however, agents who are regularly unable to determine what a moral theory requires from them can hardly be called experts at applying it.[2]

From a rule utilitarian standpoint, in contrast, the prospects for practical moral expertise are somewhat better. Actually finding the authoritative set of moral rules might still be beyond beings with merely human cognitive capacities, but we may be able to come close, i.e., to formulate rules that at least approximate the authoritative ones. In fact, there is a need to distinguish here between what might be considered higher and lower forms of rule-utilitarian practical moral expertise. The latter would only require knowledge of a reasonably good set of moral rules, a reasonable approximation of the authoritative set. The former would require the ability to formulate such rules oneself, including improving on rules that one had been taught. This higher sort of practical expertise would require knowledge of the outcomes that performing various kinds of actions in various kinds of situations will typically produce; this would at the very least require extensive empirical knowledge, including especially knowledge of human psychology and sociology.

Yet having said this, we now need to revisit the possibility of act-utilitarian practical moral expertise. The discussion above implied that from an act-utilitarian standpoint mere humans could not have more than an inconsequential amount of practical expertise. Strictly speaking, this is indeed the case; from the act utilitarian perspective genuine practical expertise would consist of the ability to recognize utility-maximizing actions, and no recognizably-human agent will have any real proficiency at this. However, the rule utilitarian's moral rules bear some resemblance to the sophisticated act utilitarian's summary rules, and there is a kind of act-utilitarian practical expertise that may be within the grasp of mere humans, namely expertise in the domain of summary rules. Much of what was said about rule-utilitarian practical moral expertise also applies to this second kind of act-utilitarian expertise. It may be unlikely that mere humans could find the ideal set of summary rules, but perhaps we can come close. We can distinguish between higher and lower forms of this kind of expertise, with the latter requiring only knowledge of a good set of these rules and the former requiring the ability to formulate such rules oneself—which would in turn require extensive empirical knowledge, especially of human nature. While mere humans might be able to attain this second, summary-rule-based kind of practical moral expertise, from the act-utilitarian perspective it is clearly inferior to the first. This is because even an ideal set of summary rules would still be a very poor guide to the requirements of act-utilitarian morality.[3]

4. MILL'S UTILITARIANISM

The vast literature on Mill's utilitarianism is an indication of its uniqueness and of the interest that it holds for philosophers. The best place to begin a survey of its most distinctive aspects may be with his way of arguing for utilitarianism. Mill's case for utilitarianism can be seen as having two parts. First, he asserts that morality must necessarily be teleological or consequentialist in nature. In his essay "Bentham" he writes that:

> Whether happiness be or be not the end to which morality should be referred—that it be referred to an end of some sort, and not left in the dominion of vague feeling or inexplicable internal conviction, that it be made a matter of reason and calculation, and not merely of sentiment, is essential to the very idea of moral philosophy; is, in fact, what renders argument or discussion on moral questions possible. That the morality of actions depends on the consequences which they tend to produce, is the doctrine of rational persons of all schools; that the good or evil of those consequences is measured solely by pleasure or pain, is all of the doctrine of the school of utility, which is peculiar to it (1838, p. 111).

Mill sees no way that one could deny consequentialism without affirming an intuitionist moral epistemology, one that says the conscience is veridical, i.e., that our moral feelings are a reliable guide to our duties. He criticizes intuitionism on two fronts: he contends both that it is false and that it is dangerous. It is false because the feeling of duty is demonstrably a product of mental associations; what a person feels obligated to do depends on what experiences he or she has had.[4] It is dangerous because it is conservative. It teaches people to regard the very fact that they hold a moral stance as conclusive proof of its truth, and hence it implies that they have no need to reconsider any of their views.[5]

The second part of Mill's case for utilitarianism is his notorious "proof that is not a proof" of the principle of utility. Mill equates well-being with happiness, and he conceives of happiness hedonistically; the long passage quoted in the last paragraph suggests this, and Mill is even more explicit when he writes in *Utilitarianism* that "By happiness is intended pleasure, and the absence of pain; by unhappiness, pain, and the privation of pleasure" (1861, p. 210). The point of the proof is to show that happiness is the only thing desirable as an end. Mill makes the well-known assertion that the only evidence that something is desirable as an end is that people do so desire it, and then goes on to argue that not only is happiness desired as an end but that it is the only thing we *can* desire as an end. This last claim rests on an account of the origin of desires that Mill takes to follow from his associationist theory of human psychology. If only happiness is desired as an end, Mill reasons, then it must be the *telos* of morality:

> If the opinion which I have now stated is psychologically true—if human nature is so constituted as to desire nothing which is not either a part of happiness or a means of happiness, we can have no other proof, and we require no other, that these are the only things desirable. If so, happiness is the sole end of human action, and the promotion of it the test by which to judge of all human conduct; from whence it necessarily follows that it must be the criterion of morality, since a part is included in the whole (1861, p. 237).

Clearly Mill's case for utilitarianism is open to question at a number of points. Even if moral teleology is true, this is surely not as obvious and straightforward as Mill takes it to be. He seems to be making an appeal to an undefended intuition about morality here, despite his anti-intuitionistic stance. And Mill's proof has long been derided as a shoddy piece of argumentation, riddled with "howlers"—errors in reasoning so egregious as to make one howl with laughter. In Mill's defense, however, moral philosophers continue to struggle to find alternatives to grounding their theories on intuitions about morality. Furthermore, if the proposition that morality must have a teleological structure does have the status of a mere intuition in Mill's thought, then at least the proposition's abstractness entails that his case for utilitarianism is not subject to many of the objections that he and other critics have leveled against intuitionism. Finally, recent scholarship on the proof contains at least as much favorable commentary as critical.[6] Few people would say that the proof is entirely unproblematic, but those who are willing to read it charitably generally find it sensible even if not ultimately successful.

When it comes to Mill's theory of utilitarianism itself, many of its most unique features stem from Mill's views about human psychology. More specifically, they stem from his rejection of certain aspects of a Benthamite view of human psychology that Mill considers overly simplistic. For example, while both Bentham and Mill are hedonists, Mill's hedonism looks very different from that of the earlier writer. For Bentham, the value to a person of the pleasure contained in an experience depends strictly upon the pleasure's quantity, where this can be understood as the product of the pleasure's intensity and duration.[7] Mill, in contrast, holds that some kinds of pleasures are of higher quality than others, inasmuch as they contribute more to a person's happiness. The so-called "informed preference test," by means of which we ascertain whether one kind of pleasure is of higher quality than another, consists simply of consulting the preferences of people who have experienced both and hence are competent to judge between them. Mill says that if a majority of the competent judges prefer one kind of pleasure over the other then the former is of higher quality (although he does not seem to think that the vote will ever be a close one). He asserts that this test will "assign to the pleasures of the intellect, of the feelings and imagination, and of the moral sentiments, a much higher value as pleasures than to those of mere sensation" (1861, p. 211). While much about Mill's doctrine of the higher pleasures is obscure, it is at least clear that he believes that a life that is rich in these pleasures will be happier than one from which they are absent, even if the latter includes a much greater quantity of "beastly" sensual pleasures.

Another respect in which Mill believes that Bentham's view of human psychology is too crude is its failure to pay enough attention to the crucial notion of character. By Mill's definition, a person's character consists of his "habits of mind and heart," along with the habits of his will or faculty of choice (1833, p. 7; cf. Carlisle, 1991, pp. 1ff.). In other words, his character is made up of the ways of thinking, of feeling, and of choosing what to do that have become, for him, a matter of habit. A concern with what people's characters are like and how they are formed is central to Mill's moral, social, and political philosophy, as a number of recent interpreters have recognized (e.g., Morales, 1996; Eisenbach, 1999; Ball, 2000; and Berger, 1984). The effects that an action has on the character of the agent are often among its most important consequences, Mill holds, although Bentham had failed to recognize this:

> In so far as Bentham's adoption of the principle of utility induced him to fix his attention upon the consequences of actions as the consideration determining their

> morality, so far he was indisputably in the right path: though to go far in it without wandering, there was needed a greater knowledge of the formation of character, and of the consequences of actions upon the agent's own frame of mind, than Bentham possessed. His want of power to estimate this class of consequences, together with his want of the degree of modest deference which, from those who have not competent experience of their own, is due to the experience of others on that part of the subject, greatly limit the value of his speculations on questions of practical ethics (Mill, 1838, pp. 111-112; see also Mill, 1833, pp. 7-8).

Nothing has been said so far about where Mill falls on the act-rule utilitarian divide. Although non-specialists may find this remarkable, the fundamental question of whether Mill is an act utilitarian or a rule utilitarian is a matter of some contention among Mill scholars.[8] There are passages in which he sounds like he is in one camp, and passages in which he seems to be in the other. It is entirely fair to say that he should have been clearer about this. Of course it would be anachronistic to expect him to use the *terminology* of 'act utilitarianism' and 'rule utilitarianism'; these terms, coined by Richard Brandt, didn't appear in the literature until 1959 (Brandt, 1959, pp. 380, 396). But the conceptual difference between these theories is sufficiently obvious for it to be reasonable to expect Mill to have appreciated it and to have done a better job of disambiguating his position.[9]

What is clear is that Mill believes that, when we are trying to determine what morality requires from us, moral rules that are distinct from the "greatest happiness principle" (the general requirement to maximize utility) should play a large role in our thinking. In his view, people ordinarily should not be performing utility calculations when they are choosing how to act; they should instead only be careful to comply with a set of "secondary" moral rules:

> There is no difficulty in proving any ethical standard whatever to work ill, if we suppose universal idiocy to be conjoined with it; but on any hypothesis short of that, mankind must by this time have acquired positive beliefs as to the effects of some actions on their happiness; and the beliefs which have thus come down are the rules of morality for the multitude, and for the philosopher until he has succeeded in finding better. . . . The corollaries from the principle of utility, like the precepts of every practical art, admit of indefinite improvement, and, in a progressive state of the human mind, their improvement is perpetually going on. . . . But to consider the rules of morality as improvable, is one thing; to pass over the intermediate generalisations entirely, and endeavour to test each individual action directly by the first principle, is another (Mill, 1861, p. 224).

This passage shows that Mill thinks the rules of conventional morality—"the rules of morality for the multitude"—are a reasonable first approximation of the ideal set of secondary rules. Even if people have seldom reasoned about morality explicitly in utilitarian terms, he maintains, they can hardly have been indifferent to their own happiness, and utilitarian considerations must have implicitly been a major determinant of what moral rules won acceptance (Mill, 1861, p. 207). And since we have available a reasonably good set of secondary rules, the only time that a person facing a moral quandary should have to reason in explicitly utilitarian terms is when she faces a situation in which two of these rules give conflicting guidance. Mill makes this point in the course of responding to an objection against utilitarianism, namely that the theory lends itself to self-serving rationalization:

> There exists no moral system under which there do not arise unequivocal cases of conflicting obligation. . . . They are overcome practically, with greater or with less success, according to the intellect and virtue of the individual; but it can hardly be pretended that any one will be the less qualified for dealing with them, from possessing an ultimate standard to which conflicting rights and duties can be referred. If utility is the ultimate source of moral obligations, utility may be invoked to decide between them when their demands are incompatible. . . . We must remember that only in these cases of conflict between secondary principles is it requisite that first principles should be appealed to. There is no case of moral obligation in which some secondary principle is not involved; and if only one, there can seldom be any real doubt which one it is, in the mind of any person by whom the principle itself is recognised (1861, pp. 225-226).

Of course, the fact that Mill talks about secondary moral rules does not establish that he is a rule utilitarian, for we have seen that sophisticated act utilitarians also believe that agents should ordinarily decide what morality requires from them by reference to a set of rules distinct from the greatest happiness principle. The passages quoted here, however, do not help us to decide whether Mill is an act or rule utilitarian. Nor, unfortunately, is the rest of his corpus overly helpful in this regard. What we can say with confidence is that Mill affirms the following points: That ordinarily we should try to decide what it would be morally right or wrong for us to do by reference to a set of "secondary" moral rules; that the rules of ordinary morality are a good first approximation of the set of secondary rules that we should employ, although there is room for improvement; and that the only time a person who must actually decide what to do—someone in the heat of the moment, so to speak—should ever engage in explicitly utilitarian reasoning is if the secondary rules she accepts give her conflicting guidance about what someone in her situation should do. Whether we take Mill to be an act or a rule utilitarian makes a difference to how we will understand all of these claims, at a certain level of detail, and we will see momentarily that it will subtly affect our understanding of the role of moral expertise in his utilitarianism as well. (It also makes a major difference to how we interpret some elements of his thought that are not directly relevant to his views on moral expertise; for instance, it is much easier to reconcile his "liberty principle" with his utilitarianism if you think that he is a rule rather than an act utilitarian (1859b, p. 223.)

5. A MILLIAN ACCOUNT OF MORAL EXPERTISE

We are now in a position to bring all of the preceding material together and develop a Millian account of moral expertise. Most of this discussion will center on practical moral expertise, but Mill's views about theoretical moral expertise deserve at least a mention. Mill would think that there are two attributes that a person would need to possess in order to deserve to be considered a theoretical moral expert, over and above general philosophical ability, and both are somewhat idiosyncratic. The first is a commitment to the possibility of resolving moral questions through the application of reason, and for Mill the test of whether someone has such a commitment is whether he or she is a consequentialist. The second is a grasp of the psychology of desire, since

Mill believes that it is only through understanding human ends that one can ascertain what the *telos* of morality must be. (Of course, Mill will believe that any who possesses a full measure of theoretical moral expertise will be a utilitarian, although he does not hold the converse. Mill would not even place Bentham in the first rank of theoretical moral experts. He describes Bentham's efforts to refute the proponents of competing moral views as question-begging, and concludes that, when it comes to abstract philosophical debates, "it has for the most part, even when he was most completely in the right, been reserved for others to *prove* him so" (1833, p. 6)).

Given the importance of secondary moral rules in Mill's thought it is clear that he is going to understand practical moral expertise in terms of such rules. We saw above that both act and rule utilitarians have a place in their theories for rules distinct from the requirement to maximize utility, and that both can give an account of practical moral expertise in terms of them. On both of these accounts, it is possible to distinguish between higher and lower forms of such expertise, with the lower consisting merely of acquaintance with a decent set of such rules, and the higher consisting of the ability to improve upon the set that is conventionally accepted. This is precisely the distinction to which Mill alludes when he distinguishes between "the multitude", who accept the rules of conventional morality (which he takes to reflect popular ideas about what sorts of conduct are most conducive to human happiness), and "the philosopher" who is in a position to improve upon these rules.

From either an act or a rule utilitarian perspective, the primary requirement for a person to claim the higher of these forms of moral expertise is empirical knowledge, particularly knowledge of human nature. From the Millian perspective, an understanding of the higher pleasures and of human character are especially important to the moral reformer. Knowledge of the higher pleasures is relatively unproblematic, because it requires only the experience of such pleasures plus perhaps the ability to reflect on the experience. This sort of knowledge is available to any person of developed faculties, and Mill himself certainly feels confident in his ability to say which pleasures are the higher ones.

Matters stand somewhat differently when it comes to knowledge of human character. There are two different types of knowledge about character that Mill would believe someone would need to possess in order to be competent to improve upon conventional morality; he or she would have to know what character traits or habits are most desirable, and how they can be instilled in people. Mill sees the question of what habits are most desirable as one that must be answered on utilitarian grounds; it is a question of what type of character yields the most happiness, both for the individual whose character it is and for everyone whom that individual affects. He does not think that there is one character type that is best for everyone, in these terms. Individuals differ in their basic constitutions, and in particular in their abilities and susceptibilities to pleasure and pain. In consequence, the precise set of habits that are best for one person will not necessarily be best for someone else. This is why Mill asserts that "free scope should be given to varieties of character," and it is one reason that he attaches so much importance to self-development or self-cultivation (1859b, p. 261). As a rule, no one has more information about the unique features of an individual's constitution than the individual him- or herself. Nevertheless, Mill does believe that there are a number of traits that should be a part of every person's character, including energy or vigor, courage, and even cleanliness.[10] There are also numerous traits that he insists should have no place in any person's character; for example:

> Cruelty of disposition; malice and ill-nature; that most anti-social and odious of all passions, envy; dissimulation and insincerity, irascibility on insufficient cause, and resentment disproportioned to the provocation; the love of domineering over others; the desire to engross more than one's share of advantages (the *pleonexia* of the Greeks); the pride which derives gratification from the abasement of others; the egotism which thinks self and its concerns more important than everything else, and decides all doubtful questions in its own favour;—these are moral vices, and constitute a bad and odious moral character (1859b, p. 279).

Knowledge of the means by which desirable character traits can be inculcated presents greater difficulties. While we have seen that he criticizes Bentham for lacking an adequate "knowledge of the formation of character," Mill is painfully aware of the limitations of his own knowledge in this domain—and those of everyone else's knowledge as well, he thinks. In his first major work, *A System of Logic*, which first appeared in 1843, Mill outlines a science of human character that he proposed to call "ethology":

> The name is perhaps etymologically applicable to the entire science of our mental and moral nature; but if, as is usual and convenient, we employ the name Psychology for the science of the elementary laws of mind, Ethology will serve for the ulterior science which determines the kind of character produced in conformity to those general laws, by any set of circumstances, physical and moral. According to this definition, Ethology is the science which corresponds to the act of education, in the widest sense of term, including the formation of national or collective character as well as individual (1973, p. 869).

This science, he adds, "is still to be created" (1973, pp. 872-873). As he writes these words, Mill intends to undertake the first systematic investigation of the laws of ethology. In 1859 he remarks (perhaps with some frustration) that he has never yet felt prepared to write on this subject, although he indicates that he still plans to do so in the future, "at least in the form of some Essays" (1859a, p. 645). Unfortunately, however, Mill never does feel prepared, and the essays remain unwritten (Bain, 1882, pp. 78-79). In the absence of real scientific laws of ethology, in his view, our knowledge of character formation is limited to mere "empirical laws." Empirical laws are generalizations drawn from observation which, because they do not refer to underlying causes, are valid only in circumstances that are more or less identical to those in which the observations were made. Such empirical laws are of little use in predicting what influences will have what effects on the characters of members of future generations, which would be important information for someone who was seeking to improve the rules of conventional morality from a utilitarian perspective. Mill would therefore have to think that this imposes a limit upon how much of the higher form of practical moral expertise anyone could claim, until the science of ethology is fairly well advanced.

So, in sum, from a Millian perspective practical moral expertise must be understood in terms of secondary moral rules. The lower form of such expertise consists in knowledge of a set of such rules that are at least decent, from a utilitarian perspective, and Mill thinks that the rules of conventional morality qualify as such. The higher form consists in the knowledge necessary to improve upon conventional morality. First and foremost this requires knowledge of human nature, and Mill would think that this would need to include knowledge of the higher pleasures, of what character traits are most desirable, and of how these traits can be instilled. The somewhat esoteric question that we cannot answer, without being able to establish definitively whether Mill is an act or a rule utilitarian, is whether he thinks that genuine practical moral expertise is beyond the grasp of humanity. If he is an act utilitarian he will think that it is, because we have seen that from an act-utilitarian standpoint secondary rules are mere summary rules, and that the sort of practical moral expertise that has to do with mastery of summary rules is only a second-best variety. If he is a rule utilitarian, in contrast, he may well think otherwise, for human cognitive abilities may be sufficient to discover secondary rules that are a reasonable approximation to rule utilitarianism's authoritative set of moral rules.

Notice that there is nothing in this account of moral expertise that suggests that moral experts—theoretical or practical—will necessarily be virtuous. Mill is a moral externalist, which means that he does not believe that there is any intrinsic connection between moral beliefs/knowledge and moral motivation; that one knows what the morally right thing to do would be, or believes that one does, does not entail that one will feel any motivation whatsoever to act on that knowledge/belief.[11] This implies that while those with greater moral expertise might be able to lay claim to greater moral knowledge, or at least greater justification for their moral beliefs, it would be a mistake to assume that they are automatically any more virtuous—in the sense of being desirous of acting morally—than anyone else.

6. CONCLUSION

It may be fitting to conclude with a very brief look at some of the implications of this Millian view of moral expertise for the field of medical ethics. We have seen that Mill believes that practical moral expertise must be understood in terms of a set of secondary moral rules that are distinct from the requirement to maximize utility, and that he thinks that "our" conventional morality comprises a decent, if not an ideal, set of secondary rules. But our ordinary moral rules are often not adequate to resolve many questions that today arise in the field of medical ethics.

A number of the most pressing issues in the area of medical ethics do involve conflicts between the rules of our society's conventional morality. The controversies over abortion and euthanasia, for example, are products of conflicts between widely accepted moral rules that forbid doing harm, require the prevention of suffering, and permit personal autonomy. Unfortunately, our conventional morality does not include any sort of second-order priority rule that tells us how to resolve conflicts between these first-order rules. In other cases, technological developments have abruptly presented us with questions to which our conventional morality may just not have enough to say, and new rules may be needed to fill in gaps that have become apparent—if not new universal rules, which apply to all of us, then at least new rules pertaining specifically to members of the medical community. While it is not the case that our conventional morality has nothing to say that pertains to the debate over cloning, for example, one sometimes gets the sense that it does not have enough to say—that if we conduct the debate purely on the basis of these rules, we will be overlooking something important.

Given the significance of these issues, Mill would think that there is a need for "philosophers" with the higher form of practical moral expertise to take on the project of revising our conventional morality in order to remove these conflicts and lacunas—if not of its universalistic elements, which apply to all of us, then at least of what it says about the special obligations of members of the medical community. Since those who would undertake this project would require a fair amount of knowledge about contemporary medical practices and technologies, in addition to the general knowledge of human nature and human affairs required of all moral reformers, in this instance practical moral expertise requires medical expertise. In fact, because many of the most serious moral controversies that we face today as a society involve medical issues, it may well be the case that Mill would think that medical ethicists and medical practitioners are in a position to contribute as much as anyone to the improvement of conventional morality. Although this assumes, of course, that they have the proper utilitarian orientation.

NOTES

* I am grateful to Melissa Longo and Claire Morgan for comments and suggestions on earlier drafts of this paper.

1. Clearly the notion of the 'general acceptance' of a set of moral rules must be explicated, and there are different ways of doing this. One relatively straightforward approach is to say that for a set of moral rules to be generally established is for it to be universally obeyed. However, David Lyons (1965) has famously demonstrated that if rule utilitarianism is interpreted in this way the authoritative set of moral rules will be "extensionally equivalent" with act utilitarianism; it will agree completely with act utilitarianism about which actions are right and which wrong. The same does not hold true, however, if 'general acceptance' is instead interpreted in terms of "universal internalization," where what it means for a person to internalize a set of rules is for him to have a disposition to feel guilty at the thought of violating them. Since human beings cannot internalize sets of rules that contain too many elements or whose elements are too lengthy or complex, this way of understanding the theory entails that the authoritative set of rules cannot include specific rules for every conceivable situation that a person might ever encounter. It is the possibility of an incredibly "fine-grained" set of rules like this that leads to Lyons's result.

Yet even this may not do. Some remarks of Richard Brandt's point to the fact that if we think about what rules it would be best for everyone to internalize, then those rules will give us no guidance in dealing with people who have failed to internalize them (Brandt, 1988). So we may do better to interpret the general acceptance of a set of rules in terms of its internalization by the great majority of individuals.

2. A number of further arguments have been given for why no moral theory that makes cognitive demands as great as those made by act utilitarianism can be acceptable. For some of these see Singer, M., 1977; 1983; Lenman, 2000; Howard-Snyder, 1997; 1999; Miller, 2003. For defenses of act utilitarianism against the objects raised by Singer and Howard-Snyder see Temkin, 1978; Carlson, 1990; Qizilbash, 1999.

3. In the foregoing discussion I have concentrated on the "objective" or "actual consequence" versions of utilitarianism, because these have the most relevance to Mill's thought (although cf. Crisp, 1997, pp. 99-101). "Subjective" or "expected consequence" versions also exist. This distinction crosscuts the act/rule utilitarianism distinction. Subjective act utilitarianism, for example, says that the moral standing of an action

depends on how much utility the agent's potential actions would be expected to yield, at the time of action, by either the actual agent or an idealized one. William H. Shaw writes:

> As these [expected-outcome] utilitarians see it, what matters is not the actual outcome of an action, but its probable, foreseeable, or expected outcome.... Some... believe that, in determining right and wrong, it is the agent's own assignment of values and probabilities that counts... [but most] believe that the utilitarian standard is expected happiness as calculated by a reasonable and well-informed agent based on the available evidence (1999, pp. 29-30).

An "actual-agent" subjective version of act utilitarianism would make it out of place to talk about practical moral expertise, because it would imply that everyone is a practical moral expert. On the other hand, if the rightness of an action depends on the consequences that it would be expected to have by a very highly idealized agent, then subjective act utilitarianism starts to look very much like the objective version of the theory. Between these extremes, however, there could be subjective versions of act-utilitarianism according to which genuine practical moral expertise is possible for human beings.

4. Mill writes that the "moral feelings are not innate, but acquired" or "implanted"; (Mill, 1861, "Utilitarianism, p. 230). For more on this, see my (1998).

5. For example, Mill says that:

> The effect of custom, in preventing any misgiving respecting the rules of conduct which mankind impose on one another, is all the more complete because the subject is one on which it is not generally considered necessary that reasons should be given, either by one person to others or by each to himself. People are accustomed to believe, and have been encouraged in the belief by some who aspire to the character of philosophers, that their feelings, on subjects of this nature, are better than reasons, and render reasons unnecessary (Mill, 1859b, p. 220).

6. Some recent examples of generally sympathetic treatments of the proof include those by Crisp (1997, pp. 67-68); Skorupski, 1989, pp. 285-313; and Sayre-McCord, 2001.

7. Bentham's "felicific calculus" actually includes seven different elements upon which he says that the value of a pleasure depends; in addition to intensity and duration these are certainty, propinquity, fecundity, purity, and extent (Bentham, 1948, pp. 30-31). Strictly speaking, however, only intensity and duration pertain to the value of the pleasure contained in an experience. The rest have to do with the value of the experience as a whole or of the action that gave rise to the experience.

8. There is a considerable body of secondary literature on this debate, including Urmson, 1953; Mabbott, 1956; Brown, 1974; Lyons, 1994; Gaus, 1980; Gray, 1983; and Crisp, 1997.

9. This is just one of a number of different respects in which the precise nature of Mill's utilitarianism is far from clear. In large part, no doubt, this lack of clarity springs from the fact that his most extended discussion of moral theory is his 1861 essay *Utilitarianism*, which runs just over 27,000 words—between 55 and 75 pages, in most editions. Compare this with Henry Sidgwick's *The Methods of Ethics*, the seventh edition of which was published in 1907, which comes in at 509 pages (Sidgwick, 1981). In a review of John Austin's *Lectures on Jurisprudence*, published in 1832, Mill chastised authors who "should and would write a book" but who write articles for periodicals instead (Mill, 1832, p. 5). Twenty nine years later, the essays that compose *Utilitarianism* first appeared as a series of articles in *Fraser's Magazine*.

10. For example see *On Liberty*, pp. 263–4; *Considerations on Representative Government, Essays on Politics and Society*, pp. 406–12; "Nature," *Essays on Ethics, Religion and Society*, pp. 393–6.

11. Note in this connection his separate discussions of the motives for obeying utilitarianism (*Utilitarianism*, Chap. III) and of his infamous "proof" of this doctrine (*Utilitarianism*, Chap. IV).

REFERENCES

Bain, Alexander (1882). *John Stuart Mill: A Criticism with Personal Recollections*. London: Longmans, Green & Co.

Ball, Terrance (2000). 'The Formation of Character: Mill's "Ethology" Reconsidered.' *Polity* XXXIII (Fall), 25–48.

Bentham, Jeremy (1948). *Introduction to the Principles of Morals and Legislation*. New York: Hafner.

Berger, Fred (1984). *Happiness, Justice, and Freedom: The Moral and Political Philosophy of John Stuart Mill*. Berkeley: University of California Press.
Brandt, Richart (1959). *Ethical Theory*. Englewood Cliffs, NJ: Prentice-Hall.
Brandt, Richard (1988). 'Fairness to Indirect Optimific Theories in Ethics.' *Ethics*, 357–360.
Brown, D.G. (1974). 'Mill's Act Utilitarianism.' *Philosophical Quarterly* 24, 67–68.
Carlisle, Janet (1991). *John Stuart Mill and the Writing of Character*. Athens: University of Georgia Press.
Carlson, Erik (1990). 'The Oughts and Cans of Objective Consequentialism.' *Utilitas* 11, 91–96.
Crisp, Roger (1997). *Routledge Philosophy Guidebook to Mill on Utilitarianism*. London: Routledge.
Eisenbach, Eldon J. (Ed.) (1999). *Mill and the Moral Character of Liberalism*. University Park: Pennsylvania State University Press.
Gaus, Gerald F. (1980). 'Mill's Theory of Moral Rules.' *Australasian Journal of Philosophy* 58, 265–279.
Gray, John (1983). *Mill on Liberty: A Defence*, 2nd ed. London: Routledge. 28–42.
Hare, R.M. (1981). *Moral Thinking: Its Levels, Method, and Point*. Oxford: Clarendon Press.
Howard-Snyder, Frances (1997). 'The Rejection of Objective Consequentialism.' *Utilitas* 9, 241–248.
Howard-Snyder, Frances (1999). 'Response to Carlson and Qizilbash.' *Utilitas* 11, 106–111.
Lenman, James (2000). 'Consequentialism and Cluelessness.' *Philosophy and Public Affairs* 29, 342-370.
Lyons, David (1965). *Forms and Limits of Utilitarianism*. Oxford: Clarendon.
Lyons, David (1994). 'Mill's Theory of Morality.' In: D. Lyons (Ed.), *Rights, Welfare, and Mill's Moral Theory*. New York: Oxford University Press. (Originally published in *Noûs* 10 (1976), 101–120).
Mabbott, J.D. (1956). 'Interpretations of Mill's "Utilitarianism".' *Philosophical Quarterly* 6, 115–120.
Mill, John Stuart (1832). 'Austin's Lectures on Jurisprudence.' In: John M. Robson (ed.), *Essays on Equality, Laws, and Education, Collected Works of John Stuart Mill*, Vol. XXI. Toronto: University of Toronto Press, 1984.
Mill, John Stuart (1833). 'Remarks on Bentham's Philosophy.' In: John M. Robson (Ed.), *Essays on Ethics, Religion and Society, Collected Works of John Stuart Mill*, Vol. X. Toronto: Toronto University Press, 1969.
Mill, John Stuart (1838). 'Bentham.' In: John M. Robson (Ed.), *Essays on Ethics, Religion and Society, Collected Works of John Stuart Mill*, Vol. X. Toronto: Toronto University Press, 1969.
Mill, John Stuart (1859a). 'Letter to Alexander Bain (14 Nov.),' In: Francis Mineka and Dwight N. Lindley (Eds.), *The Later Letters: 1849 to 1873, Collected Works of John Stuart Mill*, Vols. XIV–XVII. Toronto: University of Toronto Press, 1972.
Mill, John Stuart (1859b). *On Liberty*. In: John M. Robson (Ed.), *Essays on Politics and Society, Collected Works of John Stuart Mill*, vols. XVIII–XIX. Toronto: University of Toronto Press, 1977.
Mill, John Stuart (1861a). *Utilitarianism*. In: John M. Robson (Ed.), *Essays on Ethics, Religion and Society, Collected Works of John Stuart Mill*, Vol. X. Toronto: Toronto University Press, 1969.
Mill, John Stuart (1861b). Considerations on Representative Government. In: John M. Robson (Ed.), *Essays on Politics and Society, Collected Works of John Stuart Mill*, vols. XVIII–XIX. Toronto: University of Toronto Press, 1977.
Mill, John Stuart (1872). *A System of Logic Ratiocinative and Inductive*. In: John M. Robson (Ed.), *Collected Works of John Stuart Mill*, Vols. VII–VIII. Toronto: University of Toronto Press, 1973.
Mill, John Stuart (1874). 'Nature.' In: John M. Robson (Ed.), *Essays on Politics and Society, Collected Works of John Stuart Mill*, vol. X. Toronto: University of Toronto Press, 1977.
Miller, Dale E. (1998). 'Internal Sanctions in Mill's Moral Psychology.' *Utilitas* 10 (1998), pp. 68–81.
Miller, Dale E. (2003). 'Actual-Consequence Act Utilitarianism and the Best Possible Humans.' *Ratio* XVI, 49–62.
Morales, Maria H. (1996). *Perfect Equality: John Stuart Mill on Well-Constituted Communities*. Lanham MA: Rowman and Littlefield.
Qizilbash, Mozaffar (1999). 'The Rejection of Objective Consequentialism.' *Utilitas* 11, 97–106.
Sayre-McCord, Geoffrey (2001). 'Mill's "Proof" of the Principle of Utility: A More Than Half-Hearted Defense.' *Social Philosophy and Policy* 18, 330–60.
Shaw, W. (1999). *Contemporary Ethics: Taking Account of Utilitarianism*. Malden, MA: Blackwell Press.
Sidgwick, Henry (1981). *The Methods of Ethics*. Indianapolis, IN: Hackett.
Singer, Marcus (1977). 'Actual Consequence Utilitarianism.' *Mind* 86, 67–77.
Singer, Marcus (1983). 'Further on Actual Consequence Utilitarianism.' *Mind* 92, 270–274.
Skorupski, John (1989). *John Stuart Mill*. London: Routledge.
Smart, J.J.C. (1973). 'An Outline of a System of Utilitarian Ethics.' In: J.J.C. Smart and B. Williams (Eds.), Utilitarianism: For and Against. Cambridge: Cambridge University Press.

Temkin, Jack (1978). 'Actual Consequence Utilitarianism: A Reply to Professor Singer.' *Mind* 87, 412–414.
Urmson, J.O. (1953). 'The Interpretation of the Moral Philosophy of J. S. Mill.' *Philosophical Quarterly* 3, 33–39.

CHAPTER 6

BEN EGGLESTON

THE INEFFABLE AND THE INCALCULABLE: G.E. MOORE ON ETHICAL EXPERTISE

1. WHAT IS ETHICAL EXPERTISE?

Although the publication of G.E. Moore's *Principia Ethica* is widely acknowledged as a momentous event in the history of ethics, it is especially significant in connection with the idea of ethical expertise, since it was hailed as a turning point in the use of reason rather than sentiment to decide ethical questions. Moore's friend Lytton Strachey wrote to him saying that his book "laid the true foundations of Ethics" and that "It is the scientific method deliberately applied, for the first time, to Reasoning. . . . I date from Oct. 1903 the beginning of the Age of Reason" (quoted in Baldwin, 1993, p. xi). In addition, John Maynard Keynes famously wrote that "I went up to Cambridge at Michaelmas 1902, and Moore's *Principia Ethica* came out at the end of my first year. . . . it was exciting, exhilarating, the beginning of a renaissance, the opening of a new heaven on a new earth" (1972, p. 435). A hundred years later, Moore's work is viewed more critically, but hardly less seriously, and few doubt that Moore did, indeed, try to impose on the field of ethics more structure and analytical rigor than was previously associated with it. In this chapter, then, I shall explain what, according to Moore, would constitute ethical expertise. I shall begin by devoting the remainder of this section to the task of interpreting the notion of ethical expertise itself.

One question about the notion of ethical expertise is whether it is primarily a matter of knowledge or practice. Although the notion of expertise obviously has more of a theoretical connotation than a pragmatic one, the subject matter of ethics is so inevitably pragmatic that some have thought it to be an essentially practical endeavor. Aristotle, for example, writes in his *Nicomachean Ethics* that "Our present inquiry does not aim, as others do, at study; for the purpose of our examination is not to know what virtue is, but to become good" (p. 1103b, ll. 27–29). But in modern ethics, the opposing view is predominant. Moore's teacher Henry Sidgwick, for example, writes in his *Methods of Ethics* that "here as in other opinions we ought to aim at nothing but truth" (p. 335) and "on reflection it is generally admitted that it cannot be good to be in error on . . . any . . . point" (p. 429). And Moore himself is quite emphatic in his claims that "The direct object of

L. M. Rasmussen (Ed.), Ethics Expertise: History, Contemporary Perspectives, and Applications, 89-102.

ethics is knowledge and not practice" (p. 71) and that "What I am concerned with is knowledge only—that we should think correctly and so far arrive at some truth, however unimportant: I do not say that such knowledge will make us more useful members of society" (p. 115). Moreover, this privileging of the theoretical over the practical is reinforced when we return to the specific notion of ethical expertise. For even if a person always did the right thing, but without giving it any thought or being able to instruct others in doing the right thing (except by example), then although we might regard her as a benefactor to her fellows, and although we might even acknowledge her as being virtuous, we would hesitate, I think, to regard her as having ethical *expertise*. For the notion of expertise, like the notion of philosophy itself, seems to invoke the notion of knowledge. I shall take as given, then, that ethical expertise is essentially a matter of knowledge, not practice.

A second question about the notion of ethical expertise concerns its subject matter, because here there are two possibilities. First, one might think that ethics is primarily concerned with the good: what objects are good, what institutions are good, what states of affairs are good, and so on. Second, one might think that ethics is primarily concerned with the right: what actions and courses of action are right, what policies and laws it would be right to enact, and so on. The difference this distinction makes is that if ethical expertise is primarily concerned with the good, then although an ethical expert may be able to judge given objects (and institutions and states of affairs and so on), her ability to guide action—to tell agents which of the acts open to them would be right, and which would be wrong—is not assured. (At least, it is not assured unless she can also forge some connection between judgments of the good and judgments of the right.) On the other hand, if ethical expertise is primarily concerned with the right, then the ethical expert is in a position to use her expertise in an explicitly action-guiding way.

Moore himself is primarily concerned with what is good. Of the six chapters of *Principia Ethica*, only one has much to say about what is right, whereas each of the other five chapters contribute significantly to Moore's account of what is good. Moreover, he explicitly says that "'good' is the notion upon which all Ethics depends" (p. 192) and that "the primary and peculiar business of Ethics" is "the determination [of] what things have intrinsic value and in what degrees" (p. 78; see also p. 54 and p. 57). But he recognizes the importance of using judgments of the good to arrive at judgments of what is right, and he endorses a specific conception of the link between them (which I explore below, in section 2). In thus acknowledging the importance of questions about what is right, Moore shows some affinity with the prevailing trend in modern ethics. Moore's predecessor John Stuart Mill, for example, writes that "It is the business of ethics to tell us what are our duties" (p. 219). More recently, John Rawls writes that

> in ethics we are attempting to find reasonable principles which, when we are given a proposed line of conduct and the situation in which it is to be carried out and the relevant interests which it affects, will enable us to determine whether or not we ought to carry it out and hold it to be just and right (p. 2).

This position is advanced yet more strongly in the following remark by Jan Narveson:

> Let us begin by recalling the primary function of ethical principles: to tell us what to do, i.e., to guide action. Whatever else an ethical principle is supposed to do, it must do that, otherwise it could not (logically) be an ethical principle at all (p. 105).

According to Shelly Kagan, the "received view" is that "normative ethics is concerned with determining which actions are right, which wrong," among other things (p. 223), and Kurt Baier explains that

> action has, for some time now, occupied the limelight in ethics because, although we have a measure of control over [such things as "behavioral dispositions, character traits, habits or feelings"], we have the most direct control over [acts]: they are things we can do or perform at will (pp. 282–283).

So it seems clear, both from the perspective of philosophical tradition and from the presumption that ethical expertise should be useful in guiding action, that ethical expertise should be construed as concerning, primarily, the rightness and wrongness of acts and courses of action.

So, to sum up: I side with Moore in regarding ethical expertise as a matter of knowledge, not practice, but I differ from Moore in emphasizing the right, not the good, as the object of that knowledge. I conclude, then, that ethical expertise shall be understood in this chapter as the ability to have knowledge about the rightness and wrongness of possible acts and courses of action.

2. CONSEQUENTIALISM

According to Moore, a question about the rightness or wrongness of some act has a simple answer: an act is right, or ethically permissible, if and only if its consequences are at least as good as the consequences of any alternative act, and an act is a duty if and only if its consequences are better than the consequences of any alternative act (p. 198). So, an act is a duty if it ranks first in the production of good consequences, and if there is a tie for first place, then every act tied for first is right, or ethically permissible. Moore, then, affirms consequentialism—"[t]he view that the value of an action derives entirely from the value of its consequences" (Blackburn, 1996, p. 77).

Two things are remarkable about Moore's endorsement of consequentialism. First, consequentialism is contrary to prominent strands in the history of ethics. As Moore writes, "The very opposite of this view has been generally prevalent in Ethics. 'The right' and 'the useful' have been supposed to be at least *capable* of conflicting with one another, and, at all events, to be essentially distinct" (p. 196). As Moore goes on to acknowledge, it is a matter of ethical common sense, for many people, that the ends do not justify the means, and many philosophers have held this as well (p. 196). They have held that complying with the categorical imperative is always right, that punishing the innocent is always wrong, and that justice must be done, though the heavens may fall. They have held, that is, that there are some means that no ends can justify. So Moore's consequentialism, simple though it is, is significant.

What is also remarkable about Moore's endorsement of consequentialism is the logical status he claims it has. For he claims that it is true in virtue of the meanings of the words and concepts in which it is expressed. That is, Moore does not settle for saying that assertions about what is right are *supported* by assertions about what is good; he claims that assertions about what is right *are* assertions about what is good:

> In asserting that the action is *the* best thing to do [a duty], we assert that it together with its consequences presents a greater sum of intrinsic value than any possible alternative (p. 76).

> In short, to assert that a certain line of conduct is, at a given time, absolutely right or obligatory, is obviously to assert that more good or less evil will exist in the world, if it be adopted than if anything else be done instead (p. 77).

And he explicitly invokes the concept of meaning in his claim that "'right' does and can mean nothing but 'cause of a good result,' and is thus identical with 'useful'" (p. 196). Given that Moore is perhaps unsurpassed among moral philosophers in the attention he pays to the distinction between the meanings of moral words and what may count as evidence for sentences containing them, we must take Moore seriously when he says, in effect, that consequentialism is true as a matter of definition. Finally, Moore dispels any worries about the propriety of this interpretation by affirming that "the assertion 'I am morally bound to perform this action' is identical with the assertion 'This action will produce the greatest possible amount of good in the Universe' " (p. 197).

Moore is almost alone among prominent consequentialists in claiming that the thesis they share is true as a matter of definition. Mill, of course, affirms a form of consequentialism—holding that "the end of human action, is necessarily also the standard of morality" (p. 214)—but gives no sign of thinking that this particular statement is true in virtue of the meanings of the words, or that non-consequentialists succumb to a verbal mistake; he seems to just think that they are making a mistake in ethical judgment. Similarly, Sidgwick affirms a form of consequentialism, but regards it as a matter of intuition, and hence synthetic, and not just a matter of analytically reflecting on the meanings of words (pp. xvii–xxiii). In fact, among the major consequentialists, only Jeremy Bentham anticipates Moore in claiming that consequentialism is true as a matter of meaning; he writes that when the words 'right' and 'ought' are interpreted in conformity to the version of consequentialism he favors (hedonistic utilitarianism), "the words *ought*, and *right* and *wrong*, and others of that stamp, have a meaning: when otherwise, they have none" (p. 13). But Moore vehemently rejects Bentham's hedonism, as we shall see in the next section, so even this similarity goes only so deep. So Moore's definitional consequentialism is significant indeed.

What grounds does Moore provide for this strikingly strong position? He begins his argument with the claim that "It is plain that when we assert that a certain action is our absolute duty, we are asserting that the performance of that action at that time is unique in respect of value" (p. 197). Now some would claim that this statement is far from obvious, and that if it is defensible, then it must be because the notion of some action's being unique in respect of value is extremely vague, and is nearly vacuous. But taking this proposition as given, Moore then surveys several possible meanings that might attach to the notion of some action's being unique in respect of value, and purports to show that none is satisfactory except the one that establishes the principle of consequentialism. But given, as just mentioned, the extreme vagueness of the notion of some action's being unique in respect of value, no assurance can be provided that the options Moore considers are exhaustive of all the possibilities. So Moore's consequentialism is likely to strike many readers more as a matter of dogma aspiring to definition than as a genuine analytic truth.

In any case, consequentialism is an essential component of Moore's approach to answering ethical questions, and it has two implications that are significant for our inquiry into ethical expertise. First, it means that we can safely impute to Moore the claim that ethical expertise requires, among other things to be discussed below, an appreciation that consequentialism is analytically true. For if a philosopher were to affirm consequentialism on other grounds, as Mill and Sidgwick do, or were to affirm as analytically true some variant of consequentialism that Moore rejects, as Bentham does, then although such a philosopher might on occasion reach ethical judgments that would match those that Moore would reach, still Moore would surely say that the other philosopher does not have ethical expertise but rather just stumbles onto the truth. For if her affirmation of consequentialism is not for the right reasons, then it is just luck that she affirms it at all, and no one whose judgments are so dependent on luck can be said to have expertise. Indeed Moore writes that "The main object of Ethics, as a systematic science, is to give correct *reasons*" for holding certain ethical judgments (p. 58). So this is the first of the three elements of the conception of ethical expertise I shall be imputing to Moore: the ability to appreciate that consequentialism is analytically true.

The second significant implication of Moore's consequentialism has to do with the further abilities that it implies are required for ethical expertise. Specifically, Moore's affirmation of consequentialism leads him to maintain that questions of the rightness and wrongness of acts are reducible to other questions:

> Whenever, therefore, we ask 'What ought we to do?' or 'What ought we to try to get?' we are asking questions which involve a correct answer to two others, completely different in kind from one another. We must know both what degree of intrinsic value different things have, and how these different things may be obtained (p. 77; see also p. 79).

In other words, we need to know which ends are good, and which means will lead to those ends. I turn to Moore's thoughts on these questions in sections 3 and 4.

3. INTUITIONISM ABOUT ENDS

Moore's view of what is required in order to identify the good is a distinctive, mysterious, and troubling part of his ethical theory. Rejecting the approach taken by many of his predecessors, both consequentialist and otherwise, Moore maintains that the notion of the good is simple and unanalyzable. Here is how he puts it:

> If I am asked 'What is good?' my answer is that good is good, and that is the end of the matter. Or if I am asked 'How is good to be defined?' my answer is that it cannot be

> defined, and that is all I have to say about it. . . . propositions about the good are all of them synthetic and never analytic. . . . nobody can foist upon us such an axiom as that 'Pleasure is the only good' or that 'The good is the desired' on the pretence that this is 'the very meaning of the word' (pp. 58–59).

In these remarks, some of which apparently refer to the views of Bentham, discussed above, we find Moore espousing a view about the good that can best be understood in contrast to the view about the right that we considered above. Whereas statements of the form "X is right" are equivalent to statements about the good (as we saw above) and statements about consequences, and thus can be analyzed into statements about the good and statements about consequences, statements of the form "X is good" succumb to no such analysis. Whereas the notion of duty can be "defined" in terms of the good (p. 198), the notion of the good is "indefinable" (p. 60).

Moore's view can be more fully understood by considering his use of two concepts that have come to be closely associated with him: the open-question argument and the naturalistic fallacy. To see the substance and force of the open-question argument, begin with a purported analysis of the notion of the good, such as the claim that the good is the pleasant. According to Moore, if this were a satisfactory analysis of the notion of the good, then the claim that something pleasant is good should strike us as a tautology, and the question of whether some particular pleasant thing is good should strike us as straightforwardly self-answering, like the question of whether some particular pleasant thing is pleasant. But, Moore's argument goes, this question is actually perfectly open: one may very reasonably ask whether some particular pleasant thing is good, without asking something akin to whether some particular pleasant thing is pleasant. And this point can be generalized, Moore says: "whatever definition is offered, it may always be asked, with significance, of the complex so defined, whether it is itself good" (p. 67).

Now Moore's point here is not to show the falsity of claims such as that the good is the pleasant. Although he does think that many such claims are false, his point here is that even if they were true, they would not be correct *analyses* of the notion of the good. At best, they would be true predications of goodness to things of particular descriptions, just as the claim that my desk is gray is a true predication of the notion of the gray to my desk. But just as such a claim would have no place in an analysis of either concept, so likewise Moore maintains that no statement about the good and properties such as pleasantness, however true such a statement may be as a matter of correct predication, has anything to offer towards an analysis of the concept of the good.

This error—the error the open-question argument exposes—is what Moore calls the naturalistic fallacy:

> It may be true that all things which are good are *also* something else. . . . And it is a fact that Ethics aims at discovering what are those other properties belonging to all things which are good. But far too many philosophers have thought that when they named those other properties they were actually defining good; that these properties, in fact, were simply not 'other,' but absolutely and entirely the same with goodness. This view I propose to call the 'naturalistic fallacy' (p. 62).

In other words, the naturalistic fallacy involves "identifying an ethical concept with a 'natural' concept, or description of the features of things in virtue of which they are supposed good or bad" (Blackburn, 1996, p. 255).

In understanding the open-question argument and the naturalistic fallacy, it is important to note that these concepts pertain not only to attempts to define the good in terms of genuinely naturalistic criteria, but also to attempts to define the good in terms of what might be called supernatural criteria or, as Moore says, "metaphysical" (p. 161) or "supersensible" (p. 162) criteria. (Indeed the central chapters of Moore's book, all meant to expose instances of the naturalistic fallacy, are called 'Naturalistic Ethics', 'Hedonism', and 'Metaphysical Ethics'.) For example, the claim that the good is what God wills may, again, be true; but it is perfectly reasonable to ask whether something that God wills is good—it is reasonable in the sense that one is not just asking a question about the meanings of the concepts. So the open-question argument shows the futility of metaphysical ethics as well as naturalistic ethics (which Moore takes to include hedonism as one of its virulent strains), and the naturalistic fallacy is committed even by those who regard ethics as a supernatural, and hence (in a sense) non-natural, area of inquiry.

One objection is frequently lodged against the open-question argument, and is worth noting here. The open-question argument rests on a comparison, or contrast, we draw between the way in which different questions strike us. "Is this particular pleasant thing good?" strikes us differently from "Is this particular pleasant thing pleasant?" The latter is obviously to be answered in the affirmative, whereas we feel less certain about how to answer the former question (even assuming that the referent of 'this particular pleasant thing' is obvious). And this, Moore claims, shows that there is a difference in the meanings of the questions, and a difference between the concept of the good and the concept of the pleasant. But, it is frequently objected, the fact that certain concepts strike us differently does not mean that they are not, in fact, identical. It is possible that although questions such as whether some particular pleasant thing is good strike us as open, this is only because we lack a complete understanding of the concept of the good. So the open-question argument, and the wide applicability that Moore claims for the notion of the naturalistic fallacy, must be accepted with some caution.

Despite these concerns, no conception of ethical expertise can reasonably be imputed to Moore without taking seriously his rejection of all analyses of the notion of the good. This notion, he writes, is "a simple, indefinable, unanalysable object of thought" (p. 72). It is, in a word, ineffable.

How, then, is anyone supposed to accurately attribute goodness to anything, if the notion itself resists analysis? How is ethical expertise even possible? Here Moore offers some guidance, but more warnings as well. The bulk of Moore's guidance comes in his espousal of a particular method for ascertaining whether something is good: in order to ascertain the value of some particular quality, such as consciousness of pleasure, "it is absolutely essential to consider each distinguishable quality *in isolation*, in order to decide what value it possesses" (p. 145). That is, "We must ask: Suppose we were conscious of pleasure only, and of nothing else, not even that we *were* conscious, would that state of things, however great the quantity, be very desirable?" (p. 147). Moore goes on to explain that his method applies to both absolute and relative judgments of goodness. In order to know what things have intrinsic value, or are good,

> it is necessary to consider what things are such that, if they existed *by themselves*, in absolute isolation, we should yet judge their existence to be good; and, in order to decide upon the relative *degrees* of value of different things, we must similarly consider what comparative value seems to attach to the isolated existence of each (p. 236).

Here is an example of the application of this method, in order to judge the relative goodness of beauty and ugliness, taken in isolation from the question of what it is like for persons to perceive them:

> Let us imagine one world exceedingly beautiful. Imagine it as beautiful as you can; put into it whatever on this earth you most admire—mountains, rivers, the sea; trees, and sunsets, stars and moon. Imagine these all combined in the most exquisite proportions, so that no one thing jars against another, but each contributes to increase the beauty of the whole. And then imagine the ugliest world you can possibly conceive. Imagine it simply one heap of filth, containing everything that is most disgusting to us, for whatever reason, and the whole, as far as may be, without one redeeming feature. . . . [S]upposing them quite apart from any possible contemplation by human beings; still . . . it [is not irrational] to hold that it is better that the beautiful world should exist, than the one which is ugly (p. 135).

The method, according to Moore, is perfectly general. One simply entertains the idea of the thing(s) to be assessed, and intuitively perceives them to be good or bad, or better or worse than certain other things.

Moore warns, however, that one cannot verify, introspectively, the accuracy of one's intuitions. He writes that it is "a very natural, though an utterly false supposition that for a thing to *be* true is the same thing as for it to be perceived or thought of in a certain way" (p. 184). But, he replies, there is nothing about the state of knowing something, even knowing it with a certain level of confidence or certitude, that distinguishes that state from one that may involve a belief that happens to be false. When someone says that she knows *p* and we believe that *p* is false, we do not need to accuse her of mistaking or misreporting the contents of her own mind; on the contrary, we can affirm that she has *p* in mind in exactly the way in which she would have *p* in mind if *p* were, in fact, true (pp. 184–185). As Moore puts it, "in every way in which it is possible to cognise a true proposition, it is also possible to cognise a false one" (p. 36).

So the method of isolation cannot be augmented by introspection on the part of the person making the ethical judgment. Moreover, Moore claims, judgments of good and bad, and of better and worse, cannot be based on anything else, because no statement of any other kind can count as evidence for any of them. Not even broad agreement on some such judgment counts as evidence for it:

> the fundamental principles of Ethics must be self-evident. But I am anxious that this expression should not be misunderstood. The expression 'self-evident' means properly that the proposition so called is evident or true, *by itself* alone; that it is not an inference from some proposition other than *itself*. The expression does not mean that the proposition is true, because it is evident to you or me or all mankind, because in other words it appears to us to be true. . . . By saying that a proposition is self-evident, we

> mean emphatically that its appearing so to us, is *not* the reason why it is true: for we mean that it has absolutely no reason (p. 193).

This last remark is of the highest significance. Precisely because the notion of good is indefinable and unanalysable, a statement that a certain thing is good is, strictly speaking, unsupportable. One either regards it as true or one does not, and it doesn't have any logical connections to other statements of which one may be confident and with reference to which one might hope to judge or confirm the statement in question.

So Moore's view that ends are known to be good by intuition is, to say the least, rather mysterious. It is also troubling, since it implies that if two people have divergent beliefs about the good, then there is no method by which this can be empirically settled. Consider Moore's judgment that "By far the most valuable things, which we can know or imagine, are certain states of consciousness, which may be roughly described as the pleasures of human intercourse and the enjoyment of beautiful objects" (p. 237). Congenial though this judgment may have been to Moore and his fellows, what can Moore say to someone who thinks friendship and the enjoyment of beauty are, while valuable, no more valuable than other pleasurable states of consciousness, regardless of whether they have to do with—to use Bentham's example—pushpin or poetry? In anticipation of this concern, Moore writes that

> Many of the judgments, which I have made in this chapter [about what things are good], will, no doubt, seem unduly arbitrary: it must be confessed that some of the attributions of intrinsic value, which have seemed to me to be true, do not display that symmetry and system which is wont to be required of philosophers (p. 270).

In response to this concern, he writes that

> We have no title whatever to assume that the truth on any subject matter will display such symmetry as we desire to see—or (to use the common vague phrase) that it will possess any particular form of 'unity.' To search for 'unity' and 'system,' at the expense of truth is not, I take it, the proper business of philosophy, however universally it may have been the practice of philosophers (p. 270).

This response, one might concede to Moore, puts the burden of proof back on any proponent of "symmetry and system".

But the concern stated above has two prongs, and Moore's response deflects only the second one. For prior to any worry about symmetry and system is a worry about arbitrariness, which Moore does not address. To distinguish this worry from the one about symmetry and system, imagine an objector who says to Moore,

> Admittedly, symmetry and system are not to be found here, despite previous philosophers' conceit that it was. But my judgments are the opposite of yours, and thus are neither more nor less systematic; or they simply diverge from yours in various ways, in ways that leave them, as a whole, neither more nor less systematic than yours. How do we know that your judgments are right and mine are wrong?

To this question Moore seems to have no answer. In fact, he himself confesses that he is less interested in defending the results of his application of this method as correct than in defending the method itself as sound (p. 271). So, although Moore's method may be sound in principle, in practice it leaves room for troubling divergence among persons' ethical judgments—even, presumably, those of experts.

As I said at the beginning of this section, Moore's view of what is required in order to identify things that are good is distinctive, mysterious, and troubling. Nevertheless, it firmly establishes another element that must be included in any conception of ethical expertise we impute to Moore: the ability to discern, by intuition, the goodness and badness, both absolute and relative, of various possible ends.

4. EMPIRICAL INVESTIGATION OF MEANS

We have examined two of the three elements of Moore's conception of ethical expertise. One is the ability to factor judgments of right and wrong into (1) judgments of good and bad and (2) judgments of cause and effect, and the other is the ability to make the requisite judgments of good and bad. The third element, of course, is the ability to make the requisite judgments of cause and effect.

This element of ethical expertise, though less mysterious than the second, is no easier to attain or to execute. On the contrary, even if we assume perfect knowledge of the absolute and relative goodness of various ends, myriad factors conspire to complicate any judgment to the effect that certain means, in the form of performing a certain act, will lead to the best available consequences. First, it is hard to know what even the short-term consequences of a given act will be. Just as Lincoln did not know what awaited him at Ford's theater, or Andrew Wiles when he sat down to work on the day he saw the proof of Fermat's last theorem, so likewise we do not know what awaits us when we make breakfast, go to work, or come home and read our mail. Moreover, the long-term consequences of a given act are even harder to predict. And yet they all matter, because they all bear on what ends, and thus what sources of value and disvalue, will be produced by the means in question.

Second, even supposing perfect prediction of the consequences of any given act, it is basically impossible to be aware of, and to entertain explicitly, each of the act options that an agent may have at a particular time. Presumably chemicals that will treat or cure certain diseases can be concocted in scientists' laboratories right now, but so can thousands of others. Who can blame a scientist for failing to think of some random concoction and for failing to investigate its effects, when there is nothing (aside from what would be revealed only *by* an investigation into its effects) to commend it to the scientist's attention? The reason we don't blame such a scientist, of course, is that this happens to everyone: many of an agent's options are never noticed by her or anyone else, and they pass from being possibilities to being unexplored dead ends. And yet these possibilities would have to be noticed in order for an agent to know that a particular act is obligatory, or even ethically permissible. So the actual consequences available to an agent in a given situation are, in practice, utterly incalculable.

Considerations such as these are, of course, familiar to consequentialists; they receive ample attention in the work of such writers as Mill and Sidgwick. And for Moore they are no less compelling. As he writes, "it follows that we never have any reason to suppose that an action is our duty: we can never be sure that any action will produce the greatest value possible" (p. 199). And a similar conclusion holds for actions that are merely right, or ethically permissible.

A further complication, unnoticed by Moore and his predecessors, arises in virtue of the fact that for Moore, as for Sidgwick, the rightness or wrongness of an act depends on its actual consequences, not its intended consequences. For this view assumes that, for every act option, there is some fact of the matter as to what the state of the world will be at every future point in time. But soon after Moore wrote *Principia Ethica*, physicists began to worry that this was not true, that reality is irreducibly probabilistic, and that there is no fact of the matter as to what the state of the world will be even moments from now, not to mention into the indefinite future. Of course, the regularities observed among large-scale phenomena are not illusory, but they mask divine dice-throwing (to recall Einstein's metaphor) going on among very small-scale phenomena. Since these small-scale phenomena are not entirely swamped by the large-scale events, but occasionally rise up and influence them (as when a bomb is set to be triggered by one of them), they matter, in principle, to the prediction of acts' consequences. What *actual* consequences are to be imputed to an act whose stream of consequences includes at least one irreducibly probabilistic event?

An obvious solution is to impute to any act, as the value of its actual consequences, the average of the values of its possible consequences, each value being weighted by the probability of the occurrence of the consequences of which it is the value. But this solution has the odd implication of sometimes imputing to an act, as the value of its actual consequences, a value that does not attach to any of the act's possible consequences. For example, an act with two sets of possible consequences, one very good and one very bad, might have imputed to it, as the value of its actual consequences, ethical neutrality—neither good nor bad. But, by hypothesis, the act's consequences will be either very good or very bad. When these are the only possibilities, saying that the act's actual consequences are ethically neutral seems like a denial of the dictum that actuality implies possibility. Clearly a better solution needs to be found.

So the third element of Moore's conception of ethical expertise—the ability to make the requisite judgments of cause and effect—exceeds the bounds of feasibility, both as an epistemic matter and, apparently, as a metaphysical matter. This third element is as demanding as the second is mysterious.

5. ETHICAL JUDGMENT IN PRACTICE

I have explained that, for Moore, ethical expertise consists of (1) the ability to factor judgments of right and wrong into (a) judgments of good and bad and (b) judgments of cause and effect, (2) the ability to use intuition to make the requisite judgments of good and bad, and (3) the ability to use empirical investigation to make the requisite judgments of cause and effect. All these abilities are needed in order for a person to know that a certain act is right or wrong.

So stated, Moore's conception of ethical expertise seems hopelessly ambitious—certainly too ambitious to be of any use in practice. But Moore's view has an admirable simplicity that helpfully removes many variables from the activity of making ethical judgments, and Moore himself is not indifferent, in his *Principia*, to the need for more practical guidance. In closing, I'll touch on each of these topics in turn.

First, although the second and the third of the three kinds of knowledge involved in Moore's conception of ethical expertise are deeply problematic, Moore's view at least relieves the would-be ethical expert of having to inquire into topics that other theorists might deem essential to making an ethical judgment. For example, since Moore is concerned only with actual consequences (rather than, e.g., intended ones), the Moorean ethical expert does not need to inquire into the agent's motives, or—if she is the agent—assure herself that she does not have ulterior motives that might compromise the ethical quality of her act. Moreover, the Moorean moral agent does not need to attribute to an act any tendency, or regard the act as being of a type that usually has certain consequences rather than others. The actual consequences of the specific act itself are all that matter, regardless of tendencies and generalizations associated with acts of that type. Finally, since Moore's consequentialism is so demanding—forbidding any act that does not maximize the good—the Moorean ethical expert can remain oblivious to the distinction between the obligatory and the supererogatory. This may be an objection to the substance of Moore's ethical theory, but it simplifies the matter of ethical expertise considerably. So while in some ways Moore's conception of ethical expertise is impossibly ambitious, in other respects it boasts an admirable simplicity.

Second, Moore is concerned, in his *Principia*, to provide more practical guidance than is implicit in the notion of ethical expertise we've attributed to him. Turning his attention from the evaluation of individual acts to the determination of ethical rules that it would be desirable to have in force throughout a society, Moore writes that "If . . . we ask what rules are or would be useful to be observed in the society in which we live, it seems possible to prove a definite utility in most of those which are in general both recognised and practised" (p. 209). Acts in accordance with these rules, Moore argues, have the merit of promoting certain means and arrangements (such as life, liberty, and security of property) which are likely to be useful, under almost any social circumstances, for the attainment of just about any ends that people think are intrinsically good. So these rules can be recommended, albeit provisionally, more or less in independence of questions about exactly what things are good and bad, as long as the answers to those questions fall within certain broad ranges (p. 207). So simple common sense, even if it falls short of genuine ethical expertise, can determine a workable and responsible public morality.

But setting aside the formulation and promulgation of rules for a public morality, what about a choice that a specific individual may face? Again, Moore's answer is striking in its simplicity. If the individual's choice is one that is not governed by common-sense morality—if, that is, common-sense morality leaves the individual some latitude as to what to do—then the individual should just do what seems to have the best consequences: "the individual should . . . guide his choice by a direct consideration of the intrinsic value or vileness of the effects which his action may produce" (p. 215). If, on the other hand, the rules of common-sense morality provide guidance sufficiently determinate to effectively reduce the individual's choice to one of compliance or noncompliance, then the individual should simply comply. In exceptional cases, of course, better consequences might result from the individual's breaking the established rule rather than complying with it. But Moore's concern is with whether the individual would ever be *justified* in *regarding* his own case as an exceptional one. "And it seems," Moore says, "that this question may be definitely answered in the negative" (p. 211). For the genuine exceptions are so rare, the facts of individual cases are so hard to ascertain, and individuals are so susceptible to misjudge cases in which they are involved, that

> It seems, then, that with regard to any rule which is *generally* useful, we may assert that it ought *always* to be observed, not on the ground that in *every* particular case it will be useful, but on the ground that in *any* particular case the probability of its being so is greater than that of our being likely to decide rightly that we have before us an instance of its disutility. In short, though we may be sure that there are cases where the rule should be broken, we can never know which those cases are, and ought, therefore, never to break it (p. 212).

In practice, then, Moore enthusiastically recommends faithful support of and adherence to the rules of common-sense morality.

It follows, then, that the very simple, but epistemically demanding, conception of ethical expertise that is implicit in Moore's work is complemented by an equally simple, but epistemically undemanding, conception of how ethical decisions should be made in everyday life. Presumably, we should see Moore's support of common-sense morality as a consequence of his own reflection on the good and on the best means of promoting it—practical advice that follows from Moore's extensive reflection on the ineffable and incalculable matters that fall within the purview of genuine ethical expertise.

REFERENCES

Aristotle (1984). *Nicomachean Ethics*. W. D. Ross (Trans.); Rev. by J. O. Urmson. In: Jonathan Barnes (Ed.), *The Complete Works of Aristotle* (pp. 1729-1867), vol. 2. Princeton: Princeton University Press.

Baier, Kurt (1995). *The Rational and the Moral Order: The Social Roots of Reason and Morality*. Chicago: Open Court Publishing Company.

Baldwin, Thomas (1993). 'Editor's Introduction.' In: G. E. Moore, *Principia Ethica* (pp. ix-xxxvii), revised ed., Thomas Baldwin (Ed.). Cambridge: Cambridge University Press.

Bentham, Jeremy (1996). *An Introduction to the Principles of Morals and Legislation*. J. H. Burns and H. L. A. Hart (Eds.). Oxford: Oxford University Press.

Blackburn, Simon (1996). *The Oxford Dictionary of Philosophy*. Oxford: Oxford University Press.

Kagan, Shelly (1992). 'The Structure of Normative Ethics.' *Philosophical Perspectives* 6, 223–242.

Keynes, John Maynard (1972). *Essays in Biography*. Vol. 10 of John Maynard Keynes, *The Collected Writings of John Maynard Keynes*. London: Macmillan Press.

Mill, John Stuart (1969). *Utilitarianism*. In: J. M. Robson (Ed.), *Essays on Ethics, Religion and Society*. Volume 10 of *Collected Works of John Stuart Mill* (pp. 203-259). Toronto: University of Toronto Press.

Moore, G. E. (1993). *Principia Ethica*, rev. ed. Thomas Baldwin (Ed.). Cambridge: Cambridge University Press.

Narveson, Jan (1967). *Morality and Utility*. Baltimore: Johns Hopkins Press.
Rawls, John (1999). 'Outline of a Decision Procedure for Ethics.' In: Samuel Freeman (Ed.), *Collected Papers* (pp. 1–19). Cambridge: Harvard University Press. Originally published in *The Philosophical Review*, 60(2) (April 1951), 177–197.
Sidgwick, Henry (1981). *The Methods of Ethics*, seventh ed. Indianapolis: Hackett Publishing Company.

CHAPTER 7

GRIFFIN TROTTER

PRAGMATISM AND ETHICAL EXPERTISE

1. INTRODUCTION

Classical American pragmatism is known widely for its anti-speculative, nuts-and-bolts realism and for its focus on concrete, practical problems. William James, the first philosopher to publicly pronounce himself a pragmatist, conceived truth as "what works" and was famously allergic to grand metaphysicians such as Hegel. Charles Sanders Peirce, anointed by James as the founder of the pragmatist movement, savaged Descartes' systematic skepticism while endorsing a revised version of Scottish common-sensism. Both thinkers conceived the meaning of ideas in terms of sensible effects.

No wonder, then, that interest in classical American pragmatism has lately surged. With all the hankering for effective, user-friendly methods in bioethics, business ethics and other practical, action-oriented realms, pragmatism is a dream come true—and certainly more appealing than the remote analytic techniques that dominated moral philosophy in the middle of the 20th Century. The contemporary practical ethicist, like the pragmatist, is interested in action. She is an expert in deciding, or (more precisely) in facilitating decisions about what to do.

In this essay, I examine the phenomenon of ethical expertise, focusing on its relevance for public deliberation in secular, democratic societies.[1] Ethical expertise is potentially useful in such contexts for mediating ethical disagreements between parties with contrasting moral visions, or between parties with congruent moral visions who disagree about certain particular cases (a moral vision being a complex of leading ideas, habits and attitudes that structure the perception of moral problems and inform the response). However, the legitimacy of secular ethical expertise is threatened by the lack of a well-substantiated ethical canon. Two basic options are available for those wishing to vindicate a notion of ethical expertise, and each of these has precedents in a version of classical American pragmatism. The epistemic option occurs in Peirce and in Josiah Royce and reflects the values of the theoretician and the inquirer. The therapeutic option, decipherable at times in James and in John Dewey, draws on the values of the surveyor and the counselor.

Each option has its distinct epistemic and political underpinnings. Epistemically, there is a polarity between nominalist (James, Dewey) and realist (Peirce, Royce) accounts of ethical ends. Politically, there is a polarity between robust (Dewey) and

L. M. Rasmussen (Ed.), Ethics Expertise: History, Contemporary Perspectives, and Applications, 103-122.

attenuated (Royce, Peirce) accounts of the state's role in cultivating good lives. I examine each of these diverging pragmatic options and conclude with a tentative nod in favor of the epistemic option. In so doing I diverge from the trend of presenting "pragmatic ethics" as a relatively unified movement. My purpose, in part, is to highlight certain tensions that exist within pragmatism. Nevertheless, I regret that there is not sufficient space for a thorough review of the similarities and common causes that bind various pragmatists.[2]

2. THE PROBLEM OF ETHICAL EXPERTISE

For the purposes of this essay, an expert is any person who is recognized to possess extensive and/or specialized knowledge of a given object. Expert knowledge may be conceptual, as in knowing the details of quantum mechanics, or it may be practical and performative, as in knowing how to insert a cardiac pacemaker. The object of knowledge may be broad, as in knowledge of the evolution of species, or it may be specialized, as in knowledge of the reproductive system of the common toad (Yoder, 1998). Social recognition is important because being an expert is playing a social role (Fletcher, 1997, p. 299). Private knowledge may be powerful or enlightening, but it does not constitute expertise.

An ethical expert, then, is a person who is recognized to possess extensive and/or specialized knowledge in some domain of morals or ethics. Significant controversy exists in secular societies about the role of ethical expertise, largely because: (1) the knowledge base in ethics seems underdeveloped (Wildes, 1997), and (2) there is fear that anointed ethical experts will wield inordinate political power (Trotter, 2002).

The latter fear presupposes the former concern. That is, if ethics sported a highly developed, comprehensive, well-differentiated and widely acknowledged set of principles, like physics, there would be little hesitance about following the counsel of ethical experts. People endeavoring to make a good decision or live a good life would seek the advice of ethical experts, with no more concerns about "moral imperialism" than bridge builders have about "mechanical imperialism" when they seek the advice of engineers. Indeed, in certain moral communities (e.g., communities of Hasidic Jews) where ethical knowledge is considered secure both in its general content and in its application, the practice of consulting ethical experts (e.g., Rabbis) is widespread and unproblematic.[3]

The problem is that most of the world's secure ethical knowledge is "secure" only insofar as it can be linked with an authoritative source (such as God or Vince Lombardi). Such linkages require an element of faith, special insight or, on occasion, gullibility. In the absence of a common faith, reason flounders and moral exhortation takes on a carminative odor.

This problem is exacerbated in secular society,[4] which defines itself by its resistance to leaps of faith. Deprived of prophets, priests, and other authoritative voices, secular discourse is full of contention about the nature of the political good and the political right. Since there is no agreement in any robust sense on an ethical canon, it is difficult to agree about who possesses "extensive and/or specialized" ethical knowledge. Knowledge, after all, requires an object.[5]

Such a privation of secular ethical expertise would be less problematic in a secular society composed of a number of robust, well-grounded moral communities with their own internal authority systems (based on their several, diverging leaps of faith). Secular authority, in such a scenario, would presumably be grounded in the permission of participating individuals and communities.[6] Permission, in turn, would be negotiated, usually taking the form of a compromise (Trotter, 2002).

But contemporary secular societies are not neatly divided into particular moral communities. Many, perhaps most, individuals in countries like the United States remain un-grounded in a single robust moral tradition. The fact that a citizen of such a country attends a Lutheran or a Roman Catholic church, for instance, does not mean that the citizen will adhere to a specific, comprehensive system of moral practices and beliefs, dictated by their church. Our citizens tend to belong to a number of overlapping and sometimes diverging moral communities. We are often morally ambivalent, not merely about what we ought to do, but about where the authority for ethically right decisions lies. The result of this ambivalence is often a kind of moral ecumenism, consisting of a patchwork of moral beliefs gleaned from various divergent sources (MacIntyre, 1981, pp. 1-22).

Moral ecumenism becomes a moral vision in its own right when the patchwork coalesces into a somewhat stable body of shared moral beliefs, with its own history, its own internal dynamic and its own characteristic orientation to competing moral visions. Like any moral vision, moral ecumenism requires an account of moral authority or expertise. The ecumenical account turns out to be unique in several respects. First, ecumenists are unable to consult divinely inspired prophets and priests, but must instead rely on reason or on a common moral culture. Hence ecumenical ethical expertise tends to reside in domains of philosophy, psychology or social anthropology (rather than theology). Second, because they draw from a large variety of traditions, moral ecumenists are inclined to believe that their moral vision stands outside of tradition and possesses a kind of trans-cultural validity. Ethical experts in ecumenical communities are typically individuals who are adroit at synthesizing various strands of secular opinion and in providing an ostensibly reason-based justification for their synthesis.[7] Third, because they so strenuously eschew divine revelation and priestly authority, ecumenists tend not to recognize the manner in which their own moral communities are like the churches of old. This misperception begets ecumenists' tendency to exempt themselves from the church-state division, leading in turn to a policy of viewing their own high priests not merely as local moral authorities (with jurisdiction over a particular ecumenical community), but also as political authorities (with jurisdiction over secular society).[8]

These aspects of ecumenical ethical expertise will become crucially important in the subsequent discussion, but for now we need to underscore the fact that ecumenists, just like other members of secular societies, must find a way of living together with persons who harbor incompatible ethical beliefs. There is a need for secular moral authority—conceived as a means of mediating ethical disputes—and with it comes the need for an account of ethical expertise. The obstacle, as we earlier observed, is the paucity of epistemically well-established moral knowledge.

3. THE SECULAR OPTIONS

Four partially overlapping solutions to this conundrum have been offered in four accounts of the knowledge base of the ethical expert. The first views the ethical expert as essentially a theoretician, with extensive general knowledge of the various competing moral theories and moral visions. The ethical expert, on this account, is not an expert on what is ethically right or good, but rather an expert in the analysis of various systems of ideas about the right or the good.[9] The second account holds that the ethical expert is an inquirer—i.e., a person with special knowledge about how to conduct an inquiry about what is right or good.[10] The inquirer, once again, is not an expert on what is ethical. She is an expert on how to think and reason about ethics. Next is the surveyor. This third kind of ethical expert is an individual with extensive knowledge of prominent (though often embedded) customs about what is ethically right and good.[11] Unlike the theoretician, who is interested in how people justify what they do, the surveyor attends primarily to what people do, regardless of how they ultimately justify it. Finally, the fourth account identifies the ethical expert as a counselor. The counselor is a person who helps individuals and communities to become ethically good, or to do ethically good things.[12] Like the "client-centered" behavioral counselor, this kind of ethical expert is unconcerned about delineating a definitive account of what is right or good. Instead, the counselor assumes that clients possess an adequate rough-and-ready knowledge of the good (or at least of what is good for them), and they focus on helping the client to articulate and realize it.

There is an obvious affinity between the ethical expert as theoretician and the ethical expert as inquirer, just as there is an affinity between the surveyor and counselor roles. While the theoretician and the inquirer are concerned, respectively, with articulating and testing hypotheses about moral truth, the surveyor and counselor are relatively unconcerned with truth, but eager, respectively, to understand the existing relation between what we value and what we do, and to smooth out this relation as much as possible.

These affinities between the possible knowledge bases of ethical experts produce two major options for ethical expertise in contemporary secular society: the theoretician-inquirer (the "epistemic option") and the surveyor-counselor (the "therapeutic option").

4. PEIRCE AND JAMES ON INQUIRY AND EXPERTISE

Both of the aforementioned options can be deciphered in classical American pragmatism. They are opposite sides of a cleft that opened in the earliest stages of the movement, then plunged ever deeper. On the therapeutic side are James and Dewey, with their emphasis on realizing ends-in-view. On the epistemic side are Peirce and Royce, who are happy enough to pursue ends-in-view, but who insist that such ends are authoritative only insofar as they can be linked to a more remote end-not-quite-in-view: the truth.

Peirce argued in 1878 that "the whole function of thought is to produce habits of action" and that "the identity of a habit depends on how it might lead us to act, not merely under such circumstances as are likely to arise, but under such as might possibly occur, no matter how improbable they may be" (Peirce, 1992, p. 131). This claim begot Peirce's pragmatic theory of meaning—holding that the meaning of an idea consists in its conceivable sensible effects (Peirce, 1992, p. 132). Bothered by possible nominalist implications of this doctrine,[13] Peirce later stressed that the determinations of a general idea are largely functions of the general character of reality itself, which is encountered but not exhausted in experience, rather than being solely effects or conventions of particular thoughts (which interpret this general character) (Peirce, 1998, p. 343).

James, on the other hand, took Peirce's pragmatic maxim in the opposite direction. In order to alleviate tiresome speculation about mere possibilities or about the conceivability of an effect, James distilled the doctrine of meaning even further (and, to Peirce's chagrin, in a decidedly nominalist direction). The meaning of an idea or concept, for James, consists in its *respective* sensible effects—that is, in the actions and experiences that actually follow if the idea is true or when the idea is believed, adopted and employed.[14] Then James added to the doctrine of meaning a doctrine of truth. "True ideas," he wrote, "are those that we can assimilate, validate, corroborate and verify" (James, 1981, p. 92).[15] Truth "happens" to an idea as it proves to be useful over time.

James' revisions are nominalist in two respects: (1) generality is regarded as a feature of individual consciousness (and hence lacks repose in a reality that is independent of any particular thinker),[16] and (2) thinkers are described as assigning general ideas to constellations of experiences (objects) whose only explicit relation is that it is expedient (for the thinkers) to group them together.[17] I will call the first of these orientations "subjective nominalism" (William of Ockham is representative) and the second "instrumental nominalism" (Thomas Hobbes is representative).

James' nominalism deeply disturbed Peirce—so much that he changed the name of his own theory to "pragmaticism" in order to differentiate himself from James (Peirce, 1998, pp. 334-335). For Peirce, inquiry aims at discerning reality's objective features. If reality is not independent of individual thinkers, or if inquirers lack experiential contact with reality's general features, then the community of inquiry is only ostensibly a community and inquiry has no object. Inquiry in such circumstances would be chaotic and ultimately pointless.

Though Peirce acknowledges that human fulfillment often hinges on the application of true ideas, and that true ideas are true by virtue of the fact that they fulfill purposes,[18] he is more concerned with long-term fulfillment (experienced within a community of inquiry as it confirms a general hypothesis) than with the immediate happiness or equanimity of individuals. Because of this focus on the long run, Peirce is skeptical about overly "practical" thinking. While insisting on the intimacy between thought and action,[19] he held that the best inquirers should be rather single-minded about finding the truth, and that they should as far as possible remain detached from ordinary concerns. Peirce writes: “True science is distinctively the study of useless things. For the useful things will get studied without the aid of scientific men. To employ these rare minds on such work is like running a steam engine by burning diamonds” (Peirce, 1955, p. 48).

Strong moral conviction, deep religious faith and concern for worldly problems are viewed by Peirce as potential sources of bias, since there is little genuine ethical knowledge upon which to base them. Practical moral thinking thus presents a potential impediment to inquiry. For Peirce, the significance of truth lies, indeed, in action. But, paradoxically, over-concern with action is the single greatest barrier to finding the truth, and hence to acting well.

This contrast between Peirce and James seems to be predicated, at least in part, on their differing temperaments. Peirce grew up in the laboratory and studied chemistry. His version of enlightened philosophy (including moral philosophy) stresses objectivity and detachment. James, on the other hand, agonized about the meaning of life and studied medicine. For him, philosophy is a quest for satisfaction. These differences produce significantly diverging notions of ethical expertise that align roughly with the diverging secular options described in the previous section. While Peirce conceives the expert as helping us find the truth (in accordance with the epistemic option), James sees the expert as helping us live satisfying lives (in accordance with the therapeutic option).

5. FROM PEIRCE AND JAMES TO DEWEY AND ROYCE

Since neither Peirce nor James spent much time on social ethics, I will turn to a couple of their successors—Josiah Royce and John Dewey—to complete my account of ethical expertise in classical American pragmatism. My assumption is that Royce works out a version of the epistemic option identified with Peirce and that Dewey constructs a version of the therapeutic option identified with James.

By now Deweyans reading this essay are probably chomping on the bit, eager to take issue with at least two characterizations of Dewey that they are sure to find inaccurate: (1) I have implied that Dewey is, like James, a nominalist, and (2) I have implied that Dewey is not substantially interested in “inquiry”. Certainly Dewey would reject both characterizations. Dewey explicitly rejected subjective nominalism (he called it “traditional nominalism”) (Dewey, 1988a, pp. 145-147), and also was not an instrumental nominalist in the strong, Hobbesian sense. Yet he is clearly more an ethical nominalist than Peirce or Royce was, because his account of ethical generals or “reals” is far less robust. For Peirce and Royce, as for other classical ethical realists,[20] ethical reals have an ultimacy, and a staying power, not found in Dewey’s reals. For the former, the ethical real is akin to the classical *telos*. As an essential cause, it eclipses or transcends human experience, even as it shapes it. As a final cause, it supervenes on our present fulfillments, but becomes fully determinate only in the infinite future. Dewey arguably also has a *summum bonum* or final cause: growth (Dewey, 1934, p. 56; Dewey, 1948, p. 177). For Dewey, as for Peirce and Royce, the *summum bonum* can be viewed as a process-consummation complex.[21] But for Dewey, moral reals (including "growth") are always "generated by existents" and must be sustained by events (Dewey, 1988a, p. 311). The possibility of transcendence is ruled out from the outset. Though Dewey insists that one feature of intelligent ends is that they are projected deeply into the future (Dewey, 1988a, pp. 303, 311), his horizons are distinctively human and he believes that the "detection and definition of nature's end is in itself barren" (Dewey, 1988a, p. 309). For Peirce and Royce, such evanescent reals are not reals at all, but at best constellations of related events. Hence, from their vantage point, Dewey is an instrumental nominalist.[22]

Regarding the second point (Dewey's interest in inquiry), Dewey is clearly very energetic about articulating a theory of inquiry. However, insofar as inquiry is characterized as a truth-directed process, the notion of inquiry will hinge to some degree on the notion of truth. My above characterization of the moral expert as inquirer borrows from a more traditional sense of truth, which Peirce and Royce accept but Dewey repudiates (just as he would probably repudiate or reconstruct the distinction between epistemic and therapeutic options). The intention is not to presuppose that Peirce and Royce have identified the correct sense of inquiry, and that Dewey is mistaken. To the contrary, I acknowledge two senses, while using only one of them at a time.

Certain pitfalls in my strategy should be acknowledged from the outset. First, the pairing of pragmatist thinkers is task-dependent and might be constructed otherwise if our concerns were different. For instance, if I were discussing the significance of the individual person within the larger community, I might pair James and Royce (who emphasize the uniqueness of individuals and the important consequences of individual choice) against Peirce and Dewey (who emphasize the manner in which individual thoughts and choices are shaped by culture and nature). But I am concerned with tracing the effects of the realism-nominalism split in pragmatism, as it pertains to the conception of ethical expertise. For this project, Peirce and Royce are clearly at one pole, and James and Dewey closer to the other.

Second, I will skip many of the internal disputes and defer to the lead thinkers. Because Dewey develops the implications of a limited ethical nominalism in social philosophy more thoroughly than James, and because he is more emulated than James in contemporary practical ethics, I will follow Dewey where they part ways. Peirce and Royce are more closely linked than James and Dewey as moral philosophers (especially at the end of their respective careers, when Royce adopted Peirce's semiotics and Peirce settled into a very Roycean moral world view). However, where they diverge, I generally will follow Royce.

Third, my engagement with classical American pragmatism is not pure. That is, I have ulterior concerns about linking my discussion with contemporary issues in practical ethics, especially bioethics (where I do most of my own work). This practical bias may come across as a concession to the therapeutic school, but in fact Peirce and Royce also believe that inquiry is always and inevitably shaped by current problems. As we shall see, their differentiating characteristic (contra James and Dewey) is that they tend to interpret immediate practical and theoretical problems in light of long-range commitments such as "discovering the truth" and "serving humanity". As I have argued in the past, the marriage of the practical and the eternal is the essence of Royce's ethical project (Trotter, 1999).

6. ROYCE ON INQUIRY AND EXPERTISE

Royce bases his account of the moral life on the struggle of the individual to cultivate a coherent and satisfying life plan. Long before the birth of bioethics, Royce identified the principle of autonomy as a "first principle of all ethical inquiry" and held that it is acknowledged, at least implicitly, "even by apparently extreme partisans of authority."[23] For Royce, this principle consists of the insight that "the ultimate moral authority for each of us is determined by our own rational will" (Royce, 1995, p. 14). Alongside this principle, Royce placed an equally fundamental social principle—"that I can never find out what my own will is by merely brooding over my natural desires, or by following my momentary caprices" (Royce, 1995, p. 14). Hence, in formulating a life plan, the individual must: (1) bring his/her own will explicitly to self-consciousness, and (2) turn outward (socially) for collaboration and help.

This turning outward is accomplished in the context of moral communities. To ameliorate the tension between the autonomy principle and the social principle, individuals must freely choose one or more compatible moral communities and devote themselves to communal ideals. Hence, loyalty, which Royce defines initially as the "willing, practical and thoroughgoing devotion to a cause," is ultimately understood as the love of a community. Each person and each life plan are unique insofar as they embody a personal ideal that (1) integrates a unique array of causes, emanating from membership in a unique array of larger and smaller moral communities, and (2) embodies a unique and tangible form of service to the chosen communities.

Ethical experts are invoked at a very early stage, as youngsters imitate and consult parents, teachers, and various other role models. As persons mature, they differentiate themselves from parents and siblings through their choice of roles and causes, and by establishing new social alliances outside the original nexus of family, church, etc. Insofar as they become well-integrated members of stable moral communities, morally mature persons generally have few difficulties in identifying other persons with authority for dispensing moral advice or helping with particular moral judgments.

The authority of the ethical expert, on Royce's view, extends only to fellow loyalists within a particular moral community. Each community will have its canons of ethical expertise, and these are hard to specify in a generic way (except, perhaps, to say that ethical expertise implies a deep familiarity with the ideals that define the given moral community). Indeed, Royce recognizes the reality of a "Great Community" of all humanity, harboring moral values that bear authority across all subordinate communities.[24] But there are no ethical experts, in any robust sense, on the values of this Great Community. Our current conceptions of moral principles applying in the Great Community are far less differentiated, for instance, than the conceptions of physical or biological laws, and hence very much up for grabs. As

Peirce comments, "if the ordinary theorems of molecular physics and of archeology are but the ghosts of beliefs, then to my mind the doctrines of the philosophers are little better than the ghosts of ghosts" (Peirce, 1998, p. 156). Royce would concur.[25]

For Royce, then, the therapeutic option is available only to ethical experts who reside within a particular, robust moral community. For members of one moral community to advise or reprimand members of a rival moral community, they would have to demonstrate a distinct coherence between the moral ideals of their community and those of the Great Community, and show that this connection is lacking in the rival community. But the lack of well-substantiated discursive knowledge about the Great Community makes this sort of demonstration implausible in most cases. In the secular context (i.e., in the context of a society where diverging moral communities seek to collaborate), ethical expertise can function only in accordance with the epistemic option. Ethical experts can summarize competing theories, and clarify terms and arguments. They can also fashion new theories and arguments, and suggest ways in which these might be tested. When therapy-style experts from the robust communities enter public dialogue, they can represent these communities as trustees or delegates. But they bear no overarching moral authority for society as a whole.

7. DEWEY ON INQUIRY AND EXPERTISE

Because Dewey is an instrumental nominalist, he is relatively unconcerned with vague or far-off ideals.[26] Like James, Dewey is keen to examine and coordinate human desires and habits as they arise in day-to-day experience. Some reengineering of desires and habits is essential, but these modifications are undertaken not for the sake of epistemic progress or service to an ideal community. They are undertaken, instead, because certain aims and habits (especially those that involve social integration and personal growth) are more rewarding in the foreseeable future, and demand that we bring counterproductive aims and habits (such as the will to dominate our enemies) to heel. The regulative ends are always distinct ends-in-view, but never final causes.

In *The Public and its Problems*, Dewey offers a threefold analysis of early Twentieth Century America (Dewey, 1984, pp. 308-315): (1) a bewildered public has lost its sense of collective identity, to the point that it is no longer genuinely a "public,"[27] (2) the public will find itself only when it achieves a sufficient moral and political consensus to promote large-scale cooperation, and (3) movement toward this end should be guided by scientific experts (including ethical experts). Many contemporary Deweyans advance similar claims about contemporary America. Let us consider each of the claims in turn.[28]

7.1. The Lost Public

For Dewey, the diminution of social ideals results from a complex of intersecting (and often colliding) governmental, corporate and private activities. Dewey laments the demise of mid-size deliberative bodies such as we found in the heyday of "town hall" democracy. If the public is to realize its potential as an intelligent social force, Dewey believes it must cultivate new forms of face-to-face association and interaction.

Yet Dewey and some of his contemporary followers seem reluctant to concede the political importance of extant deliberative bodies such as families, churches and professional associations. John Stuhr, for instance, invokes Dewey in criticizing the Republican Party for its "crucial assumption" that local communities now exist. On Stuhr's view, this assumption is "simply false" (Stuhr, 1997, p. 263).

The no-community thesis, combined with Dewey's insight that full personhood is a function of community life, implies a startling conclusion: America is essentially devoid of politically functional persons. In order to restore progressive politics, and self-rule, Stuhr holds that we must create new moral and political ideals and instill these ideals so deeply that human nature itself will appear transformed. Hence, Stuhr writes:

> Dewey's point is more radical and complex than it may seem to those who are concerned simply to redistribute income or extend economic initiatives and opportunities to those who are poorest. For Dewey, this is insufficient: We must form a new psychological and moral type of individual; we must forge a new individualism, a new human nature (Stuhr, 1997, p. 266).

7.2. The Need for Consensus

Dewey's emphasis on consensus can be articulated in terms of the experimental view of inquiry, shared roughly by all classical American pragmatists (but applied very differently by Peirce and Royce). For Dewey, the reclamation of our lost public requires large-scale cultivation of new cultural habits (such as liberal humanism and development of the social sciences). Dewey regards these habits as elaborate social hypotheses, to be tested in the laboratory of public intercourse. Their success or failure will be measured by the degree of human fulfillment, harmony and sustainable growth that they inaugurate.

In order for Dewey to conduct his social experiment, he needs to control for potential confounders. These confounders include: (1) misgivings about habits that Dewey wants to inculcate, (2) misgivings about the potency of the social sciences as tools for engineering a good society, and (3) non-scientific modes of belief fixation. The major problem for Dewey is that in order to eradicate (1) and (2) he must accept (3). That is, since the scientific evidence for Dewey's versions of humanism and social engineering is sparse at best, Dewey must employ a non-scientific mode of belief fixation in order to sell his vision of social reform. To accept Dewey's program on the basis of (or perhaps despite) current evidence from the social sciences requires a leap of faith, not a devotion to scientific method.

In fact, Dewey provides two pathways for such a leap. The first hinges on Dewey's employment of a rather non-pragmatistic quasi-transcendental argument. Dewey writes:

> Flux does not have to be created. But it does have to be directed. It has to be so controlled that it will move to some end in accordance with the principles of life, since

> life itself is development.... In any case, civilization is faced with the problem of uniting the changes that are going on into a coherent pattern of social organization. The liberal spirit is marked by its own picture of the pattern that is required: a social organization that will make possible effective liberty and opportunity for personal growth in mind and spirit of all individuals. Its present need is recognition that established material security is a prerequisite of the ends which it cherishes, so that, the basis of life being secure, individuals may actively share in the wealth of cultural resources that now exist and may contribute, each in his own way, to their further enrichment (Dewey, 1991, pp. 61-62).[29]

Dewey argues that material security (of unspecified degree), is a necessary condition for fulfillment, but is lacking in contemporary society. Liberty, he concludes, is no longer possible without a social scheme that guarantees security. And it is no longer conceivable merely as freedom from interference (Dewey 1988b, pp. 199-200). Liberty now also denotes the possession of goods and services that seem necessary for physical, intellectual and spiritual development—and the defense of liberty begets an entitlement to these goods and services. Because liberal reform is portrayed as a necessary condition for human flourishing, empirical data showing that Dewey's program produces specific outcomes is no longer required.

A second pathway to consensus about liberal humanistic values is Dewey's quasi-religious "common faith." Dewey believes that religious traditions often pose a troublesome impediment to social progress. He notes, however, that the "religious" adjective applies to "attitudes that may be taken toward every object and every proposed end or ideal" (Dewey, 1934, p. 10). Hence it is possible to direct the religious impulse toward Dewey's social reform movement. And such is precisely what Dewey recommends. On his view, special religious truths should be "surrendered" on behalf of a common faith in human progress (Dewey, 1934, p. 33). He even suggests that we give the name "God" to the natural process by which human ideals are generated, adorned, pursued and modified (Dewey, 1934, p. 26, 51).[30]

7.3. The Need for Ethical Expertise

Dewey's conception of participative democracy precludes a robust role for ethical experts—at least in the long run. Democratic values, including an understanding of inquiry and a capacity to deliberate effectively, are for everyone, not just for the experts. However, the impoverishment of public life, insofar as it is perceived by Dewey, or in the present case by some of his philosophical descendents, results in a situation where those adept at the theory and practice of inquiry form a depleted corps of capable individuals. Their leadership would seem to be essential—in the schools, courts, and policy chamber—if we are to cultivate a new generation of intelligent inquirers.

Hence, we can provisionally carve a Deweyan therapeutic version of ethical expertise. This version of expertise dovetails with Dewey's concept of expertise in the social sciences. He tends to the view that (1) sociobiologic regularities characterize human life and (2) progress occurs when we harness these regularities to enhance predictability and the integration of human energies.[31] The ethical expert is an individual who adroitly wields knowledge of the social, biological and physical sciences, in the criticism and reconstruction of social ends, and in the cultivation of means and habits that fulfill these ends. In other words, the ethical expert is the agent of "liberalism," whose task is "the mediation of social transitions" (Dewey, 1991, pp. 54-55). The specifically ethical content of this role resides in the expert's ability to (1) identify ethical values at work in ordinary social interactions, (2) trace positive and negative consequences of these values, (3) imaginatively revise these values (both as means and as ends-in-view), (4) imaginatively rehearse the revised values, employing social sciences to predict consequences, and (5) assist in the reconstruction of public morality based on intelligent hypotheses about the most plausible and useful value revisions.

In order to convert parents, schoolteachers, clergy and other traditional authorities into agents of constructive reform, there is a need for oversight by those who can interpret the big picture. Ethical experts from traditional communities must be unshackled from the restraints of superstition and thoughtless habit. Until this transformation is effected, ethical expertise—viewed in the therapeutic sense as the authority to prescribe moral and political remedies—must reside in the hands of Deweyan liberals (Dewey, 1984, pp. 362-365).[32]

8. CONTRASTING THE OPTIONS

It is impossible to articulate a doctrine of the operation of ethical expertise in secular society without making presumptions about the proper structure and function of secular society. In other words, ethical expertise in the secular context is an inherently political concept. Hence, the contrast between John Dewey's version of the therapeutic option and Josiah Royce's version of the epistemic option can be sharpened by a more explicit statement of the respective political trappings.

In the vernacular of contemporary political philosophy, one could argue that Dewey adheres to a perfectionist version of liberalism.[33] Like contemporary perfectionists (Raz, 1986), Dewey thinks that states are obligated to discern and articulate the characteristics of good living, and then to encourage good lives through legislation. As a perfectionist, Dewey is an ethical pluralist in an attenuated sense only. It is desirable, Dewey thinks, to promote a plurality of concrete lifestyles, and contrasting values, because human satisfaction is enhanced by the growth of complexity. But these values must be congruent and they must conform to the broader, liberal vision of good living.[34]

Ethical experts, when they are needed at all, are on this view social therapists.[35] They advise politicians about the best policies for promoting good living. And they counsel individuals and institutions about how to live well in their given circumstances. Because Deweyan ethical experts repudiate the notion of an ultimate ethical good or *telos*, their counsel hinges on accounts of how to revise, reconstruct and integrate certain ends-in-view. To legitimate such advice, they must foster consensus (at least among themselves). When therapeutic experts think deeply on ethical issues, scientifically survey relevant realms of human experience, examine relevant arguments, and confer earnestly among themselves, they tend to think that their views have special moral and political import.[36]

Royce, on the other hand, never offered a robust account of the good society, or even of a good life. His account of loyalty focuses on dedication to a cause, but greatly under-determines the specific characteristics of a good cause and of the kind of moral community that begets it. For the most part, Royce avoided politics. He was more interested in helping inquirers to seek the truth (an operation of epistemic expertise) than he was in rounding up proselytes for a specific vision of truth (an operation of therapeutic expertise).

Two of his rare political prescriptions are instructive. First, he held that "wise provincialism," understood as a kind of peaceful individualism of distinct geographic areas, was an asset to the greater polity (Royce, 1908). The great virtue of provincialism, on Royce's view, is that it countervails the "leveling tendency" of contemporary society, in which mass media and public education instill a monotone worldview that precludes innovation and excellence. Second, he suggested that international relations would be improved if nations agreed to underwrite their mutual agreements and contracts with funds placed under the control of an independent, non-partisan, non-profit international board of trustees. This trusteeship relation would be undertaken voluntarily by the contracting nations and would provide a kind of "international insurance" insofar as the trustees would pay damages to any parties wronged according to the terms of their respective contracts (Royce, 1914, pp. xxi-xxv).

This political advice reinforces the link between Royce's practice of epistemic ethical expertise and his endorsement of political conditions that favor the principle of permission. The recommendation for provincialism is in line with his appreciation for the benefit of multiple lines of ethical inquiry and his international insurance scheme is founded on regard for non-coercive cooperation and a balance of power. Royce's principle "social purpose" is the prerogative of citizens to fashion (usually communally) a cause and to voluntarily serve it—which is the essence of inquiry into the good and the right. The good society, in other words, is a society that allows citizens to associate freely into moral communities and to peacefully pursue moral truth, collaboratively or independently, as their convictions dictate. This freedom is promoted through the construction of checks and balances, which Royce believes are necessary because (more than Dewey) he is impressed by the immiscibility between virtue and unchecked political power.

While Royce thinks each community is likely to have its ethical experts in the therapeutic sense, the larger society will acknowledge ethical expertise primarily in the epistemic sense. Epistemic experts will contribute with: (1) knowledge of various, competing moral communities, (2) skill in representing these communities in compromise-oriented negotiations (compromise replaces the consensus of therapeutic experts as the dominant focus of political deliberation), and (3) accounts of the conditions for effective moral inquiry.

Though a comprehensive argument is far beyond the scope of this essay, there are several rather straightforward reasons for preferring Royce's version of ethical expertise to Dewey's version. First, Royce's lack of confidence in the current state of ethical and sociological knowledge is better justified than Dewey's confidence. On this point Royce could invoke Peirce's principle of fallibilism, which is the doctrine that no truth claim is epistemically certain—that justification is never fully determinate or complete and that alternative hypotheses are always possible.[37] For epistemic experts, fallibilism penetrates more deeply than it does for therapeutic experts—especially in the realm of value theory.[38]

A second reason resides in the aforementioned tension between Dewey's regard for scientific method and his non-evidence-based evangelism for his liberal "common faith." If ethical inquiry is really experimental, as both Dewey and Peirce maintain, then the success of inquiry would seem to be helped by: (1) establishing multiple loci of inquiry (i.e., multiple distinct moral communities rather than a single, enormous, integrated community), (2) allowing inquiry to play out without preemptive interference, and (3) appropriating the moral energies (i.e., the natural loyalties) of inquirers. These benefits are available to epistemic experts, but they are subverted by grand notions of therapeutic expertise.

Ultimately, the tug-of-war between epistemic and therapeutic versions of ethical expertise will be decided not merely by the experts, but also by others who nurture and digest their work. Probably there is a role for both versions. And there is reason for skepticism about the current state of ethical expertise, of either variety. How, then, are ethicists to interact with, and serve their public? Let us hope that this question is answered in a spirit of loyalty—if not loyalty to truth, then at least loyalty to causes greater than the acquisition of grants and appointments.

NOTES

1. In framing my topic this way, I confine myself to a portion of what might count as "ethical expertise"—namely the portion that functions in explicitly public (and often political) contexts within the democratic polis. This focus corresponds with my interest in bioethics and other forms of practical ethics that grapple extensively with public policy issues.
2. Some readers may be surprised that I include Josiah Royce in the inventory of American pragmatists. But this claim is hardly controversial among Royce or Peirce scholars. Peirce, traditionally viewed as the original American pragmatist, said that Royce was America's greatest pragmatist (Brent, 1993, p. 293) and that Royce was the only true American pragmatist other than Peirce himself (Brent, 1993, p. 265). Like other pragmatists, Royce repudiated the dualism between cognition and conation, described the meaning of ideas in terms of their possible or intended fulfillments, and incorporated a conception of scientific method into a broader, fallibilist theory of inquiry. Like Dewey and Peirce, Royce focused (especially late in his career) on the process of interpretation, emphasizing its social nature.
3. Of course, one does not have to belong to a religious enclave to recognize and consult ethical experts. Children routinely consult parents, for instance, because they view them as resident experts in handling situations such as bullying, toy sharing, and such. In many families, the parents' standing as ethical experts dissipates as children mature. Some might complain that I am conflating "expertise" with "authority." However, a hard and fast distinction between expertise and authority seems plausible only in the face of a hard and fast distinction between theory and action—the expert, on this view, being recognized for her vast theoretical knowledge, but not for her practical wisdom. The distinction between theory and practice may be useful on occasion (as we shall see), but is not taken for granted by the pragmatists and is not, in my opinion, a useful benchmark for delineating ethical expertise.
4. Here I use the term "secular" in the manner employed by H. Tristram Engelhardt, Jr. to denote a body of knowledge and practices that is purported to be derived from reason and human experience alone—without recourse to private or contestable sources of knowledge such as noesis—and hence thought to be readily accessible to human inquirers (Engelhardt, 2000, pp. 14-16). A prominent use of "secular" is to

denote that which is non-clerical or non-religious. This sense is elusive and less useful, in my opinion, as I believe (with John Dewey) that the "religious" stance is best understood as a unifying, pious, devoted human orientation to unseen (though not necessarily non-natural) forces that are believed to shape collective human destiny (Dewey, 1934, pp. 4, 10-11, 33). On this conception, Marxism and Dewey's proposed "common faith" are plausible programs of religious devotion. But they are also secular, in Engelhardt's sense, because they lack an explicit connection to forces outside of nature and are entirely human in genesis, orientation and scope. We might say, as many now do, that they are "secular religions". Ultimately, I prefer a third way of conceiving "secular", overlapping with one of the senses of "liberal" and one of the senses of "pluralistic", as denoting the absence of an ethical canon. "Secular", "liberal" or "pluralistic" societies, in this third sense, are those that welcome individuals and communities with diverging ethical canons. A secular society may have a public philosophy, but this philosophy will lack sufficient normative content to guide ethical inquiry or to resolve most important moral dilemmas. In essence, then, I conceive secularity as a form of neutrality with respect to conceptions of the good or the right. If perfect neutrality is impossible (as I think it is), then no society is perfectly "secular". Another implication of this conception of secularity is that certain purportedly secular moral visions, such as "secular humanism," are not secular (since they are content-rich and they exclude a large number of competing moral visions). Alas, none of the terms "secular", "liberal", or "pluralistic", are consistently used in this manner.

5. Even performative knowledge, such as knowledge of how to ride a bicycle, has its object. For performative knowledge, as in other forms of knowledge, the object is an intended (actual or ideal) consummation of the knowledge (e.g., riding a bike). Hence, "knowing" how to ride a bike, as in knowing the biomechanical features of bike riding, and "knowing" how to ride a bike, as in being able to do it, may be different species of knowledge, but they can have the same object. As Mounce observes (1997, p. 134), John Dewey tends to restrict the term "knowledge" to that which emerges in the course of inquiry (which would seemingly eliminate many forms of performative knowledge from the domain of knowledge proper). However, Dewey stresses (1988a, pp. 120-128) the consummatory nature of knowledge, and on this point other pragmatists agree. Different pragmatists differ, as we shall see, on the nature and scope of consummations.

6. H. Tristram Engelhardt characterizes the principle of permission as holding that: "Authority for actions involving others in a secular pluralist society is derived from their permission" (Engelhardt, 1996, p. 122).

7. See Engelhardt's discussion of conceptive ideology (Engelhardt, 2002, 14-15).

8. Engelhardt (2000, p. 27) has invoked the notion of "secular priests" in making a point similar to mine. His use of the term "secular" (discussed above) overlaps considerably with my use of the term "ecumenical"—as both his "secular" thinkers and my "ecumenical" thinkers purport to achieve a rational synthesis of values gleaned from diverse moral traditions.

9. The theoretician-expert has not played a very prominent role in practical ethics, for the obvious reason that practical ethics arises from a need to make practical decisions—usually quickly. K. Danner Clouser, the first bioethicist to be appointed to a medical school faculty in America, was in many respects a theoretician-expert. After more than twenty years in medical education, he wrote: "Philosophy asks probing questions and understands how to discover and work with assumptions, implications and foundations. Conceptual analysis, which is central to the doing of philosophy, has been central also to the doing of medical ethics. On the other hand, I doubt that philosophy has brought much of substance, that is, what things to value, what goals are most meaningful, what philosophies of life are most fulfilling" (Clouser, 1992).

10. Jonathan Moreno (1995) provides a detailed overview of bioethics' most prominent method of moral inquiry—namely, the procedure(s) for obtaining a valid consensus. However, expertise in consensus formation is not the sort of thing I have in mind in characterizing the inquirer-expert, since the quest for consensus is oriented mainly towards immediate prescriptions rather than towards the attainment of moral truth. Robert Veatch is closer than Moreno to the inquirer-expert (Veatch, 1981, p. 145), but still misses the mark insofar as his inquiries aim only at the political good, which he does not explicitly link to the truth, or to the right or the good as such. Neo-Peircean Cheryl Misak (2000) is a much better example of a practical ethicist who adopts an inquirer-expert model.

11. Arthur Kleinman's bioethicist as ethnographer is a sophisticated version of the surveyor-expert (Kleinman, 1995, pp. 41-67). Since the ethnographer's surveys are intensely local, his method contrasts sharply with surveyors who seek to uncover a universal morality. In this latter category are Tom Beauchamp and James Childress, who claim to have identified a universal "common morality" that exists

at a middle level between theory and concrete application. Their orientation exemplifies the surveyor, not merely because it is based on an empirical reckoning of actual values, but also because it dismisses the analysis of theoretical justifications for this common morality. On their account, it is an empirical fact that "all persons in all cultures who are *serious about moral conduct* do accept the demands of the common morality" (emphasis theirs) (Beauchamp and Childress, 2001, p. 4). They go on, however, to claim that this non-normative fact has normative implications. Since they spend most of their time working out the normative implications of their proposed common morality, rather than in interrogating various personal and institutional moral agents in order to discover the substance of the common morality, Beauchamp and Childress end up being more like counselor-experts than like surveyor-experts.

12. Casuists, such as Albert Jonsen and Stephen Toulmin (1988), are in many respects a prototype for the counselor-expert. The contemporary casuist allows each community, profession or practice to supply its own concrete moral standard (consisting of an array of paradigm cases), and then helps individuals and groups to deliberate about difficult cases by comparing them to applicable paradigm cases.

13. Nominalism, on Peirce's view, is the doctrine that denies the reality of generals (or "thirds" in Peirce's system). Peirce conceives nominalism as entailing that common nouns "represent in their generality nothing in the real things, but are mere conveniences for speaking of many things at once, or at most necessities of thought" (Peirce, 1992, xxiv). Likewise, general ideas, for the nominalist, are mere constellations of thoughts or experiences, grouped together for expedience.

14. James (1981, p. 26) holds that the pragmatic method is the business of interpreting "each notion by tracing its *respective practical consequences*" (italics mine). He writes: "What difference would it practically make to any one [sic] if this notion rather than that notion were true? If no practical difference whatever can be traced, then the alternatives mean practically the same thing, and all dispute is idle."

15. "Verification" should be understood in a very broad sense (including indirect forms of verification). James writes: "The overwhelming majority of our true ideas admit of no direct or face-to-face verification" (James, 1981, p. 97). He also claims that many of our beliefs count as true, not because we have verified them, but because of their verifiability. As Mounce points out (Mounce, 1997, p. 48), this claim (that verifiability without verification is sufficient for truth) is incompatible with James' radical empiricism. Mere verifiability is not an identifiable element in the stream of consciousness.

16. By confining the meaning of an idea to its respective sensible effects, James identifies the idea with a series of discrete thoughts or moments in consciousness. James' various accounts of the meaning of ideas always trace the trajectory of the ideas within an individual consciousness. Ideas are shared or social, in James' opinion, in a comparative sense (when people think they have experienced similar effects from similar thoughts) or insofar as thoughts and sensible effects are conterminous. The sharing of general ideas does not imply that these ideas are independent of particular thinkers, however. Since James holds that generality is a form of aggregation, the deletion of the thoughts of any particular thinker will change the aggregate and thus transform the meaning of the general concept. See William James, "How Two Minds Can Know One Thing" and "A World of Pure Experience" (James, 1977, pp. 194-214, 227-232).

17. An "expedient" grouping is a grouping that serves some immediate purpose or end-in-view (if we expanded "expedient" to cover the fulfillment of a vague but ultimate purpose, then the criterion would no longer fit as part of the conception of nominalism).

18. Peirce's doctrine of fulfillment as both the criterion and the definition of truth qualifies him as an "internalist" in the sense described by Hilary Putnam (1981, pp. 49-74). Internalism rejects the thesis that truth is constituted by proper correspondence between words or thought-signs and un-interpreted objects. To be an object, for Peirce, is to be interpreted. However, Peirce would take issue with several aspects of Putnam's description of the internalist perspective, specifically Putnam's claims that: (1) objects do not exist independently of conceptual schemes (Putnam, 1981, p. 52), and (2) knowledge can have no source of input that isn't shaped by our concepts (Putnam, 1981, p. 54). In distinction to Putnam, Peirce recognizes sub-conceptual ways of apprehending (and interpreting) an object (Peirce, 1998, p. 4) and also recognizes sources of epistemic input that transcend human experience (Peirce, 1998, pp. 439-440; Raposa, 1989, pp. 50-52).

19. Peirce wrote in "How to Make Our Ideas Clear" that "the production of belief is the sole function of thought" (Peirce, 1992, p. 127) and that the "essence of belief is the establishment of habit, and different beliefs are distinguished by the different modes of action to which they give rise" (Peirce, 1992, pp. 129-30).

20. For the purposes of this essay, "realism" will be defined merely as the rejection of subjective and instrumental forms of nominalism. Realism in this sense is epistemic (pertaining to our knowledge of generals or universals) rather than metaphysical (implying the existence of a mind-independent reality).

Neither Royce nor Peirce is a metaphysical realist. They hold that generals are real and independent of the thoughts of any individual person, but not that they are independent of mind.

21. Dewey and other pragmatists should be differentiated from utilitarians and crude instrumentalists – from the former, because they reject the notion that human fulfillment is "a dialectic abstraction of a feature of concrete goods" (Dewey, 1988a, p. 323), and from the crude instrumentalists because they reject a simple ends-means distinction.

22. Another way of getting at Dewey's nominalism is through the doctrine of continuity, accepted by both Dewey and Peirce. If all things and events are continuous, then they ultimately all hang together as a morally ordered and coherent unity, or they do not. If the former, then there is an ultimate moral truth (embodied in the ordering principle) and this is real. If not, then moral truth, and moral reals, are transiently useful fictions at best. Peirceans should challenge the integrity of Dewey's fallibilism on similar grounds. In what sense is it possible to commit a genuine moral error when moral truth is fluctuant (as it is in Dewey)? Dewey sometimes seems to regard belief in transcendent or ultimate truths as the enemy of fallibilism, probably because he frequently conflates belief that there is an ultimate moral truth (absolutism) with insistence that this truth is already known or grasped in its essential elements (dogmatism). This confusion has become so prominent in popular discourse that absolutism is now taken as a virtual synonym for dogmatism. In any case, fallibilism's greatest enemy may be extreme moral contextualism, which takes the sting out of moral errors by curtailing their life span.

23. Royce uses Augustine as an example (Royce, 1995, p. 14), noting that on Augustine's account the human will remains unstable and restless until it finds repose in the will of God. Hence, Royce believes that even for authoritarian Augustine, the ultimate warrant for God's authority resides in the human will (which God has created in the image of His divine will).

24. Peirce also claims there is a logical necessity for "interest in an indefinite community, recognition of the possibility of this interest being made supreme, and hope in the unlimited continuance of the intellectual activity" (Peirce, 1992, p. 150). On this basis, Peirce claims that loyalty to a universal community is a prerequisite to inquiry. "Logic," he writes, "is rooted in the social principle" (Peirce, 1992, p. 149).

25. Royce's dedication to the Great Community did not prevent him from recognizing that loyalty is embodied primarily in the mundane circumstances of ordinary living. Nor did it keep him from realizing his own ignorance about human destiny and about the communities of the future. Royce wrote: "A newspaper once asked me to contribute to a so-called symposium whose problem was this: What characteristics will the ideal man of the future possess? As I only knew about the ideal future man this, that when he comes, he will, as in him lies, adequately attend to his own business, I felt unable to contribute anything original to the proposed discussion" (Royce, 1908, pp. 154-55).

26. Dewey writes: "In classic Greek thought, the perception of ends was simply an esthetic contemplation of the forms of objects in which natural processes were completed. In most modern thought, it is an arbitrary creation of private mental operations guided by personal desire, the theoretical alternative being that they are finite copies of the fulfilled intentions of an infinite mind. In empirical fact, they are projections of possible consequences; they are ends-in-view" (Dewey, 1988a, p. 86). Peirce would have a lot to say about this passage, including the following. (1) The objects of inquiry are never fully ends-in-view (determinate objects), but also simultaneously ends-still-coming-into-view (dynamical objects). If our ends were clearly in view, we would not need to inquire about them. As a psychologist, Dewey recognizes this fact, but as a social reformer he sometimes forgets it (Mounce, 1997, pp. 139-143). (2) The aesthetic contemplation of ultimate ends is entirely appropriate—since it is the best way of understanding them (Peirce, 1998, p. 440). Dewey's prescriptions for growth and integration seem themselves based on aesthetic contemplation. (3) Teleological operations of personal desire need not be arbitrary or private. They can plausibly also be determinations of an infinite mind, guiding thought toward an ultimate end. In the latter case (as with Aristotle) the formal cause melds with the efficient and final causes.

27. According to Dewey (Dewey, 1984, p. 257), the "public" arises when those "indirectly and seriously affected for good or for evil form a group distinctive enough to require recognition and a name." The public is lost or eclipsed insofar as we are ineffective at perceiving the important (often unintended) consequences of non-public associations (such as friendships, business partnerships, etc.) and at recognizing and representing the affected groups.

28. The analysis in this section follows my analysis in an earlier essay (Trotter, 2001).

29. In the cited passage Dewey sounds like a transcendental moralist—claiming in essence that material security is a transcendental condition for the possibility of fulfillment in a liberal society as it is currently

envisaged. There is, in fact, strong symmetry between this Deweyan argument and H. Tristram Engelhardt's well known transcendental argument. Engelhardt defines secular morality as the activity of peaceably resolving disagreements between moral strangers, then argues that adherence to the principle of permission is a necessary condition for the possibility of a secular morality (Engelhardt, 1996, pp. 70, 94-97). Dewey characterizes liberal democracy as the integrated social pursuit of "effective liberty" and personal growth, then argues that material security is a necessary condition for the possibility of liberal democracy, given current historical circumstances. The major difference between Dewey's quasi-transcendental argument and Engelhardt's neo-transcendental argument is that Dewey tells us that his conception of liberal democracy is historically contingent. It is not clear, however, that mentioning the fact that one's dogmas are historically contingent will do anything to mitigate the intensity with which those dogmas are maintained (especially when one of the dogmas is the claim that all claims are historically contingent).

30. Though Dewey is certainly correct in recognizing that atheistic humanism potentially supports most of the emotional and volitional determinations of theistic religions, he seems relatively oblivious to a major negative implication of this connection. If secular humanism is a religion, then it is subject to the church-state division and cannot be invoked as the public morality of the secular state.

31. Dewey believes that "our accumulated knowledge of man and his ways, furnished by anthropology, history, sociology and psychology, is vast, even though it be sparse in comparison with our knowledge of physical nature." This confidence in the present potential of the social sciences (Dewey was a critic of the "dualistic psychology" that held sway in many circles during his life, but his thought cohered closely with that of Mead and other prominent social psychologists) contrasts sharply with Peirce's less enthusiastic estimation. Because of his confidence in the social sciences, Dewey laments the infrequency with which they are appropriated in the development of public policy. He writes: "We are habituated to the idea that every discovery in physical knowledge signifies, sooner or later, a change in the processes of production… There is next to nothing of the same sort with respect to knowledge of man and human affairs" (Dewey, 2000, p. 52).

32. Against therapeutic ethical experts of a non-liberal humanist stripe, Dewey writes: "The attempt to decide by law that the legends of a primitive Hebrew people regarding the genesis of man are more authoritative than the results of scientific inquiry might be cited as a typical example of the sort of thing which is bound to happen when the accepted doctrine is that a public organized for political purposes, rather than experts guided by specialized inquiry, is the final umpire and arbiter of issues" (Dewey, 1984, pp. 312-313). Dewey rarely equivocates about the need to remove divine revelation, ancient tradition and unscientific reflection from the repertoire of the ethical expert.

33. The claim that Dewey is a "perfectionist" will inevitably remain arguable because the conception of perfectionism is so slippery. As Patrick Neal observes (1997, pp. 4, 35), the term was devised by egalitarian liberals (such as Rawls and Dworkin) to contrast their own view with the view that "the purpose of the state is to affirm and aid its citizens in seeking the realization of some idea of the good." What is sometimes perplexing is understanding why such thinkers believe that they are not advocating a political implementation of an idea of the good while holding that thinkers as diverse as Nietzsche and Aristotle are advocating one. In any case, if we stipulate that the state's idea of the good must be static, then Dewey is not technically a perfectionist (though we may wonder how dynamic his conception of the good really is). If we don't, then Dewey seems to count as a perfectionist. For an extended discussion of Dewey's perfectionism from a sympathetic scholar, see Savage, 2002, pp. 5-6, 130-132, 140-143, and 164-169.

34. Dewey's version of liberalism includes not only his three explicit "enduring values" ([1] liberty as freedom from oppression, [2] liberty as the opportunity to develop inherent capacities, and [3] the utilization of "free intelligence" in inquiry), but also his claims about the meaning and value of "development," about the necessity of large-scale cooperation, and about the falsity of theological claims (Dewey, 2000, pp. 37-93).

35. Richard Momeyer (2002) is a good example of a thinker outside pragmatism who subscribes to the therapeutic option. Like Dewey, he promotes a particular version of ethical inquiry—namely one that begins with a survey of current practices—and aims for an imaginative integration of various conflicting values.

36. The policy advisory function, and its connection with the cultivation of consensus, is treated elegantly by Moreno (1995). Though I have been critical of the hunger for consensus and the tendency to inflict a weak consensus on dissenters (Trotter, 2002), consensus formation is often salient and contemporary Deweyans like Moreno have, in my opinion, often made critical contributions to public

policy. My opposition to therapeutic notions of ethical expertise, here and elsewhere, is a matter of degree. Certainly there are instances of the operation of politically legitimate therapeutic moral expertise, but these instances are limited in number and scope, and the attempt to multiply them can be dangerous.

37. Peirce writes: "We never can be absolutely sure of anything, nor can we with any probability ascertain the exact value of any measure or general ratio" (Peirce, 1955, p. 8). Peirce is not saying that we cannot *feel* certain (certainty in a psychological sense). To the contrary, his doctrine is that our knowledge is never totally justified, that we cannot *be* certain (certainty in an epistemic sense). Nevertheless, Peirce thinks that refractoriness to psychological certainty is a desirable consequence of fallibilism, writing that "no blight can so surely arrest all intellectual growth as the blight of cocksureness" (Peirce, 1955, p. 4).

38. H.O. Mounce observes (1997, p. 138) that Dewey's basic beliefs form an essentially impervious "apperceptive mass" that Dewey is not willing to depose, revise or reconstruct. See endnote #22 for commentary on Dewey's version of fallibilism.

REFERENCES

Beauchamp, T.L. & Childress, J.F. (2001). *Principles of Biomedical Ethics*, 5th ed. New York: Oxford University Press.

Brent, J. (1998). *Charles Sanders Peirce* (revised and enlarged ed.). Bloomington and Indianapolis: Indiana University Press.

Clouser, K.D. (1992). 'What Bioethics Brought to Philosophy.' (A paper read at the Birth of Bioethics Conference, Seattle, Washington, September 23-24). Quoted in: A.R. Jonsen, *The Birth of Bioethics* (p. 80). New York: Oxford University Press, 1998.

Dewey, J. (1934). *A Common Faith*. New Haven: Yale University Press.

Dewey, J. (1948). *Reconstruction in Philosophy*. Boston: Beacon.

Dewey, J. (1984). *The Public and its Problems*. In: *John Dewey: The Later Works, 1925-1953: Volume 2: 1925-1927*, J.A. Boydston (Ed.). Carbondale: Southern Illinois University Press.

Dewey, J. (1988a). *John Dewey: The Later Works, 1925-1953: Volume 1: 1925: Experience and Nature* (paperback), J.A. Boydston (Ed.). Carbondale: Southern Illinois University Press.

Dewey, J. (1988b). *John Dewey: The Later Works, 1925-1953: Volume 4: 1929: The Quest for Certainty* (paperback), J.A. Boydston (Ed.). Carbondale: Southern Illinois University Press.

Dewey, J. (1991). *Liberalism and Social Action*. New York: Prometheus.

Engelhardt, H.T. (1996). *The Foundations of Bioethics*, 2nd Ed. New York: Oxford University Press.

Engelhardt, H.T. (2000). *The Foundations of Christian Bioethics*. Exton: Swets & Zeitlinger.

Engelhardt, H.T. (2002). 'Consensus Formation: The Creation of an Ideology.' *Cambridge Quarterly of Healthcare Ethics* 11, 7-16.

Fletcher, J.C. (1997). 'Bioethics in a Legal Forum: Confessions of an 'Expert' Witness.' *Journal of Medicine and Philosophy*, 22, 297-324.

James, W. (1977). *The Writings of William James: A Comprehensive Edition*, J.J. McDermott (Ed.). Chicago: University of Chicago Press.

James, W. (1981 [1907]). *Pragmatism*. Indianapolis: Hackett.

Jonsen, A.R. and Toulmin, S. (1988). *The Abuse of Casuistry*. Berkeley: University of California Press.

Kleinman, A. (1995). *Writing at the Margin: Discourse Between Anthropology and Medicine*. Berkeley: University of California Press.

MacIntyre, A. (1981). *After Virtue*. Notre Dame, Indiana: University of Notre Dame Press.

Misak, C. (2000). *Truth, Politics, Morality*. New York: Routledge.

Momeyer, R.W. (2002). 'What Conception of Moral Truth Works in Bioethics?' *Journal of Medicine and Philosophy* 27, 403-416.

Moreno, J.D. (1995). *Deciding Together: Bioethics and Moral Consensus*. New York: Oxford University Press.

Mounce, H.O. (1997). *The Two Pragmatisms: From Peirce to Rorty*. New York: Routledge.

Neal, P. (1997). *Liberalism and Its Discontents*. Washington Square, NY: New York University Press.

Peirce, C.S. (1955). *Philosophical Writings of Peirce*, J. Buchler (Ed.). New York: Dover.

Peirce, C.S. (1992). *The Essential Peirce: Selected Philosophical Writings*, Vol. 1 (1867-1893), N. Houser and C. Kloesel (Eds.). Bloomington: Indiana University Press.

Peirce, C.S. (1998). *The Essential Peirce: Selected Philosophical Writings*, Vol. 2 (1893-1913), Peirce Edition Project (Ed.). Bloomington: Indiana University Press.

Putnam, H. (1981). *Reason, Truth and History*. New York: Cambridge University Press.

Raz, J. (1986). *The Morality of Freedom*. Oxford: Clarendon Press.

Royce, J. (1908). *Race Questions and Other American Problems*. New York: Macmillan.

Royce, J. (1914). *War and Insurance*. New York: Macmillan.

Royce, J. (1995). *The Philosophy of Loyalty*. Nashville: Vanderbilt University Press.

Savage, D.M. (2002). *John Dewey's Liberalism*. Carbondale and Edwardsville: Southern Illinois University Press.

Stuhr, J.J. (1997). *Genealogical Pragmatism: Philosophy, Experience and Community*. Albany: State University of New York Press.

Trotter, G. (1997). *The Loyal Physician: Roycean Ethics and the Practice of Medicine*. Nashville: Vanderbilt University Press.

Trotter, G. (1999). 'The Medical Covenant: A Roycean Perspective.' In: G. McGee (Ed.), *Pragmatic Bioethics* (pp. 84-96). Nashville: Vanderbilt University Press.

Trotter, G. (2001). 'Pragmatism, Bioethics and the Grand American Social Experiment.' *American Journal of Bioethics*, 2 (1), *In Focus* article at http://bioethics.net.

Trotter, G. (2002). 'Bioethics and Healthcare Reform: A Whig Response to Weak Consensus.' *Cambridge Quarterly of Healthcare Ethics*, 11, 37-51.

Veatch, R.M. (1981). *A Theory of Medical Ethics*. New York: Basic.

Wildes, K.W. (1997). 'Healthy Skepticism: The Emperor Has Very Few Clothes.' *Journal of Medicine and Philosophy* 22, 365-371.

Yoder, S.D. (1998). 'The Nature of Ethical Expertise.' *Hastings Center Report*, 28 (6), 11-19.

SECTION II: CONTEMPORARY PERSPECTIVES

CHAPTER 8

MARY ANN G. CUTTER

EXPERT MORAL CHOICE IN MEDICINE: A STUDY OF UNCERTAINTY AND LOCALITY

1. INTRODUCTION

In this piece, I argue for a "localized" account of moral expertise in medicine. While I reject the view that there is a universal or absolute account of moral truth available to women and men, I also reject the view that there is no moral truth. The concept I wish to develop is one of a "localized" account of moral knowledge, one that grounds an interplay among facts, values, and culture in moral knowledge claims. By this I mean that description, values, and socio-political influences enter into how we judge clinical problems. On this view, neither is it the case that anything goes nor that there is one and only one answer to questions concerning moral probity in medicine. As a consequence, a "local" notion of expertise in medicine is most defensible. This view has implications for how we understand bioethical expertise, how we train bioethical experts, and what we take to be the warrants of bioethical expert claims.

2. RISE OF THE BIOETHICAL EXPERT

Professional bioethics developed out of movements during the latter third of the twentieth century (Jonsen, 1998). These movements involved granting greater rights, privileges, and protections to patients, research subjects, women, prisoners, and consumers. For sure, events in Nazi Germany in the 1940s and their after-effects contributed greatly to increasing awareness of the lack of protections for individual rights and welfare of those participating in medical activities. In addition, dramatic advances in medical technology, such as the development of pharmaceutical agents and surgical techniques, raised a host of moral questions that required careful attention. Discussions about the beginning of life (*Roe v. Wade*, 1973), the definition of death (President's Commission, 1981), and the limits of the use of technology (Jonas, 1984 [1979]; Reiser, 1978) became commonplace. Numerous legal and policy changes were implemented in the 1960s and the years that followed to address these and related problems. At the same time, a large body of literature (e.g., *Principles of Biomedical Ethics* [1979], *Hastings Center Report*

L. M. Rasmussen (Ed.), Ethics Expertise: History, Contemporary Perspectives, and Applications, 125-137.

[begun in the 1980s], *Journal of Medicine and Philosophy* [begun in the 1970s], and the *Philosophy and Medicine Book Series* [begun in the 1970s]) emerged that formed the basis in the late twentieth century for thinking about what a bioethicist does and should do. This coincided with a rise of academic programs (e.g., Georgetown University's Kennedy Institute of Ethics, The Hastings Center) to serve the training of bioethicists.

In the bioethical literature (e.g., *Principles of Biomedical Ethics*), there are clear alliances to the thinking of Aristotle (1985), St. Thomas Aquinas (1948), Immanuel Kant (1985 [1785]), and John Stuart Mill (2002 [1861]) in providing the basis for expert moral decision-making. This is in part because such thinkers significantly influenced the development of Western thought on moral matters. Aristotle's appeal to virtue, Aquinas's analysis of duty to the Creator, Kant's development of the concept of respect for persons, and Mill's careful account of utility are some of the most influential accounts of the character of moral decision-making. These texts constitute what one would call a "canon" in Western philosophical ethics as well as in Western biomedical ethics.

According to Smith (1963), American moral philosophy may be seen as an extension of the teachings of the canon for three reasons. First, in American moral philosophy, thinking is primarily an activity in response to a concrete situation and this activity is aimed at solving problems. Second, ideas and theories must make a difference in the conduct of people who hold them. Third, the Earth can be civilized and obstacles to progress overcome by the application of knowledge. To these characteristics, Jonsen (1998, p. 71) would add: American moral philosophy has traditionally been respectful of religious feeling, enthusiastic about democracy, and convinced of the power of individual freedom. It has also been, in its epistemological foundations, naturalistic and intuitionist. That is, moral notions arise from sentiments innate in human nature, grasped first by intuition, and then cultivated by reason and experience, which is used in turn to analyze and recommend what ought to be done.

One of the marks of the canon and its expression in America is the message that moral questions are answerable through a principle-based framework. The moral resources in the West derive from Athens and Jerusalem in the form of Greek philosophy and Judeo-Christian faith. Up until the Enlightenment, those seeking moral counsel could go only to the clergy. Moral direction was almost always connected with faith and the acceptance of religious authority, which is based on the mystical phenomena of divine prophesy and revelation. With the Enlightenment comes the attempt to ground moral theory on human reason alone, during which time secular moralists canonized the ethics of Aristotle, Kant, Mill, and others. In the twentieth century, professionals trained in this canon are asked to speak their collective mind whenever there are public-policy, clinical, or legal questions.

The rise of the bioethical expert is supported by the view that moral experts possess special knowledge by virtue of which they can speak with authority on moral matters. This story is not new. In fourth-century Athens, there were teachers called "sophists" who claimed to be experts on ethics. For rather substantial fees, the sophists offered to impart their moral knowledge to the children of wealthy Athenians. Though he vehemently denied the charges, Socrates was accused of being one of this kind of ethical expert. In the trial of Socrates, as described by Plato (1985), the oracle at Delphi proclaimed that there was no one wiser than Socrates. Confused by this announcement, Socrates attempted to find someone wiser than he. He cross-examined representatives from a wide span of Athenian society. According to Plato, Socrates eventually came to understand that he was the wisest person in Athens by virtue of the fact that he alone knew that he was ignorant. Socrates concluded that the ethical experts of his time were imposters, a lesson that has relevance in the twenty-first century discussions of the status of bioethicists.

3. THE QUESTION OF CERTAINTY

Socrates was right; uncertainty pervades human knowledge and, in particular, human moral knowledge. One might note that there are different senses of uncertainty that arise in the process of moral decision-making. There is the sense of uncertainty that refers to a lack of facts or evidence. Humans are not omnipotent and therefore lack full knowledge at any one time. They lack facts because of their limited perspective and because some facts are as-of-yet unknown. In addition, there is a sense of uncertainty when values are involved, for values are not universally shared by individuals. As Osborne (2001) and others (Cutter, 2003) report, clinical conditions worldwide are anything but univocal given the role culture plays in their construction.

There is also an uncertainty that comes from not having access to an ultimate rational justification. All moral reasoning is based on assumptions that in the end cannot be justified against competing assumptions. Typically, bioethicists are given cases to analyze and judge. Take, for example, the case of an elderly person with terminal cancer who wants to stop all treatment. Should she be allowed to, or should her depression be regarded as rendering her incompetent to make the decisions? The conclusion drawn will depend not only on how the expert understands depression as a biopsychological phenomenon, but also on how right and wrong are understood. A neo-Kantian might argue that the woman's choice ought to be respected insofar as it is autonomously made. A Utilitarian might surmise that the stakeholders would best be served if the woman were not allowed to discontinue treatment because there are greater goods to be achieved (e.g., life is better than death, suffering is a positive value). Here, some calculation regarding what best maximizes benefits and minimizes harms is required. A Divine Law theorist would claim that refusing treatment is no different from suicide, which is against God's will (Aquinas, 1948) and thus morally prohibited. The first expert begins with the assumption that there is nothing more important than preserving human autonomy, the second focuses on maximizing benefits and minimizing harms for the relevant parties, and the third tracks the moral vectors from the Bible.

In each case, an additional premise must be employed to justify the selection of the prior premise required to advance the argument, and so on. The rational assertion that utility (or autonomy or God's will) is the overriding value requires an additional premise justifying why this is the case, and so on, *ad infinitum*. One cannot appeal to utility (or autonomy or God's will) to determine the correct ranking of benefits and burdens (or rights or God's plan). One must appeal to some other claim to justify why the world is this way, and so on. The task is formidable when one is committed to the rational justification of moral claims (Engelhardt, 1996).

For many, the complex and sometimes uncertain character of moral decision-making is not easy to accept, especially when one is in the business of making concepts clear (as in the case of philosophers, scientists, and clinicians) or the position of patient or advocate of a patient wanting answers. Part of the difficulty we face is that we have inherited traditions of thought (e.g., the Enlightenment) that promise answers, and clear ones at that (Sassower, 1993). One of the main points forwarded by this essay is that we must begin to accept that moral decision-making, by the expert and otherwise, is fraught with uncertainty.

4. WHY RELATIVISM IS NOT AN OPTION

It may worry some that relativism appears to be compatible with the position advanced here. If this is so, then the so-called bioethical expert is a fraud. Such a conclusion is in error, as the following illustrates.

Relativism is the belief that knowledge is determined by specific qualities of the observer, including age, race, gender, and cultural conditioning. For example, the Sophists (e.g., Protagoras [481-411 B.C.]; Curd, 1996) claim that place of birth, family habits, personal abilities and preferences, religious training, age and so forth control an individual's belief, values, and preferences. Based on this view, we need only accept what seems true at the moment according to our culture. More extreme Sophists claim that even within the same culture, individuals have their own beliefs.

Two basic variants of relativism are distinguishable: cultural and individual. *Cultural* relativism is the belief that all values are determined by the ideas, concepts, and beliefs humans in community live by. Humans in community tend to assume similar perspectives, and likes and dislikes, in order to live harmoniously together. Monthly magazines that feature home, dress, art, and music fashions illustrate such trends well. Humans from *different* communities often diverge in perspective, likes and dislikes, in order to respond appropriately to their environment, to distinguish themselves from each other, and to give themselves an identity. A Siberian culture will likely differ from a tropically-based one given the different environmental influences and demands placed upon it.

Individual relativism carries the logic of cultural relativism to a more refined conclusion. It holds that even in the same place and time, good and bad, and right and wrong, are relative to the unique experiences and preferences of the individual. There is no unbiased way to say that one standard is better than another, for the standard used to make a given claim is in itself the reflection of a preference, *ad infinitum*. No matter how far back we push "ultimate" reasons, they always reduce to individual preferences, likes or dislikes, nothing more.

Relativism is an attractive position because it seems to account for a culture's diversity and commitment to tolerance. Unfortunately, these advantages cannot be achieved because relativism fails on two counts, namely, conceptual and practical. *Conceptually* speaking, relativism advances the view that "truth is relative." If this claim is taken seriously, then it backfires on itself: if it is true, it must be false. If truth is relative, then the claim itself is relative, thereby undermining its ability to establish anything. Correspondingly, we can be certain of nothing. Here's the problem: relativism commits one to the thesis that one and the same proposition can be true in one system and false in another. This is equivalent to endorsing and condemning one and the same state of affairs, which is a problem of self-refutation.

Practically speaking, relativism fails to take into account what common sense tells us: that we do act as if there is understanding or certainty in our world. We rely on math and science to build our world and are often quite successful in doing so. We do think and act as if things matter. We vote, crusade for causes, make sacrifices for various ideals, and hold each other responsible for personal actions. Rationally and practically speaking, then, there is really no such thing as a consistent relativist, despite postmodern persuasions.

5. "LOCAL" MORAL MEDICAL CHOICES

If moral decision-making is neither inflexible nor relative, then what is it? I suggest that it is "localized." It is localized first in terms of what and how it describes. Moral decision-making in medicine involves, in part, judging a clinical problem. How the clinical problem is understood is key to any judgment. The facts of this clinical problem may vary depending on the observer. AIDS, for example, can be correlated with genetic, immunological, anatomic, and social variables, depending on whether one is a molecular biologist, immunologist, oncologist, or public health official. The construal will depend upon the particular clinician's appraisal of which etiological variables are most amenable to the framework the observer brings to bear on the observation. For example, a molecular biologist will likely focus on the genetic make-up of the cell and its vulnerability to HIV invasion and control. An immunologist will probably decide that the major factor in AIDS is the T-cell count where an increased number correlates with HIV-associated signs and symptoms. The public health official will likely decide that the basic variables in AIDS are elements of a lifestyle that include unprotected sex or use of unclean needles. Much depends on how the facts are nested.

In addition, moral decision-making in medicine is localized in terms of what and how it values. Clinical problems (e.g., AIDS) serve as treatment warrants. Since we understand clinical problems in terms of responding to patient complaints, it follows that clinical problems are constructed against a background of value presuppositions. Simply put, functional values tell us what ideals of activity are proper to an organism (e.g., a functioning immune system). Aesthetic values tell us what ideals of form and grace are worthy of achievement (e.g., a body without wasting and infection). Instrumental values tell us how to get from means to ends (e.g., boosting the immune system, which results in health), where the end is the maximization of benefit and the minimization of harm. Since all may not share in the same view regarding what ideals are proper to be achieved, ethical values are at stake as well. (The foregoing indicates the more obvious sources of attention for bioethicists.) In the case of AIDS, the recognition of patient rights may favor anonymous AIDS testing in order to prevent discrimination. Diagnosing, predicting, and treating, while taking proper regard of patients, requires acknowledging the range and specific expressions of values embodied in how we understand and undertake clinical problems.

Finally, moral decision-making in medicine is localized in terms of cultural forces. We judge clinical problems not only in terms of the relation of the condition to the organism, but in terms of the relation of the organism to the environment (Reznek, 1987, p. 69). As an example, AIDS is considered by some not only a medical condition but a spiritual one (Nash, 2000). Dyslexia is not a disease in a pre-literate environment, but it is in a literate one. Even within certain clinical categories, seemingly universal afflictions can differ profoundly from culture to culture so as to encourage the developers of the Diagnostic and Statistical Manual of Behavioral Disorders IV to introduce a new category of psychiatric illnesses known as "culture-bound syndromes" (American Psychiatric Association, 1994, pp. 845-849). Examples include anorexia nervosa, a diagnosis in a resource-rich culture (e.g., the United States) that places positive value on thin body image, and *taijin kyofusho*, a syndrome found in Japan that involves intense fear that one's body is displeasing, embarrassing, or offensive to other people. Generally put, culture-bound syndrome comprises several kinds of illness or affliction that do not have one-to-one correspondence with a disorder recognized by Western, allopathic nosologies. Nevertheless, they do appear to account for signs and symptoms brought into the clinic, and therefore gain credence as a clinical problem.

A local account of clinical problems allows for a range of judgments. On the one hand, agreement may be widespread. One might think here of a long-standing patient-physician relationship in which both parties (1) understand the nature of the patient's problem, (2) wish to achieve a specific therapeutic good (in a functional, aesthetic, and instrumental sense), and (3) respect each other (in a normative sense). Such are the kinds of relationships patients typically seek and bioethicists hope for. On the other hand, disagreement may abound. One might think here of consensus groups in which experts from diverse clinical and allied specialties are brought together to decide on what ought to be done: (1) There is not a univocal understanding of what constitutes the problem, (2) there are conflicts regarding which goods are at stake (functional, instrumental, and aesthetic), and (3) there are disagreements between the principles of respecting persons (by maintaining confidentiality of medical records) and serving the public good (by disclosing information for the public good). Early attempts to classify AIDS in order to treat the condition more effectively feature many of these differences (Cutter, 2003, Ch. 2).

6. WHY LOCALIZATION IS NOT RELATIVISM

Although we may lack a singular interpretation of a moral problem in medicine, relativism is not an option. There are constraints or limits on interpretations of a moral problem. There are similarities or common grounds. The constraints on interpretation in clinical medicine are the reality of patient suffering or pain, the methods humans use to know a clinical problem, the actions that can be taken, the ways suffering or pain are evaluated, and the influences of cultural forces.

To begin with, and despite the influences of laboratory medicine (Foucault, 1973 [1963]), patient suffering must remain the focus of moral decision-making in medicine. As a clinical science, medicine studies the human being not simply as an objective entity (e.g., a body, an organ, a tissue, a cell), but as a being who has purpose, values, consciousness, reflection, and self-determination (Pellegrino and Thomasma, 1981, p. 23). Medicine does and will fail if it neglects to consider what patients bring to the clinic. Although interpretations of what specifically constitutes suffering may differ, medicine faces the global challenge of addressing patient desires to prevent or ameliorate what, biopsychologically speaking, hurts.

The method that medicine uses to diagnose, predict, and treat clinical problem is not relative. As a clinical science, medicine brings theory to explain and interpret observations associated with the clinical problem. It proposes an explanation of third-person observations and first-person reports based on durable knowledge. This involves hypothesis formation and testing. In contemporary Western medicine, problems are usually stated in the form of a need for a causal account for the observed events. Constant and regular temporal association of events supports the proposal of a causal relationship but clinical medicine requires further analysis. Clinicians use statistical methods to consider large sets of evidence to determine whether their regular and constant association occurs more frequently than chance would predict. When they do occur more often than chance, clinical investigators say that they suspect a causal relation. What results is the possibility of enduring, but dynamic, causal relations, the kind of which will vary in statistical degree of certainty depending on the evidence.

There are good reasons for viewing this clinical method as clustered around therapy: as enterprises of treatment, enterprises in the prevention of the need for therapy, studies of the biological and/or psychological bases for the occasions for therapy, or studies of the efficacy of therapy. Therapy is what ties medicine to the clinic, to the practical endeavor of responding to patient complaints.

This relation between understanding and treating clinical problems is also one between knowing and valuing. Some facts in medicine are more important than others not because they help us know more truly but because they help us treat more effectively. In this way, "true" knowledge in medicine is in the service of effective and efficient action. Put another way, there are no such things as neutral, naked, or bare facts in medicine (Fleck, 1979 [1935]). Facts always appear interpreted within the embrace of evaluative frameworks, whether or not these frameworks are formally or informally developed as scientific accounts. These evaluative frameworks are often stated in terms of utilitarian values, or cost-benefit calculations, but they also appear in terms of non-utilitarian considerations, such as claims regarding patient autonomy or sanctity of life.

Of course, the knowledge, actions, and values in medicine will vary depending on resource availability, societal rules, patient requests, etc. Put another way, socio-political influences are unavoidable in medicine. Medicine can no longer be seen as an isolated endeavor. It is inextricably tied into the economics, laws, and customs of a culture. It relies on government (e.g., federal), institutional (e.g., insurance), and private (e.g., fee-for-service) funding. It is influenced by the laws of a land (e.g., those regarding criminal and civil matters) as well as cultural (e.g., ideals regarding beauty, ability, and function) and religious (e.g., views aligned with the Divine Will) forces.

Despite variations in how clinical problems are known and valued, there is *much* that is shared within and across communities of medical professionals and those that judge their actions. A local account of moral choice in medicine is not relative.

7. THE ROLE OF NEGOTIATION

Medical professionals are members of particular communities that focus on particular sets of facts and values and know which ones are important and why. These facts and values inevitably lead to recipes for action, which incorporate implicit, and at times explicit, value judgments about proper trade-offs between costs and benefits. The character of proper trade-offs will often be negotiated without claiming that *the* answer can be discovered. In this way, moral decision-making in medicine is in part a negotiated reality.

A negotiation model of moral decision-making in medicine offers a schema for assessing various claims regarding the ways in which we understand and undertake moral decision-making. Negotiation (L. *negotiatio*, from *negotiari*, to carry on business) as a procedural expression of respect is often the only way to resolve disputes where there is disagreement about what it is proper to do (Engelhardt, 1996). In the law, "to negotiate" means to conduct communication or dialogue with a view toward reaching a settlement or agreement. It is the *process* of arranging the terms and conditions of a transaction (e.g., contract) (Black, 1979, p. 934), within the constraints of a given world.

Negotiation in medicine takes on either a formal or informal character. Formal discussions regarding moral decision-making are found in the literature and in special organized forums such as conferences and committees. Recall, for example, the works of such bodies as the President's Commission for the Study of Ethical Problems in Medicine and Biomedical and Behavioral Research (1981) on defining death and the President's Bioethics Commission (2004) on genetic engineering. This kind of formal negotiation in science and medicine is seen as well on committees that create and/or revise clinical classifications in specialty clinical groups (Cutter, 1992). Discussions regarding what constitutes moral decision-making take on as well an informal character and are found in the offices and laboratories of clinicians, scientists, and health policy makers as well as media roundtables across the world. Discussions concerning what to focus on, how to name, whether to fund, and what to do typically begin on an informal note and often go hand-in-hand with formal discussions.

This conclusion does not doom us to a hopeless relativity with regard to moral judgments in medicine. Clinical communities can and do agree on a great many explanatory accounts because they are bound by shared views of what constitutes pain, the constraints of scientific methodology, limited options for interventions, and similar notions of what makes human biological and psychological life pleasant versus unpleasant, good versus bad, and right versus wrong. Indeed, all is not relative.

8. IMPLICATIONS

There are numerous implications of this view for bioethical expertise, including recognizing that (1) bioethicists are at best geographers of facts, values, and socio-political influences, (2) interdisciplinary training for bioethicists is crucial, and (3) the development of a bioethical establishment is to be discouraged.

To return to an earlier discussion, bioethicists evolved out of an interest to assist health care professionals in thinking through value conflicts in an age of extraordinary scientific, technological, and social development. They are trained to illuminate conflicting moral interests. They can rehearse the reasons for holding one moral position as opposed to another and can often provide the best arguments for their moral positions. But for all their acumen in catching fallacious moral reasoning, bioethicists do not enjoy a privileged relation to moral truth. A privileged status assumes that ethicists have access to the additional premises that must be employed to justify the selection of prior premises, and so on. As previously argued, humans do not have rational access to ultimate justifications. It follows that the bioethicist is to be seen as mapping out the terrain of the facts, values, and socio-political influences that frame a certain debate. In the case of whether a partner of a patient diagnosed with AIDS should be told about the diagnosis, the bioethicist can assist in identifying the facts (e.g., data regarding HIV transmission), values (e.g., concerns regarding patient autonomy and welfare), and socio-political (e.g., legal, economic) influences. He or she can assist as well in clarifying the conflicts at stake in the case (e.g., patient autonomy or right vs. partner welfare) and the options for resolving the conflicts. In the end, although the answer to the moral quandary may not be achievable, there most likely will be better answers than others in terms of the criteria employed.

Given that moral decision-making is nested in facts, values, and socio-political forces, it becomes important that bioethicists are trained in an interdisciplinary fashion. What it means to be interdisciplinary, to traverse disciplines, given the institutional constitution and methodological and epistemological cores of academic disciplines, is an important focus of discussion. In order to be interdisciplinary, some (but not all) methodological differences among disciplinary approaches may be subordinated when there is a common project and expectation for progress (Ceccarelli, 1995). Evolutionary biologists and molecular geneticists, for example, find shared interests in understanding how Darwin's principles of natural selection operate at the molecular level within large populations (Fuller, 1995). Geneticists and ethicists find shared interest in thinking through the extent to which humans are determined (Lewontin, 1991). Philosophers and medical professionals create common ground in discussions of human psychosomatic existence, often resulting in novel ways to understand human existence, ways that are enriched by thinking in the humanities *and* the sciences (Khushf, 1997). In this way, interdisciplinary studies are marked by an attempt to bridge seemingly disparate ways of knowing (Gardner, 1993) and doing, which is one of the functions of bioethical consultation.

In successful interdisciplinary discourse, there must at least be a shared sense of vision, values, and expected outcomes in the process of working together.

Participants must agree upon the general purposes and value of the discussion and seek to achieve certain outcomes. As an example, health care providers and bioethicists may agree to think through the grounds for limiting access to critical care medicine that are shared by Roman Catholic clinicians and provide a consensus statement (Engelhardt and Cherry, 2002, pp. 35ff). For a second example, parties may wish to discuss the possibility of a global bioethics (Tong, 2000; Po-wah, 2002), highlighting the importance of shared visions of the nature and role of bioethics in the new millennium in order to arrive at some conclusions. For a third example, by the bedside, parties will come together to isolate and clarify competing facts and value claims for purposes of providing the best and right care for the patient. In this way, bioethics, as an interdisciplinary endeavor, provides assistance in thinking through issues that bridge science and the humanities.

An interdisciplinary discipline offers the advantage of remaining critical in its investigations and views because it remains open to new ideas from fellow disciplines as well as ones that join the debate. To use an analogy from science, an explanation is strengthened when there are multiple lines of evidence. Different lines of evidence for the same hypothesis reduce the likelihood that we have ignored confounding factors or that we spuriously concluded that our hypothesis predicts the observable data. Given that bioethical reflection necessarily involves consideration of the clinical facts, values, and socio-political influences of the case, it will be important that it remain open to new knowledge and ways of construing a debate or conflict. Transitions from one view to another will not be arbitrary. One moves from one explanation to the other if the second frees one from the contradictions and ad hoc assumptions integral to the first, allowing one to see things more clearly and more insightfully, and to approach new problems with greater clarity and insight (Lakatos, 1970).

This leads us to another concern, one that relates to the decreasing degree of bioethics' interdisciplinary character: bioethics may itself be in danger of becoming an establishment. Increasingly, bioethicists are hired by large institutions such as medical schools and hospitals. These are the only entities that can afford to pay such persons a wage large enough to live on. (For sure, those trained in "bioethics" are found in college and university settings, typically in philosophy departments, but they may not be as beholden to those at the local hospital because their major source of compensation comes from elsewhere.) Bioethicists employed by medical centers and hospitals are typically those called to consult on a case involving a dispute with a patient's family. In this role, there may be considerable pressure for the ethicist to come to a moral conclusion consonant with that of the employer, including those employed such as the health care professionals, who usually are the first to provide the bioethicist information. Such situations contribute to the rise of an establishment of bioethicists who tend to favor the institution's point of view and fail to consider the multiple factors that enter into a bioethical conflict.

This is not to argue that an institution that defines itself in a particular way cannot hire those who assist in speaking on its own behalf. A Roman Catholic hospital, for example, would likely hire a Roman Catholic bioethicist to sit on their ethics committees and participate in bioethical consultations. Given that the majority of the patients at the hospital might well accept the fact that they are in a

Catholic hospital for services and may well accept the teachings of the Catholic Church, there is no problem in having a bioethicist who, in the case of a patient's request for an abortion, does not take into consideration the arguments in favor of abortion. Rather, the problem arises when the bioethicist thinks he or she is not beholden to an institution or point of view. There is no such thing as a neutral bioethicist. The minimal requirement of the bioethicist will be to acknowledge the lack of neutrality. A more complete response is to acknowledge the forces that influence the consultation and map the options for resolution with care.

Add to this problem the rise of bioethical "speech", which may in the end be less than accessible to the untrained person. Disciplines tend to speak to themselves, alienating those who have no training and history in the fields. The advantage of this, of course, is that the discipline may become more specialized and thereby push its discussions and research in directions that would not be possible without specialization. Science (e.g., genetics) serves as an example here. The disadvantage is that the discipline may cease to serve the very people it sets out to serve because such people can no longer participate in the discussions. Contemporary feminist discussions have received this critique because they have failed to take into account the experience of colored women; it has traditionally been middle-and upper class white women who have contributed to the field (Rothenberg, 2003). The same may be said of bioethical discussion. Discussions that frequent the bioethical literature regarding cosmetic surgeries and the permissible boundaries of infertility services (including cloning) serve primarily the upper class. Where are the bioethical warriors about access to basic medical care, including prenatal care, vision and dental care, nursing home care, and affordable prescription drugs? In the end, we may increasingly be wary of the future of the bioethical consultant on grounds that bioethics is increasingly becoming an institution.

9. CONCLUSION

The contemporary world is replete with "experts". In the consumer arena, we have little difficulty questioning the "expert" in terms of background, training, actual talent, and expectations. We do this when buying vacuums; why not do this when assessing the very people who gather around the hospital bed or committee table to make life-and-death decisions? This essay is written in the spirit of generating reflection on what constitutes legitimate bioethical expertise, training, and claims. This is not to undermine the profession but rather to encourage critical thinking in and about it. What I have suggested is that bioethical expertise requires an awareness of the local character of knowledge and value claims, the need for interdisciplinary education, and the perils of support of the establishment.

REFERENCES

American Psychiatric Association (1994). *Diagnostic and Statistical Manual of Mental Disorders*. 4th ed. Washington, D.C.: American Psychiatric Association.

Aquinas, St. Thomas (1948). *Summa Theologica*. Trans. Fathers of the English Dominican Province. Benzinger Brothers, Inc.

Aristotle (1985). *Nicomachean Ethics*. Trans. T. Irwin. Indiana: Hackett Publishing Company.

Beauchamp, Tom L. and Childress, James F. (1979). *Principles of Biomedical Ethics*. New York: Oxford University Press.

Black, Henry Campbell (1979). *Black's Law Dictionary*. 5th ed. Minnesota: West Publishing Company.

Ceccarelli, Leah (1995). 'Rhetoric of Interdisciplinary Scientific Discourse: Textual Criticism of Dobzhansky's *Genetics and the Origins of Species*.' *Social Epistemology 9, 2,* 91-111.

Curd, Patricia (Ed.) (1996). *A Presocratic Reader*. Indiana: Hackett Publishing Company.

Cutter, Mary Ann G. (1992). 'Value Presuppositions in Diagnosis: A Case Study of Cervical Cancer.' In: J.L. Peset, et al. (Eds.), *The Ethics of Diagnosis* (pp. 163-170). Holland: Kluwer Academic Publishers.

Cutter, Mary Ann G. (2003). *Reframing Disease Contextually*. Holland: Kluwer Publishing Company.

Engelhardt, H. Tristram, Jr. (1996). *The Foundations of Bioethics*. 2nd ed. New York: Oxford University Press.

Engelhardt, H. Tristram, Jr. and Cherry, Mark (2002). *Allocating Scarce Resources: Roman Catholic Perspectives*. Washington, D.C.: Georgetown University Press.

Fleck, Ludwik (1979 [1935]). *Genesis and Development of a Scientific Fact*. T.J. Trenn and R.K. Merton (Eds.). F. Bradley and T.J. Trenn (Trans.). Chicago: University of Chicago Press.

Foucault, Michel (1973 [1963]). *The Birth of the Clinic: An Archeology of Medical Perception*. A.M. Sheridan Smith (Trans). New York: Pantheon Books.

Fuller, Steve (1995). 'Interdisciplinary Rhetoric: Lessons for Both Rhetoric and Rhetoricians.' *Social Epistemology 9, 2,* 201-204.

Gardner, Howard (1993). *Multiple Intelligences: The Theory in Practice*. New York: Basic Books.

Jonas, Hans (1984 [1979]). *The Imperative of Responsibility: In Search of an Ethics for the Technological Age*. Chicago: University of Chicago Press.

Jonsen, Albert R. (1998). *The Birth of Bioethics*. New York: Oxford University Press.

Kant, Immanuel (1985 [1785]). *Foundations of the Metaphysics of Morals*. New York: Macmillan Publishing Company.

Khushf, George (1997). 'Why Bioethics Needs Philosophy of Medicine: Some Implications of Reflection on Concepts of Health and Disease.' *Theoretical Medicine and Bioethics 18,* 145-159.

Kierkegaard, Søren (1980). *Sickness Unto Death: A Christian Psychological Exposition for Upbuilding and Awakening*. H.V. Hong and E.H. Hong (Trans. and Eds.). New Jersey: Princeton University Press.

Lakatos, Imre (1970). 'Falsification and the Methodology of Scientific Research Programmes.' *Criticisms and the Growth of Knowledge*. I. Lakatos and A. Musgrave (Eds.). England: Cambridge University Press, pp. 91-125.

Lewontin, Richard (1991). *Biology as Ideology: The Doctrine of DNA*. New York: Harper Perennial.

Mill, John Stuart (2002 [1861]). *Utilitarianism*. Indiana: Hackett Publishing Company.

Nash, Mark (2000). 'The Simple Solution to the AIDS Epidemic.' *The Philadelphia Trumpet, July,* 11-13.

Osborne, Lawrence (2001). 'Royal Disturbances.' *The New York Times Magazine*, 98-102.

Plato (1985). *Euthyphro; Apology; Crito; Phaedo*. F.J. Church (Trans), with an introduction by R.D. Cumming. New York: Macmillan.

Po-wah, Julia Tao Lai (Ed.) (2002). *Cross-Cultural Perspectives on the (Im)Possibility of Global Bioethics*. Holland: Kluwer Academic Publishers.

President's Bioethics Commission (2004). 'Advising the President on Ethical Issues Related to Advances in Biomedical Science and Technology.' *The President's Council on Bioethics* [On-line]. Available: <http://www.bioethics.gov>

President's Commission for the Study of Ethical Problems in Medicine and Biomedical and Behavioral Research (1981). *Defining Death: Medical, Legal, and Ethical Issues in the Determination of Death*. Washington, D.C.: U.S. Government Printing Office.

Reiser, Stanley (1978). *Medicine and the Rise of Technology*. England: Cambridge University Press.

Reznek, Lawrie (1987). *The Nature of Disease*. London: Routledge and Kegan Paul.

Roe v. Wade, 1973.

Rothenberg, Paula (2003). *Reace, Class, and Gender in the United States*. 6th ed. New York: Worth Publisher.

Sassower, Raphael (1993). *Knowledge Without Expertise: On the Status of Scientists.* New York: State University of New York.
Smith, John Edwin (1963). *The Spirit of American Philosophy.* New York: Oxford University Press.
Tong, Rosemarie, with G. Anderson and A. Santos (eds.) (2002). *Globalizing Feminist Bioethics: Cross-cultural Perspectives.* Colorado: Westview Press.
Wittgenstein, Ludwig (1985). *Philosophical Investigations.* New York: Macmillan.

CHAPTER 9

CORINNA DELKESKAMP-HAYES

SOCIETAL CONSENSUS AND THE PROBLEM OF CONSENT: REFOCUSING THE PROBLEM OF ETHICS EXPERTISE IN LIBERAL DEMOCRACIES

The practice, in modern Western democracies, of engaging and offering ethics expertise for legislative, judicative and societal decision-making has come under critical scrutiny. This essay questions the direction of that scrutiny. While the literature concentrates on the issue of *offering* expertise, this essay argues that the deeper difficulty lies with *demanding* such expertise. That difficulty arises in the context of the political self-understanding embodied (especially, even though not exclusively) in the more socially oriented Western democracies: Such democracies derive the legitimacy of political power from that power's moral[1] function for protecting human rights (in an encompassing, positive sense of the term). This derivation establishes, so this essay argues, the normative horizon within which ethics expertise must be understood: Insofar as such expertise is sought for implementing what a supposedly general societal consensus seems to impose, it disregards that principle of consent, which the truly universal obligation of respect for autonomy imposes. The demand for such expertise in such political societies thus highlights the clash between democracies' (contingently substantial) moral claims and their (inherently formal) liberal commitment. The essay closes with some remarks about why attempts to repudiate that clash are doomed. It will thus establish at least *prima facie* evidence for the conclusion that ethics expertise in liberal democracies merely serves to provide an ethical veneer for political and societal injustice.

The literature predominantly disregards this political context, restricting the relevant critical concerns to three questions:

(1) Should ethics expertise merely lay out alternative solutions for given problems, thus merely educating the decision makers,[2] or should such expertise advocate one "best" solution?[3] (I.e., should their service be descriptive or prescriptive?)

(2) Can question (1) be resolved in view of a decision concerning the nature of ethics? That is, must one assume the existence of only one canonical "ethic" or morality for all humans, or should one settle for ethical pluralism?

(3) How, if one opts for advocating, and furthermore assumes that the purpose of offering expertise is to bring an existing societal consensus to bear on the issue to be

L. M. Rasmussen (Ed.), Ethics Expertise: History, Contemporary Perspectives, and Applications, 139-163.

determined (Kipnis, 1997, pp. 330, 339), can philosophers' professional and personal integrity be maintained, either
a) if their own philosophical position differs from the socially accepted one? Or
b) if none of the available academically acceptable positions coincides with that socially accepted one? Or
c) if they must fear that openly disclosing the philosophical basis for and likely consequences of their recommendation (thus remaining true to their scholarly mission) would jeopardize decision-makers' willingness to follow what they recommend (Kamm, 1990, pp. 348, 358)?

In a first step, I shall show why these questions do not capture the problem of ethics expertise at its deepest level, namely (as it will turn out) its involvement with the democratic procedure as a morally legitimizing principle of its own. In a second step, I shall explore the difficulties attending that principle in the context of a liberal (consent-oriented) society that is also committed to realizing a (consensus-based) common good. A third and final step will suggest reasons for the impossibility of rendering commitment to such a good compatible with the liberal commitment to individuals' right to normative self-determination.

1. UNDOING THE RECEIVED ACCOUNTS OF THE PROBLEM OF ETHICS EXPERTISE

On closer examination, the educative and advocative function of ethics experts (first question, above) are not even sufficiently separable to warrant creating a preference problem in the first place. On the one hand, even advocates of experts' advocating agree that these experts should be professional philosophers. Only training in the history of and literature about the relevant ethical theory alternatives, after all, renders their judgement "informed" (Agich and Spielman, 1997, p. 387).[4] The "descriptive function" of ethics expertise is thus incorporated into experts' very "prescriptions". On the other hand, even advocates of experts' merely educating role concede that philosophers, rather than sociologists or psychologists, are chosen as ethics experts, precisely because they usually take their scholarly commitment to include posing the validity question, thus taking a personal stand in the matters under discussion (Cf. Birnbacher, 2002, p. 105).

To be sure, some philosophers understand their professional task as exclusively analytical and critical. For them, professional competence reduces to technical training in logic, historical exegesis and conceptual analysis. But these philosophers are also human beings (and supposedly qualified to provide ethical guidance). Accordingly, they cannot shun taking a moral stand (Moreno, 1991, p. 52, Benjamin, 1990, pp. 376ff). Such personal positions will inevitably influence the manner in which they present the relevant ethical alternatives (even if the purpose is mere education) (Momeyer, 1990, pp. 396ff; Childress, 1996, p. 214).[5] Moreover, regardless of such scholars' efforts to clarify the merely (as they would understand it) personal nature of their bias, and to discount that influence, still the very fact that they were chosen as "experts" will endow their viewpoint with special authority and persuasive compellingness. In the end, the effect of their purportedly educative testimony will thus be no different from that of a committed (and also scholarly) advocate-expert.

Some authors have tried to salvage an at least interpretive contrast between ethics experts' educative and advocative roles by relating that contrast to different views of the nature of ethics (cf. second question). Advocating advocacy might reflect belief in only one true ethical account, which is ("objective", and therefore) valid for all human beings. Advocating education, by contrast, might reflect belief in an irreducible plurality of (merely "subjective") ethical accounts, so that ethical decisions involve a choice. Experts' educational guidance here has the task of rendering such essentially contingent choices merely more "enlightened".

But even for ethics experts' interpretation of what their services amount to, this difference in meta-ethical positions (of monism/objectivism or pluralism/subjectivism) is at bottom irrelevant. Even if ethical monists advocate for what they consider "right", they can only hope that decision makers will act accordingly. Similarly, ethical pluralists, after having discharged their educative duty, must acquiesce in decision makers' implementing what to the decision-makers themselves seems "right" (Fletcher, 1997, pp. 311ff; Agich and Spielman, 1997, pp. 387ff). Even theoretical pluralists, in other words, must discount their pluralism when it comes to endorsing clinical responsibility hierarchies, due process and democratic procedures. Just as monist advocates whose advice is disregarded, so pluralist educators will be able to make sense of having offered their services only if they, on some second-order ethical level, accept the democratic process. It is this process, after all, which authorizes decision makers either to take their personal pick (given that they were elected or hired with a view towards their typical way of picking) or to choose in a way concordant with the normative commitments of those who elected or hired them for that purpose. Both interpretive positions thus acknowledge that such choices acquire a "created general validity" for the political society involved (Moreno, 1991, pp. 54ff).

There is a further interpretive dimension that surfaces in the discussion of ethics experts' educative or advocating role. Clearly, ethical monism or pluralism will determine the expert's role only if a decision has already been made concerning the function of ethical theories *vis à vis* practice. Only if ethical theories ground such practice, so that one can deduce the relevant solutions from the theoretical principles, will pluralism and monism necessarily be correlated with experts' restriction to educating or their authority also to advocate. Proponents of the latter position argue that the un-tenability of foundationalism (as defended by proponents of the former position) is not sufficient to repudiate the legitimacy of advocating (Agich and Spielman, 1997, p. 384). After all, so the argument goes, such a strictly foundational view (often associated with rationalism) about the function of ethical theories is usually not endorsed.[6] Most authors (Moreno, 1991, is a typical example) are aware of the difficulties involved in bridging the gap between philosophical theory and real life. Ethical theories are frequently conceived as mere instruments for reconstructing given courses of action.[7] Properly analyzing and evaluating different courses of action may require engaging different ethical theories. Such theories may offer more or less helpful conceptual tools for morally orienting any given course of action. Given this further dimension, could philosophy's function *vis à vis* practice provide a conceptual resource for salvaging the relevance of ethics experts' educational or advocating role? Could, in other words, their respectively pluralist or monist acceptance of the democratic procedure as an ultimately right-making condition amount to something different, depending on whether they were foundationalists or instrumentalists? Could a monist foundationalist's acceptance of decision makers' not following his advice (i.e. deciding "wrongly") impose a more resigned, or more cynical self-understanding than that of a foundational pluralist or an instrumentalist?

In order to answer this question, we must take a closer look at how, in the context of an acceptance of democratic procedures as "right-making", any one conceptual "instrument's" helpfulness is to be evaluated. Such an evaluation presupposes a point of reference. In the given context, that point of reference must be that societally accepted way of proceeding in certain types of situations, which at the same time underlies the democratic authorization of decision makers, and which is re-interpreted (and sometimes modified) through their ways of using that authority. We can call this "accepted way" the "societal consensus". Not only while practically engaged in providing their time-pressure-hampered consulting services to those who solicited them, but also when critically reflecting on the ethical acceptability of those services, ethics experts acknowledge the (quasi-Aristotelian) imperative that they should be, as Fletcher put it (1997, p. 299) "responsive to the … values and laws of the communities, states, and nations" of their environment, and thus obliged to maintain an inside-perspective check on their critical acumen as ethicists.[8] Instrumentalist ethics expertise thus involves philosophically reconstructing that consensus in such a way that it can be acknowledged as adequate by decision makers, and at the same time allows one to optimally accommodate that solution to the practical problem at hand, which is recommended as "right" on the basis of that reconstruction. This understanding, of course, also accommodates the fact that particular "application problems" in ethical theory may modify what gets applied. It makes room for a creative contribution of ethics expertise to the theories it engages.

This same understanding of instrumentalism in ethics expertise, moreover, also throws light on foundationalism in ethics expertise. Unlike with instrumentalism, of course, we shall have to consider both foundationalism's pluralist and monist versions: As for foundational pluralist ethics experts, they can construe decision makers' eventual choices as contributing to the further development (or evolution, or change) of that same societal consensus, which also had guided the democratic votes (ultimately) underlying, or authorizing, those particular decision makers' authority. As for foundational monists, their willingness to expose their expertise to the risk of decision makers' discretionary disregard must be assumed to rest on their willingness to modify their position in terms of an epistemological skepticism: the one ethical truth that (for them) grounds all practice must be admitted to escape present thinkers' adequate grasp.[9] Foundational monists' acceptance of the democratic procedure thus implies their recognizing that procedure's heuristic function, and the evolving societal consensus as the path to ethical truth. In other words, that procedure, in experimenting (as it were) with marginally different normative consenses (as these surface with changing democratic majorities) is seen by such monists as gradually working out an ever-clearer general understanding of that ultimately valid ethical theory, the existence of which these monists affirm (Momeyer, 1990, pp. 404, 405ff). Whereas monists, whether foundationally or instrumentally oriented, will understand the democratic process as promoting a society's moral progress, pluralists will merely look for an unfolding of interesting changes, making up a narrative that shapes that society's normative identity. But then the contrast that was supposed to exist either between monism and pluralism, or between instrumentalism and foundationalism, has turned out to be irrelevant, when viewed in the context of each position's more basic endorsement of the right-making character of the democratic procedure. Accordingly, these different interpretive ethical-theory-contexts, in terms of which ethics expertise's educational or advocative function could be conceived, make no practical difference. All contributors to the discussion about ethics expertise turn out to endorse the societal consensus as its democratically authorized point-of-reference.

It is precisely this acknowledged framework which also underlies the integrity problems for philosophical ethics experts raised in the third question posed above. Yet it has become clear that already the very commitment to serve as ethics expert imposes a willingness to relativize those very personal, or even professional opinions, which had triggered those integrity problems. Once their expertise is understood as integral to the way in which a democratically authorized decision process shapes a society's normative consensus, that expertise's content can be subordinated to its function. Rather than primarily worrying about what an expert recommends, and attending to the educational or advocative intention, meta-ethical orientation (in terms of monism or pluralism), or practical import (in terms of foundationalism or instrumentalism) of their service, and construing any "problems" for their self-understanding in those terms, one should attend to the – as we may call it – "performative" aspect of that service: the fact that the involvement of ethics expertise, irrespective of any of the above particulars, and irrespective of whether decision makers agree with that expertise or not, endorses the ethical rightness of that consensus, which authorizes and guides their eventual decisions.

Such a changed perspective has two significant consequences. On the one hand, it allows the integration of experts' seemingly disparate activities into one coherent account. Ethics experts, after all, are chosen not merely for their academic proficiency. As the example of their clinical engagement reveals, "bio-ethicists" must also possess experience with clinical settings and professional standards. They are quite generally invited for their ability to trace the ethical dimension of what often merely appear to be just technical or managerial problems, to draw out the ethical implications of any one profession's self-understanding for such dimensions, and to offer their skill in analyzing ethical arguments and in resolving ethical conflicts, as these arise in professional practice, policy, or legal proceedings. They must, in other words, sensitivize decision makers to the concerns of those affected by the decisions, and mediate discussions for mutual information, inspiration and the normative agreement among those affected themselves. Especially with regard to that latter task, experts not only orient their service towards an already existing societal consensus, to the affirmation and preservation of which they implicitly contribute. Rather, these experts also explicitly work out or restore a threatened or absent consensus. On the other hand (to come back to the second consequence), their participation in the process through which such consensus is brought to bear (in democratically authorized manners) on societal decision making suggests to their clients as well as to the public that these decisions are ethically acceptable. Over and above the – as it were – pragmatically right-making function of the democratic procedure and the societal consensus it both presupposes and ratifies, moral philosophers set their professional validity seal on the resulting decisions. It is this implication of their cooperation, so the following section argues, which will disclose a more serious problem than what is commonly perceived.

2. UNCOVERING THE POLITICAL ROOT OF THE PROBLEM

If moral philosophers thus accept, further, and justify what the invisible hand of democratic majority generation discloses, democracy cannot be understood in a "mindless" sense as the (institution-mediated) self-government of egoists. If a society's normative consensus results from merely tactical alliances, securing electoral support for merely private aspirations, then the provision of ethics expertise would indeed present a threat to philosophers' professional integrity. If their willingness to cooperate in such endeavors is to be plausible, democracy must be understood (at the very least) in a republican manner: The public's voting behavior must be driven by an encompassing vision of a common good. In pluralist societies, such visions must in turn rest on (real or virtual) discursive processes. Through these, the divergent normative commitments are brought together for mutual enrichment and differentiation, suggested policies are reflected in terms of divergent perspectives, and solutions are designed so as maximally to accommodate the interests and concerns of all those who will be affected. (Obviously, those ethicists who believe in humanity's moral progress will be able to "explain" that progress by reference to such discourse.) It is, moreover, precisely this discursive underpinning of what is then exposed to the test of democratic decision making, which can be found reflected in the additional sensitivising and mediating functions of ethics expertise, as these were discussed at the end of the first section. Similarly, on the content-level of experts' service, the societal consensus which they were obligated to invoke and restore can itself be understood as resulting from such discursive agreements. After all, the discursive agreements from which that consensus is derived can be taken to guarantee that liberal democracies' formal commitment to the principle of consent is preserved by whatever material commitments their quest for common good implementation might impose on such democracies' constituents. In that sense, the discursive presuppositions of ethics experts' normative point of reference as well as the (virtual or real) discourses which their sensitivizing and mediation function envisage, can be seen as safeguarding the harmony between liberal democracies' twofold moral commitments.

Yet if such processes of discursive reflection among the various moral stakeholders within a pluralist society already by themselves provide a legitimizing ground for the exercise of legislative, policy, adjudicative or professional authority (and discretion in decision making) within such a society, why should it be ethicists (irrespective of their different ethical and multi-dimension meta-ethical orientations), who are to imprint an additional ethical quality certificate on the resulting practical decisions? Why could the advising, sensitivizing or mediating experts not be psychologists, sociologists, cultural anthropologists, social workers, or communication specialists? If the republican character of modern democracies imposes on members a regard for the common good, and if the discursive process is by itself sufficient to "digest" the irreducibly divergent visions guiding that regard into one generally endorsable vision, why could one not settle for some merely procedural supervision of that digestive task? The additional engagement of ethics expertise makes sense only if the democratic process (as based on discursive agreements), while perhaps necessary to secure, and be compatible with, the ethical acceptability of what it endorses, is still not sufficient to guarantee that acceptability.

The need for such additional moral legitimization becomes obvious if one attends to the conditions under which the (real, virtual, or invoked) discourses preparing democratic decision making take place. The model underlying the discourse ethics developed by Habermas is voluntary cooperation. The democratic discourse (within different levels of political and societal institutions), by contrast, prepares legislative, adjudicative, professional, and policy proposals, which are forcefully implemented even on those who have not participated in the discursive process. Moreover, even the participants' engagement is not un-constrainedly voluntary, but accepted as the only chance to influence what would eventually affect everyone anyway. Their cooperation is situated within a framework of (ultimately) political compulsion. Given the liberal commitment morally to justify the use of political power, the involvement of ethics experts as participants in the relevant procedures (for determining the way in which that power is used) thus becomes understandable. This participation is to establish an ethical quality certificate designed to elicit that virtual consent among the governed on which modern government ultimately depends (Wildes, 1997, p. 369).

All of this implies that ethics expertise (*pace* Agich and Spielman (1997, p. 394) essentially differs from expertise in carpentry or plumbing, and that the critical scrutiny to which the former is (rightly) exposed in the literature should be extended to its political context. The decisive question then is: Can ethics expertise fulfill the legitimizing function for the sake of which it is engaged? This question concerns not only the macro-level of legislation. Ethics expertise permeates even (to take an example from the health care context) the remotest institutional review boards and hospital ethics committees: Even if professional institutions are (in principle) at liberty to grant decision making authority on the basis of professional (and managerial) excellence, nevertheless, to the extent that these institutions are integrated into systems of state-sponsored funding and insurance schemes, majority-backed political priorities will influence these grantings. Moreover, insofar as even such professional institutions embody democratic decision-making elements, they not only partake of the putative legitimacy-bonus of the political system, but also require ethical support for the compulsion embodied in that system.

A closer look at liberal democracy's moral claims reveals that the very possibility of such support faces a difficulty. Such democracies derive their moral legitimacy (or the legitimacy of their use of political power) from their commitment to granting freedom rights. These encompass the right to normative self-determination. Ethical diversity among the constituents of such polities is thus accepted. Its flourishing even provides evidence of a polity's liberal character. Some types of ethics expertise are directed, in a narrowly restricted sense, towards merely implementing that liberal principle.[10] They partake of the legitimacy to which they contribute.[11] Most cases of expertise discussed in the literature, however, concern content-rich moral solutions to value-impregnated practical problems. They thus concern decisions which will affect those whose own substantial moral commitments would have imposed different (or even contrary) solutions. In such cases, and regardless of how remote (and procedurally mediated) experts' involvement may be for the final outcome, it legitimizes the use of force against the non-consenting.[12] It morally justifies forcefully interfering with those whose right to moral self-determination was supposed to have been respected.[13]

How could one defend such involvement? There are some obvious strategies. One may suppose that the common good, the obligatory regard for which qualifies a democracy as a republic, is not entirely empty. That good may accommodate some universally valid norms or values, about which all ethics experts can reasonably agree. Of course such universal validity, in the face of an accepted moral diversity, will distinguish only a meager "core".[14] Its validity, if it is to be universal, must be grounded rationally.[15] Let us disregard, for the moment, the question of what such a supposition could finally accomplish. This will be addressed in the next paragraph. For the present, let us further suppose, by proceeding in the (very thinly conceived) Kantian spirit that permeates at least continental Europeans' political self-understanding, that all humans, in spite of their different rational abilities or accomplishments, are morally obligated to acknowledge what their (potential) rationality dictates. Refusing to acknowledge that obligation would then be tantamount to forgoing one's respectability as a human being, and thus one's very entitlement to that normative self-determination which one might invoke against any generally enforced moral vision. Accordingly, subjecting a democracy's constituents to the necessity of forming mutual agreements concerning how such core values (which by themselves are already rationally compelling) are to be implemented could not be charged with using force against the non-consenting (in any relevant sense). Ethics experts' advising, sensitivizing and mediating such agreements, even though all of this would guide and justify imposing merely particular ways of implementing such core values and norms on all, would not constitute a violation of the latter's right to self-determination.[16] Quite on the contrary, that right would, from the very start, have been conceived to be limited with regard to such indispensable core values. Imposing these values even on those not accepting their compelling-ness would constitute a merely educational support towards the development of greater realized human dignity (on its way to either moral progress or cultural sophistication). (The ethical pluralists discussed in the first section of this essay would have to restrict their pluralism to the fringe areas of ethics, thus leaving such a supposed common core intact.)[17]

Of course, the services ethics experts actually offer are not restricted to endorsing and (somehow) implementing (or protecting) such universally accepted core values (such as "health care for all"). The case studies addressed in the literature revolve around quite particular ethical solutions, incorporating quite specific positions concerning singular practical problems. To whatever extent "core values" might be universally accepted in theoretical discourse, the actual moral disagreements separating members of different moral orientations will sometimes suggest even directly opposing solutions to practical moral problems.[18] In regard to such real-life differences, that general agreement on the theoretical level reveals itself as merely notional.[19] Insofar as ethics experts guide the use of political power in concrete situations, their service once again turns out to legitimize the use of force against the (now relevantly) non-consenting. The latter's own moral commitments are thereby disrespected.

Could this involvement be justified (second defense strategy) by reference to a further, implicit meta-level agreement? All members of modern democracies will admit, so it may be claimed, the fairness of the democratic procedure. This procedure, after all, sometimes gives preponderance to one set of normative principles, and sometimes, with changing majorities, to others. Even those against whom force is used in a way that disagrees with their more specific moral preferences can still consent to the rules of the game. They can work at doing better at the next election. One might argue in addition that the liberal principle itself, on which the obligation to grant normative autonomy rests, endorses public self-government, which in turn unavoidably sometimes favors some, at other times, other, specific value preferences. In a democracy that principle imposes the (formal) need to secure the consent of those against whom force will be used not (so this strategy argues) merely by itself, but always in the context of the need to endorse the normative validity of whatever (materially specified) societal consensus voting majorities may generate. Even if ethics experts, considered as philosophers, may maintain some critical distance from what is generally believed, still their engagement for guiding (sensitivising, mediating) societal decision-making was found to impose a countervening obligation to acknowledge such consensus (see for example Kamm, 1990, pp. 357ff, 361), always provided it was developed according to the rules of an intrinsically fair decision making procedure.[20]

On closer look however, this (second) defense of ethics experts' legitimizing function extends to only those moral minorities who have a real chance of winning power at the next election. It leaves unresolved the issue of "intrinsic" moral minorities. To be sure, the discursive processes, which filter the normative programs proposed for electoral endorsement, are designed to secure maximal inter-group acceptability. The societal consensus, which provides normative orientation for the experts' performance, can thus be specified as presupposing as well as encouraging such efforts at securing inclusiveness. Yet there is ample evidence in the literature on ethics expertise for the claim that even such expertly mediated discursive processes in the end result in consensus statements, which affirm only the view of the participating majority.[21] Diverging minority voices in such committees, if they are separately published at all, regularly get neglected by the decision makers (or decision making bodies), [22] who are concerned about not compromising their ability to act by attending to too many small groups' detailed reservations.[23] Moreover and more seriously, from the very start only those moral minority groups are included in such discursive consensus developments, whose moral commitments would not proscribe the very process of inter-ethos (or, as this often has religious motives, inter-faith) agreement. There exist communities for whom the very project of common consensus formation with its implied requirement to (in the end) settle for compromise positions, counts as morally compromising.[24] Surely the former, by entering the discursive process, had at least (if grudgingly) accepted the rules of the game that were implicitly enforced on them. They share in the responsibility for the consequences. But with regard to the latter more "fundamentalist" non-mainstream groups, the democratic procedure, even if mediated by discursive processes, in effect openly disrespects members' right to normative self-determination.

A further (third strategy) attempt at vindicating liberal democracies' moral legitimacy, and the appropriateness of ethics experts' involvement in the attendant disrespect, again engages the notion of rationality. Philosophically speaking, of course, the number of believers in a certain system of thought is insignificant for evaluating that system's truth. Plato always emphasized that the wise are few. Nevertheless it is often supposed that unwillingness to enter into the discursive dialogue with adherents of other such systems, to expose one's faith to criticism and give an account of its foundations is a sign of irrationality. This lack of rational 'civility' then disqualifies (Kamm, 1990, pp. 364ff)[25] such believers from the circle of those who ought (morally) to be respected.[26] The controversy about such disqualification often revolves around the problem of tolerance. It is then argued that no political society needs to extend tolerance (and specifically, a tolerance that expresses itself in refraining from intrusion into people's normative self-determination) to those who do not offer a corresponding tolerance to members of other moral creeds.[27] But this is a side-track. Some moral communities may happily concede that others' moral commitments are fine for them, and still insist that their own members should not be exposed to the outcome of compromise formation among those others. Members of such communities may simply claim that discursive arguments (or, if they are Platonists or Aristotelians, discursive arguments with crowds who lack proper enlightenment and education) are not the way in which their own position can be adequately presented or appreciated. For them, talking on that level with those of a discursive mind-set (or discursing with the un-inspired many) is simply pointless. Prominent secular examples for such a position are feminists. Feminists are often quite tolerant *vis à vis* the ways in which men (unfortunately just simply cannot help but) think and behave. They insist however, not only that such thinking and behaving would be inappropriate for them, but also that the very act of entering into dialogue with men tends to impose rules of evidence and criteria for consistency, which foreclose any in-depth mutual understanding and fruitful exchange.[28] Similar problems are encountered by the more traditional religions, which have not undergone the Scholastic transformation characteristic of Western, post-Enlightenment Christianities. Implicitly to disqualify members of such moral communities from human dignity-constitutive rationality, and thus from entitlement to normative self-determination, is to play a dominant societal (material value) consensus against modern democracies' liberal commitment to secure consent. It constitutes a glaring *petitio principii* in favor of discursively framed mainstream creeds.[29]

If there is to remain any hope for ethically defending ethics experts' involvement in such disrespect, it must give practical salience to those "core values" or "core norms" which were (somewhat boldly) stipulated above as being rationally compelling, and thus (at least on the level of theory) universally obligatory. For that purpose, one would have to establish an argument according to which, in addition to that universal validity in theory, these values or norms must be realized or implemented politically.[30] Only in that case, after all, could one discount the attendant normative diversity losses, or the normative disrespect costs, as being morally unavoidable.[31] Suppose it could be shown that health care, or social support for those unable to sustain themselves, must be secured through the state. In that case, not only the technicalities, but also the moral priorities of such securing would have to be settled by democratic majorities. Through the respectively particular manner of such settling, these majorities would determine the respectively particular (and admittedly partial) cultural identity of their politically constituted society. That identity, in other words, would comprise not only a formal element of commitment to liberty. It would also limit that formal commitment in view of a material preference concerning how to realize the "common good". Moral minorities, and especially their more divergent, off-mainstream varieties, would simply have to settle for the constraints, which any common-good-providing political environment inevitably imposes on them.

In the third and final section of this essay I shall offer some considerations which render the possibility of providing such an argument unlikely.

3. EXPLORING THE UNLIKELINESS OF A TENABLE SOLUTION

In this final discussion I shall disregard the question of how those stipulated "core values" or "core norms", which underlie the supposed "common good", can be established as rationally compelling. Even if they could, the decisive question here concerns their political realization. Even very off-beat minority moral communities, after all, may concede (even if not always on discursively rational grounds) that one ought to help those in need, and that the sick must be medically treated. But they will deny that such services should be offered by a state, and thus in a manner that eventually disrespects those minorities' normative freedom.

Let us from the very start concede (somewhat boldly) that health care and social services can be secured efficiently only when supported by the political power resources of the state. Relegating such charitable undertakings to the private sector (even if that sector were supported by strong moral communities) might imply accepting (so we might further concede) significant efficiency losses. Still, from the perspective of some minority groups, such losses are less threatening to their moral concerns than the policy infringements, which come with government involvement. Such groups will be morally committed to prefer a less "civilized" environment (where certain goods and services can be less taken for granted).

Which political philosophy should be invoked? Which of the many candidates could provide an answer to the question of political common good enforcement, without from the very start discounting (at least) those more off-mainstream (fundamentalist) minorities, the normative self-determination of which the liberal commitment of modern Western democracies cannot help to endorse? This second condition from the very start excludes utilitarian, as well as communitarian accounts of the state.[32] We are left with deontological theories.

The most promising candidate was developed by Immanuel Kant.[33] His philosophical account of the state legitimizes the latter's use of political power even against those who actually do not consent. It does so by reference to what is morally obligatory. His theory thus promises to justify such power in precisely that way which renders the involvement of ethics expertise for guiding that use appropriate. This essay cannot offer an even remotely adequate sketch of his theory. It can merely indicate the general direction of an argument that would establish why even Kant's theory does not yield the desired justification for the claim that a society's (materially specified) common good must be implemented through political force, and that therefore the engagement of ethics experts for directing that implementing was legitimate.

The truly rational core of Kant's (in fact very involved) moral theory of the state roughly amounts to this: Political power is morally justified to the extent that it implements the state's central function for making morality possible. How did Kant get there, and where does it lead him? Kant appeals to all humans' experienced commitment to morality. It is from this commitment that persons derive their human dignity. Morality, as governed by the laws of practical reason,[34] imposes obedience to the "categorical imperative". This imperative imposes respect for persons' autonomy. Persons realize their autonomy through the means of their bodies and their property. Insofar as persons are committed to rational morality and hence endowed with dignity, they are entitled to those means. Hence, morally respecting persons' autonomy amounts to leaving un-invaded their bodily integrity (and *a fortiori* their life) as well as their property, except where such invasions are covered by the persons' consent. Outside of a state, property lines are contested. Hence, outside of a state the morally required autonomy respect cannot be offered. Hence the state is meta-morally indispensable, and its use of power for that purpose morally legitimate.

It will not be necessary in the present context to discuss the question of how the use of penal force is justified on that account. (I have offered an account of this issue in Delkeskamp-Hayes, 2005.) It is more important to return to the issue of ethics expertise, especially when it is engaged for questions regarding the political implementation of a (substantially determined) common good. If political power in a Kantian state is morally justified (exclusively) insofar as that state renders morality possible, in what sense could the use of such power for realizing such a common good be understood as a way of rendering morality possible? Surely realizing a common good might be understood as a way of behaving morally. But behaving morally differs from behaving in a way that renders others' behaving morally possible. Should we conclude that by forcefully extracting taxes from its members for the sake of helping those in need the state indeed "makes it possible" for these members to act morally? But then "making morality possible" would have been used in two senses: The first, primary one, concerned providing the conditions (property laws) for the possibility of persons freely obeying the moral imperative. This second, hybrid one, concerns applying a force that constrains persons into realizing a moral good. Could we conclude that Kant's view of the moral justification of political power accommodates both senses? Whereas the first imposes respect for personal property, the second depends on violating that respect. Kant's account would become inconsistent, were one to surmise such a twofold justification.

Undeniably, Kant himself posited an obligation for the state to care for those in need. He was careful, however, not to permit this obligation to ratify corresponding rights on the side of those in need. He was also careful to keep such charitable support minimal: one must not encourage a "laziness" among the receivers that would thrive on invading other persons' property. This, after all, would repudiate the state's ability to realize its moral function. Kant offers no criteria by which such minimal-ness (and the permissible extent to which the state could yet violate its members' property for the purpose) might be evaluated. Understandably therefore, Kant's present-day European social democratic grandchildren have invoked psychological and sociological findings concerning what is included in "being in need" and "being unable to sustain oneself". Social services that support psychological stability, societal inclusion and recognition have become accepted as being included (at least theoretically, and during economically good – or foreign-policy-challenged – days) in the state's charitable obligations. And since being exposed to "charity" has come to be understood as also hurting human dignity, those obligations have been supplemented by corresponding positive rights on the receiver's side. As a result, even if ethics experts today were engaged for problems connected with the moral effect of state-funded support (which they are not), it would be difficult for them to apply Kant's maxim of not encouraging "laziness". They would find it equally difficult to determine the point at which the state's social-service/common-good-geared invasion of individual property comes into conflict with the state's primary meta-moral obligation to secure autonomy respect with regard to individuals' property.

There is obviously something lacking in Kant's account of the state's obligation to charity. Perhaps his attempt to include beneficence under what is required by the categorical imperative (rather than restricting beneficence to the theory of virtues, which remain in the private realm) can supply the needed explanation. Again, within the limits of this essay we cannot fully lay out the problems with which that attempt burdens Kant's theory. The basic difficulty concerns the fact that his account of persons' moral dignity, which provided the ground for their entitlement to respect, was linked with their "noumenal" or rational nature. Only on this level of regarding humans, after all, could that moral freedom be posited, which provides the theoretical condition for the possibility of morality in the strict sense. In what concerns humans' empirical reality, they must be considered as causally conditioned. Accordingly, the rational world, of which humans are taken to partake (in a way which accounts for their holding themselves and one another morally accountable), functions as the exclusive ground of normative validity. To be sure, humans' phenomenal self, as linked with their embodiment and egoistic needs and passions, is always presupposed as given, and as the field against which the demands of morality must be fought out. Yet that empirical self has no normative significance of its own.[35] Thus Kant has, strictly speaking, foreclosed the route on which he could have considered helping those in (bodily, empirical, phenomenal) need normatively relevant in any immediate sense.[36]

But then a properly rational Kantian state ought to have remained committed to securing the conditions for the possibility of autonomy respect exclusively. It should have restricted its "common good" orientation to implementing the liberal principle. Such a (truly minimal) state would have allowed for ethics expertise only insofar as it guides the way in which that liberal principle is implemented.[37] But such a state would, in particular, not be authorized to disrespect the autonomy rights of moral minorities for the purpose of implementing any further, substantial "common good".

This very thin sketch of Kant's political theory is surely not sufficient as an argument for repudiating the moral legitimacy of ethics expertise, once it was engaged in liberal democracies for guiding the political implementation of substantial common goods (as for example education and health care). Such an argument would require more space. But this sketch is sufficient to shed serious doubt on the moral legitimacy of ethics experts' role. They, after all, morally guide, and thereby implicitly justify, the democratic state's infringing on individuals' normative autonomy. That this happens for the sake of realizing some "common good" is no excuse, as long as the goodness of that good is not commonly appreciated. Moreover, moral diversity in modern pluralist societies is generally proclaimed to be an important resource (be that for moral progress or for cultural richness). It is usually supposed that ethics experts' "sensitivizing" decision makers for the concerns of those varied moral communities who will be affected by their decisions, and ethics experts' mediating between such affected groups, serves to protect that diversity. Yet since even these functions were found always merely to support an existing "societal consensus" or to ratify majority preferences, ethics expertise in such societies instead contributes to the leveling out of moral differences and to the trivializing of moral disagreements.

There exist viable alternatives. The least "invasive" one is to impose a moratorium on those decisions concerning which not all those who will be affected can agree. But this solution will be unavailable in many cases, where more or less immediate action is required. Even more importantly, imposing a moratorium usually rests on the assumption that better chances for a truly universal agreement can be expected in some foreseeable future. It thus again presupposes a problematic faith in the foreseeable consequences of a universally unifying ethical progress. An entirely different response to ethical pluralism involves renouncing the project of common action altogether. Occasionally even bioethics experts recommend that a patient, whose end-of-life choices cannot be reconciled (after all have honestly tried) with the particular ethical orientation of the institution in charge of caring for her, should not be forced to submit to any ever-so-skillfully-mediated consensus solution, but that she should be moved to another, more compatible hospital.[38] Generalizing this model, the project of regulating (as well as common-good-endowing) the lives of people who are divided by irredeemably different moral commitments could be trusted to different sets of regulatory agencies, or diverse kinds of regulatory, policy or even legal frameworks. Just as Hardwig reminds those on the demand side of technological and scientific expertise that problems which cannot be settled by sound argument might require that one refrain from engaging in common action altogether (1994, pp. 96ff), so a diversification of regulatory institutions that responded to the normative diversity within a given state would permit that state to refrain from society-wide common action, while still leaving space for community-based regulation and common good implementation.[39] The assumption that the liberal state, which grounds its moral authority first and foremost in its commitment to respect for autonomy, and thus on the toleration of moral difference, should nevertheless subject that difference to its democratic predominance and harmonization procedures, and to enforce their outcome on all, thus appears to be ethically indefensible. We are left with the conclusion that the practice of engaging and providing ethics expertise in such a state indeed merely secures an ethical veneer for the continuing self-destruction of its liberal commitments.

NOTES

1. I shall use the terms "ethical" and "moral" synonymously for "normatively regulated". I shall reserve the term "ethics" (along with "ethical theory" or "moral theory") for the scholarly reflection about such regulation.

2. Fletcher renders this same contrast as a one between process-oriented and pro-active consulting (1997, p. 311).

Frequently, a general consensus in the field is invoked according to which ethics experts, at least when they appear in court, should educate, not advocate (see Singer, 1972, p. 115; Wildes, 1997, p. 367; Kipnis, 1997, p. 325; Jonsen, 1980, p. 158; Weinstein, 1994, pp. 63ff; Siep, 2003, p.127; and, disapprovingly, Agich and Spielman, 1997, pp. 381 ff). Two major reasons are offered. First, credentials for those engaged as experts include an academic proficiency, which requires facing a diversity of philosophical approaches and conclusions. While these can clearly be evaluated in terms of general cognitive criteria (Birnbacher, 2002, p. 102), controversy about their intrinsic merits will remain unresolvable (Kipnis, 1997, p. 340). Since "expertise about an ethical issue and experience in ethics" must be distinguished from one another (Fletcher, 1997, p. 317), mere academic knowledge does not qualify an expert for the former, i.e., to ethically guide others' practical decisions (cf. Agich and Spielman, 1997, p. 386). Secondly, particular cases (even outside of courts) are sometimes so morally complex, that even any one ethicist must recognize that there are different ways of resolving them (Fletcher, 1997, p. 311).

3. On reasons for opposing the educational restriction of experts' role see Momeyer, 1990, p. 391, or Menzel, 1990, p. 419.

A seemingly different way of understanding the major bone of contention is presented by Illhardt, who opposes "compromise-oriented" to "polarizing" ethics counseling (1999, p. 262). Yet as he develops these concepts, it becomes clear that both are merely varieties of normative advocacy, their difference reducing to that between a principled theoretical, and a prudential, mediation-oriented way of advocating. A similar conclusion can be drawn for Siep's distinction between external and internal counseling: both involve prescriptive language.

4. Kipnis invokes for that position the *Canadian Health Care Ethics Consultant* with its demand for "judgments that plainly take into account the diversity of options, the associated values and consequences" (1997, p. 341). In a complementary sense, Siep (2002, p. 95) wants ethics experts to combine the educational task with a personal assessment of what they consider "best".

5. A good illustration for this influence is provided by Fletcher's own – otherwise admirably self-critical – example concerning his having provided ethics expertise with regard to life-support for an anencephalic baby. Fletcher endorses experts' obligation to attend to the ethical commitments and interests of all affected parties. Yet he himself turns out to be unable, irrespective of years of experience in court negotiations, to truly enter into the mother's frame of mind (as she "irrationally" insisted on such support), to take seriously the perspective of committed vitalists who publicly supported that mother (1997, p. 314), or to consider the technological interventions with such a baby as anything but "harmful" (1997, p. 306, see also Note 9).

6. As far as rationalism is concerned, Benjamin (1990, p. 383) is a good example.

7. Thus Agich and Spielman (1997, pp. 392ff) argue that the profession of an ethics expert, while requiring a philosophical education, occupies a separate realm of "phronesis", or prudential reasoning (cf. also Jonsen, 1980, p. 163; Illhardt, 1999, pp. 267ff; Jonsen and Toulmin, 1988, pp. 63 ff). Such reasoning, while also forming a subject of philosophical speculation (Aristotle is frequently quoted in this regard) is yet accessible, for any given field, only to those with experience in that field. Such experience in turn is taken to reveal, first, that different philosophical theories or conceptual foundations do not determine differences in ethical decision outcome (Jonsen and Toulmin, 1988, pp. 18ff; Kipnis, 1997, pp. 340, 330, note 4; Agich and Spielman, 1997, p. 392). These differences are merely *epiphenomena* of, or merely engage different ways of conceptually reconstruing and accounting for, such practical wisdom. Such experience therefore, secondly, pushes scholarly philosophy into the background, where various construals of philosophical theory serve a merely instrumental purpose for the ethical conduct of life and the practice of expertise. (Some problems can be clarified with the help of *this* philosophy, others with *that*.) Accordingly, in practical contexts philosophies, as theoretical idealizations, present merely "strategic choices" (Fletcher, 1997, p. 298), or, with a slightly darker twist of sarcasm, "disingenuous rationalization(s)" (Kipnis, 1997, p. 327).

For a critique of such positions, however, see Birnbacher (2002, p. 113) and Fuchs (1997, p. 4). The latter rightly points out that Beauchamp and Childress' attempt to render disagreement on the level of philosophical theory practically irrelevant by recurring to middle level principles is bound to fail in knife-edge cases, where the differences between theoretical understandings of those principles revive the very philosophical dissensus those principles were to overcome. (Oddly, Fuchs disregards the practical non-irrelevance of different theoretical interpretations he thus conceded when endorsing (op.cit. p. 15) ten Have's critique of Engelhardt, whose restriction of a necessary general consensus to respect for autonomy is motivated by precisely that non-irrelevance).

8. Only Siep seems to be aware of the danger that such expertise may result in "immunizing one's culture" (2002, p. 95).

Failure to look beyond the ethical *status quo* is certainly tempting if one limits one's reflection to unproblematic cases. Had Kipnis, for instance, addressed the issue of nutrition and hydration withdrawal from a patient (who had been non-communicative for years) not in a secular but in a religious institution, and the issue of Jehovah's witness parents' refusal of blood transfusion for their child or Jehovah's witness soldiers' cooperation in DNA banking not with regard to experimental but to therapeutic and preventive care settings, he might have had to face a serious moral diversity that calls into question his (and the Canadian handbook's) faith in the one "ethically best" solution that also maintains "the societal consensus".

9. Even though "prudential ethical monists" insist that for each practical problem only one optimal ethical solution exists, they are aware of the possibility of, and indeed overwhelming evidence of, disagreement concerning which one it is, or even which one is better than all the other proposed ones, for any given situation. Even such confessed anti-skeptics, thus, have no other recourse but to rely on the "test of good reasoning", on "appropriate critical analysis" (Agich and Spielman, 1997, p. 386), on the legal system's safeguards (p. 388), rules of evidence (p. 389), and thus – so we must conclude - on politically procedural solutions of compromise or consensus formation, due process and – ultimately – voting.

Because of this practical agreement, it is sometimes hard even to classify an author as pluralist or monist. In spite of his invocation of pluralism two sentences prior, Fletcher claims that "Ethics, law, and public policy evolve together dialectically in open and democratic societies in a process analogous to a mutual and interactive conversation" (1997, p. 316), thus suggesting one single "best" ethical theory to be in the societal making. Similarly, Siep derives the need for ethics expertise from the development of new medical technologies, as these give rise to ethical problems that cannot adequately be addressed through legal and political procedures (2003, p. 124). He thus suggests that ethics might supply the otherwise un-available valid solution. But then he again refers the decision about that solution back to legal and political procedures.

10. Kipnis' essay (1997) is a good example. A similar (if less obvious) case is Siep (2003, pp. 128 ff). He develops three maxims for the work of experts in (the German equivalent of) IRBs: 1. that one should not centralize such commissions regionally, nationally or even internationally (as some have demanded in the interest of enhancing the efficiency of counseling output), because the situation-ethical ingredients to the process of counseling would thereby be lost; 2. that researchers should seek patient subjects' consent through personal conversation; 3. that researchers' and generally medicine's interest in speedy publication must be weighed against the pharmaceutical industry's interest in secrecy to protect

intellectual property rights. With these maxims, he addresses conditions under which individual and communal contract partners can effectively develop a consensus on voluntary common undertakings that adequately preserve all partners' autonomy rights.

11. This holds even if the question of how, in any given case, normative self-determination should be secured, must be settled – ultimately – by decision makers' *fiat*.

12. Characteristically, the literature uses the term "ethically acceptable" in three different senses. First (as in Weinstein, 1994; Birnbacher, 2002; or Fletcher, 1997, p. 311), it refers to the quality of justifications offered, as a matter of academic excellence, and without regard for the consent of those who are affected by the decisions thus justified. Here conflicting solutions can be admitted, each of which is "ethically acceptable". Second, the term refers to endorsing the value consensus held within society (at least) in general ((McConnel, 1984, p. 250). Or finally (as in Agich and Spielman, 1997), its meaning addresses the fact that consent has been reached (or is assumed) among all concerned.

The problem involved with glossing over the difference especially between the second and third senses (or with confusing consensus with consent) seems to be profoundly ignored by most authors on ethics expertise. Some present their own endeavors as experts, when they were merely advocating that laws must be clarified in order to better protect individuals' autonomy, thus preserving society's liberal principle, with qualms and hesitations which make sense only in view of the substantial-value-pluralism problem (see for example Yarborough, 1997, pp. 347ff). In a kindred discernment-lacking spirit, Jonsen interprets the progressing consensus concerning the impact of autonomy rights in new medical contexts as progressing consensus concerning value implementation (Jonsen, 1980, p. 167). He approvingly quotes a case study (by Gustafson on a demented patient's refusal of dementia-corrective surgery), which conflates the issue of the patient's competence to consent with that of a person's social obligations (pp. 170ff). Wildes is one of the few who distinguishes the two classes of issues. He recognizes the unproblematic legitimacy of expertise in view of the liberal permission principle exclusively (1997, p. 370.) Moreover, it is in an essay about an "Ethics of Expertise" (pointedly not addressing expertise in ethics but about technical or scientific facts), that one finds the pertinent warning: "often an expert's expertise is applied to those who have not appealed to the expert. They may well not even agree that this opinion represents a form of expertise. Indeed expert opinion alters many people's lives without their knowledge, much less consent" (Hardwig, 1994, p. 86). But even this author, when discussing sources of experts' self-deception (pp. 88ff, 93), and thus of circumstances that attenuate the problem of such lack of consent, concentrates only on dishonesty and greed. He does not attend to experts' own particular ideological or value orientations.

13. Whereas in the case of the United States, the commitment to liberty in the Preamble of the Constitution is merely joined to a commitment to a "general welfare" which is left strategically undetermined, continental European constitutions invoke more substantial normative goals, which impose explicit limits on individual liberty. In these latter cases therefore, there exists a tension between what at given times is accepted as the canonical interpretation of those goals and the fact that individuals may wish to determine their moral identities in terms which are incompatible with that interpretation. Whereas for the United States, ethics expertise in political contexts raises a principal problem *vis à vis* their basic liberal commitment (or raises the problem of whether "general welfare" has surreptitiously assumed a positive meaning), in the continental European context such expertise merely highlights a problem that is already implied in liberal states' substantially normative set-up.

A good illustration for the way in which liberal and common-good commitments are muddled in European accounts of bioethics and bio-law is provided by Campagna. He emphasizes (rightly) that the state's infringement on individuals' liberty must be "covered" by their consent. Yet in spite of his awareness of ethical dissensus in modern societies, he remains unencumbered about the fact that the German constitution endorses not only the state's obligation to render its members' exercise of autonomy compatible, but also to protect certain substantial values (2000, p. 296). Accordingly, his treatment of "conflicts" which the state must safeguard against (p. 293) extends not only to conflicts over members' freedom spheres but also to substantial moral disagreement (i.e. about what should count as "abuses" of freedom, or as failing to act "rationally" or "responsibly", pp. 304ff). His understanding of the state's task of protecting the "culture of basic rights" thus encompasses both respect for autonomy rights and some value-impregnated "framework" which is claimed to "sustain" everyone's freedom (p. 299). From such an account, vigorously illiberal conclusions can be derived concerning the need to couple the granting of negative rights with the imposing of corresponding responsibilities (p. 298).

14. Such core moral norms or values are usually framed in terms of positive rights, to which members of modern Western democracies are believed to be entitled over and above their basic right to freedom. On

the level of specific implementation, it is the legal system, which is often seen to incorporate the normative core of a political society's commitments. Thus Fletcher claims that in a "pluralist and multicultural society", "the legal forum is an important arena" for clarifying "the scope of ethical principles, norms, and procedures and their relation to law and public policy" (1997, p. 316).

A good example of the impossibility of understanding some authors' positions, except by recourse to commitment to such core norms or values, even where explicit confirmation is lacking, can be found in an essay by Sharpe and Pellegrino (1997). As pluralists, they oppose experts taking a normative position in court. Courts should follow their own rules of due process in subsuming cases under laws and regulations. Moral and legal issues should be kept separate (p. 378). Even the relevance of moral considerations in legal disputes is repeatedly questioned (pp. 373, 376, 378). And even where the law refers to standards of medicine's professional integrity, that integrity is described as being so unclear that the state should better determine its own interest in matters relating to such integrity.

So far their presentation is straightforward. Yet the authors also affirm a pedagogical role of experts in court, and their opposition to advocating is strengthened by the fear that advocating will compromise that pedagogical function. They offer two major reasons for this pedagogical role: first, ethics experts know about ethical practice in clinical and other professional settings; second, they are skilled in analyzing arguments with an ethical dimension (p. 374). This is strange. First, how can professional ethical practice, as it underlies a professional integrity that was described as "unclear", get clarified in court so as to serve a legal purpose? Perhaps there are clear partitions within an unclear professional ethos? Or perhaps exposing its un-clarity helps the legal purposes? The authors seem not to question the fact that courts refer to such ethical practice in spite of its (pluralism-imbued) contested nature. Second, even though they had questioned the legal relevance of the ethical dimension of legal problems, nevertheless the authors also admit a relationship between ethics and the law, which they describe as both dynamic and unsettled. How could ethics experts be expected to address such a relationship, given the fact that ethical pluralism will surely aggravate that dynamism and unsettledness? In their concluding remarks, the authors affirm a moral underpinning to the law, and the necessity of adding an ethical dimension to legislative political deal making. So all of a sudden, the significance of a separate ethical orientation is affirmed, which was denied before on grounds of ethical pluralism.

Such vacillating can only be explained by supposing that the authors' ability adequately to determine the political power implications of the moral pluralism they have so aptly recognized is compromised by their desire also to endorse some determinate ethical commitments underlying the state's laws and regulations. There is plainly an awareness of the tension between that recognition and this endorsement: In issues of legal development, the authors want ethics experts to represent all of a society's ethical communities or traditions when engaging in pedagogical (analytic) as well as normative deliberations in panels or committees. Not surprisingly however, it remains unclear in what sense such panels, should they comprise advocates of mutually exclusive ethical orientations, could be hoped to produce "the" ethical dimension "so often ignored in the political process of expedient deal making."

15. In his discussion of ethics expertise in medical contexts, Siep, following Vieth and Quante, distinguishes between externalist and internalist ethical positions (2002). This distinction in one sense parallels what I would call a distinction between a rationalist and a "culturalist" account of core values. Siep's "externalism", to be sure, is not restricted to rationalism in the historical sense of the term. It extends to utilitarian or discourse-ethical senses. Yet all of these positions are characterized by their claim to universal validity. My use of "rationalist" similarly intends that meta-level: It refers to ethical theories that ground universal claims, regardless of whether these are addressed to rational beings, sentient beings (with higher animals as classmates), or those who are able to participate in public discourses. In each of these cases, exclusion from the class of such beings is tantamount to exclusion from the class that is entitled to respect. Neither internalist (in Siep's terms) nor culturalist (in my terms) ethical positions, on the other hand, establish such universalist claims. The validity of normative judgments here is restricted to the group that shares a common geographical area, ethnic progeny, history or tradition, and that confesses its commitment to the norms involved. Sometimes the core values appealed to in defending ethics expertise are indeed construed in such culturalist terms for members of a particular polity (which is thus construed in communitarian terms). With regard to that polity, these values circumscribe a sort of majority-backed "Leitkultur" over and against diverging moral minority milieus, thus from the very start repudiating the polity's liberal commitment.

16. A typical example of such a position can be found in Braun (2000, p. 410). In her presentation, the (positive along with negative) "basic rights", which the modern state's commitment to both the principle of liberty and to such core values implies, along with the practice of "democratic discourse" (as the

principle of realized self-government) appear as the two legitimizing grounds for modern Western political societies.

17. Interestingly, whereas discussions about particular societal problems, especially the pressing bio-medical kinds, often revolve around positions (usually in some natural law context), which affirm such rational groundedness, and affirm it not only with regard to a minimal core but also with regard to specific issues, in the literature on ethics expertise these positions seem under-represented. Understandably, those who realize that there is a problem with ethics expertise are also reticent about the practical relevance of such supposedly universally shared "rationality".

18. Agich and Spielman (1997, pp. 390ff) recognize that the "demand for an objectively valid theoretical justification" for ethical decisions is a "philosophical conceit". Nevertheless, they themselves in some (perhaps sub-theoretical, intuitive) way endorse such objectivity. They argue that the strict separation of facts from values, which renders the quest for objectivity in ethics illusory and gives rise to skepticism, must be overcome. Thus, human suffering (a fact) is supposed to "cry out" to fellow humans (thus constituting a disvalue that implies a call for help). But on closer look matters are less clear-cut. In order for such a fact-value link to serve its norm-grounding purpose, the account to be taken of the manner of such suffering must be carefully qualified. One must explain, for example, why some specifically non-Judeo-Christian cultures (e.g., Buddhism) do not recognize such an obligation. One must also take seriously that some states of human suffering either belong to a beneficial warning system, or are self-inflicted in ways that defy efforts at helping. (Imagine sympathetic nurses intercepting a suffering marathon runner within the last 200 yards of the goal!)

19. The effective normative pluralism in modern liberal societies, which thus reduces any supposedly rational agreement concerning general core values or norms to a matter of empty words, is well appreciated by Menzel (1990, pp. 412ff). Against those who fail to take that ineradicable pluralism seriously, the most persuasive refutation is offered by Engelhardt (2000, pp. 28ff).

20. Where the literature on ethics expertise does in fact address experts' political influence, it refers only to those ethicists who fail to underwrite that latter obligation. Their influence (as endorsing merely their personal or professional position) is therefore seen to disrupt the legitimacy-constituting role of the democratic process (*Enquete Kommission* of the German Bundestag, 2002, p. 401). Especially with regard to the European Bioethics Convention of Oviedo (Council of Europe, 1996), but also with regard to European Union research and policy projects and the German Chancellor's national ethics council (Bahners, 2001), political criticism of ethics expertise highlights merely the secret power of expert elites which is exercised behind the back of the European member democracies' public (see for example Paul, 1998, pp. 68ff). Alternatively, that criticism castigates the subservience of ethics counseling to the promotion of particular (industry) interests (Höfling, 1998, p. 82; Rehmann, 1990, pp. 33ff). See also Braun, whose critique of the dominant category of rationality does not argue for a reduced competence of the state to implement substantial moral norms, but, on the contrary, for protecting the state's competence in this regard from the discourse-ethically grounded dominance of ethics experts (2000, pp. 423ff).

All such criticisms miss the underlying problem on two accounts. First, as Agich and Spielman (1997, pp. 387ff, 397ff) rightly emphasize, even non-consensus-endorsing ethical advice will be fed into the democratically authorized decision making process. Second, these criticisms fail to question the legitimizing function of that process itself, or to question ethics experts' service for attaching an additional "ethical seal" on its outcome.

21. It is sometimes argued that societal ethical conflicts "often" recur on the level of individual moral uncertainty. Consensus formation in society is therefore extolled as also helping individuals to reach their needed practical decisions (Fuchs, 1997, pp. 124, 132). But it is only the majority portion of post-modern societies, or those whose links with normative communities are weakened, who might conceivably profit from illumination through ethics experts' consensus or compromise formation. The appeal to the need to counteract "moral dam-breaking" (implied in Fuchs' presentation of ethics experts as educators of the people) through such compromise formation in effect endorses further weakening the moral commitments of those minorities which still adhere to their un-compromising moral principles. Justifying societal compromise formation in ethics thus amounts to justifying the societally promoted weakening of ethics.

Fuchs' other line of defense engages the need to preserve the conditions for the possibility of communal cohesion within a diversely oriented society. Durkheim, Parsons, Honneth and also Habermas are invoked for the claim that only a society-widely endorsed core set of values can secure a polity's internal peace (Fuchs, 1997, p. 12). Yet the issue at stake here is much more basically the peace between a polity and its minority members.

22. As Fuchs' (1997, p. 127) thorough survey of international ethics committee work makes plain, and contrary to the hopes he still seems to place in dissensus statements, even the critical reasons advanced by minority members, and included in the resulting (mainstream) statements, were regularly disregarded by political decision makers (p. 80). (Cf. the parallel experience reported by Saretzki, 1998, p. 93). An obvious motive (and one that plays way below the level where rationality criteria are applied) lies in the limited political power of (at least the more peaceful) minorities in democracies (cf. Gottschalk, N. & Elstner, M., 1997, p. 164).

Accordingly, Siep restricts the role of ethicists in pluralist societies to rendering decision makers more aware of the complexities involved (2002, p. 94), and rendering expert panels' agreement on common consensus statements more difficult (p. 96). The problem posed by ethicists' participation in such sometimes tragic choices is thus considered – as it were – *aufgehoben* in each participant's intensified ethical struggle. (A similar view is endorsed by McConnel, 1984, pp. 207ff, Bayertz and Runtenberg, 1997, p. 109))

23. Childress shows in a case study concerning research on fetal tissue how the attempt of a committee majority to accommodate the minority's concerns led to a weakening of the ethical claims advanced for the final decision which repudiated the ethical impact of the consultation (1996, p. 226) and in the end resulted in the minority's opinion to prevail (pp. 233ff). Thus one can appreciate Benjamin's remark that advisory panels, if they are to secure any corporate public influence at all, are well advised to sweep their members' differences under the rug (1990, p. 385), or Mieth's wry comment: "in the context of ethical counseling offered on the political level there is no room for political naiveté" (1997, p. 217).

A critique of the discourse-ethical presupposition of such committee work is provided in Bayertz, 1991, p. 232 and in Deutscher Bundestag, 2002, pp. 403ff.

24. Practical decision making through discursive processes that include members of different normative orientations functions on the presupposition that these members are thereby placed in a position where they can discursively learn from one another. Such mutual learning (and competing in the heuristic game of winning majority support) presupposes, of course, a general willingness to acknowledge that each of the differing ethical positions may contain some partial truth, which, once integrated into the completed account, will partake of the latter's validity for all. Mutual tolerance among adherents of differing ethical creeds must imply some at least potential presumption in favor of one's ethical opponents. In addition, there must be some common sense that majority support is what "counts" in the course of ethics competition. Ethics experts' endorsement of such a view is thus itself a mainstream culture phenomenon. By contrast, those who take their moral commitments seriously and also happen to be outnumbered will more likely find themselves dissatisfied. Many off-mainstream members of modern societies in a characteristically pre-modern sense claim to be oriented towards an absolute ethical truth. Communities advancing such claims often (even though not exclusively) derive the authority for deciding which areas of human action are open to cross-community consensus-formation from a transcendent source. Participation in cross-societal normative compromise formation within the pointedly secular environments offered by liberal democracies today would require that such communities separate the moral values they endorse from the spiritual contexts of those values. They would have to frame their values in secularly accessible language and, when addressing their members "back home" in terms of the understanding thus developed, in effect advocate restricting that spiritual context to the privacy of every one's inner life. Such participation, thus, would threaten those communities' (at least spiritual) integrity.

A helpful example of committed moralists who, even though they do not belong to any pre-modern faith tradition, still refuse to participate in a societal process of consensus-formation is presented in the report on the "discourse project" concerning gene technology in Lower Saxony (Saretzki, 1998, p. 86).

25. Kamm's demand that minority groups, precisely on account of their inferior numbers (1990, pp. 367), should be willing to compromise more, rests on Scanlon's definition of "reasonable". Kamm interprets that definition as what no one can reject who consents to finding principles all can agree on, or what no one can reject "given a desire to reach unforced agreement", where that agreement not merely concerns the compatibility of everyone's freedom with everybody else's, but also substantial values making up a society's "common good". Her maxim that "it is morally wrong to force compliance" (p. 360) is obviously restricted to minorities.

In a similar spirit, Agich and Spielman (1997, p. 383) quote (with only practical, not however in-principle reservations; see their note 1) Delgado and McAllen's vote for discounting "Machiavellian, Nietzschean, or other moral skeptics, radical existentialists, and ethical egoists" and for allowing only those whose views are "generally consonant with both the considered moral judgments of persons in our society and the moral norms incorporated into law". Such restrictions clearly make sense in situations with a practical urgency of reaching moral agreements (see Engelhardt's pointed analysis, 2000, pp. 29ff.). But when imposed on the ethical reflection concerning ethics expertise, these restrictions endorse the question-begging nature of the "ethical quality seal" which ethics experts publicly imprint on the results of their services. The fact of (selectively) reached agreement (consensus) is taken to prove the existence of (general) agreement (consent).

Likewise, for Jonsen the fact that medical ethics consultation "works" already establishes the correctness of his claim that there exists always one "most probably good" solution to ethical problems. The fact that mainstream ethicists agree with other mainstream ethicists about mainstream ethics is taken to render irrelevant the extant criticism and "offence to many", which Jonsen himself concedes (1980, p. 167).

26. Birnbacher (2002, p. 102) can be taken to support a meta-theoretical, Kamm a prudential (1990, p. 367) account of such a canonical commitment to rationality. This commitment at the same time imposes a secular public reason censorship on what is recognized as normatively relevant. The narrowing down of plurality issues to academic philosophy (always excluding the "radicals"), and the reduction of even the "standard" ethical theories to conceptual resources for sophistry (see note 7) renders the authors on expertise oblivious to the plurality of non-academic world-views, ideologies and religions in modern societies.

An exception seems to be Jonsen. He does acknowledge world-view diversity, but claims that the method of casuist "elucidation" (1980, p. 160) of normative principles transforms such diversity into normative agreement. Yet the evidence he offers concerns only detail-opinion diversity within a (Roman Catholic or Jewish) context, where "an overarching background of values and beliefs" (op.cit., p. 164) is still intact. He considers not the real (Christianity/Deism/Atheism) diversity of modern societies, which he all too euphemistically characterizes as lacking "explicit" common ethical principles (p. 164). The problem, so we must insist against Jonsen, is not that such principles are implicit and only wait for their being explicated for general acceptance. The problem is whether anything universally normative is even accessible to humans' reasonable grasp.

27. Once any position-taking in philosophy is relativized by the recognition that there are "reasonable differences of opinion", one's own normative commitment degenerates into a matter of mere personal preference (Benjamin, 1990, p. 385), which can be discounted without threat to one's philosophical integrity (p. 386). A "serious moral view" thus is nothing which one is willing to act upon (let alone die for), but one "for which reasons could be given" (Kamm, 1990, p. 364). Understandably, such an attitude must render the idea appalling that moral commitment might be taken seriously in a practical sense, and that someone committed to a position would refuse exposing it to compromise. For tolerance-committed academics, strong moral principles must appear uncivilized.

Similarly, once moral commitments at bottom reduce to aesthetic or utility preferences, the political imposition of a common good will be seen as presenting no serious threat to commitment to freedom rights. With human self-fulfillment as the ultimate normative ground, the ability to sympathize with a religious or secular pro-lifer's unhappiness about the fact that his tax money supports abortion and stem cell research is reduced. It will be argued that the superior happiness of seeing one's daughter's receiving expert surgical care, and at a moderate cost to the family, should outweigh that unhappiness. Likewise, a scientologist's unhappiness about his inability to advertise his healing services in the medical yellow pages, and a Jehovah's Witness' unhappiness of having his child exposed to enforced blood transfusion will be argued to be outbalanced by the therapeutic success of the existing health care system.

When members of pre-modern faiths refuse to accept such utilitarian balancing (Fuchs, 1997, p. 123) and to tolerate it (in a positive, welcoming sense that goes much beyond peaceful suffering), they challenge what is perceived by the mainstream to be the very basis of civility. Members of "fundamentalist" communities are therefore often seen as simply not qualifying for that societal respect which would render using un-consented societal regulatory force against them illegitimate. Their existence undercuts liberalism's constitutive, but now material value consensus in view of the significance of "protecting" the resources for unhindered, moment-to-moment exercise of individual autonomy against lasting and binding moral commitments. The liberal society's own countervening commitment to that consensus thus even imposes efforts at "integrating" such deviant communities into the mainstream. Marriage laws, family policies, public education, and public health care all present inroads for de-composing pre-modern traditions' normative integrity. Mainstream ethics expertise plays a central role in justifying such de-composing

28. A critique of the dominant reduction of "rationality" to mainstream deontology-tampered utilitarianism and a defense of moral emotions (even outside of religious contexts) can be found in Braun (2000, pp. 415ff).

29. Such disqualification, of course, also deprives both those (ethical monists) who believe in the democratic procedure as a means to further the moral progress of humanity, and those (ethical pluralists) who believe in that procedure's contribution to an interesting unfolding of humans' potential for cultural achievement, of the more interesting resources for either critical testing or narrative depth.

30. In an un-typically explicit way this presupposition recurs throughout Benjamin's article (1990, especially pp. 381ff, 385)

31 .Such an argument would have to be fleshed out into a comprehensive framework-ethic, which could impose limits on the normative pluralism a liberal polity must respect.

32. The same holds for hypothetical contract theories, which provide an ignorance-veiled, and thus supposedly non-partial, utilitarian framework for the supposedly rational endorsement of both negative human rights and entitlements to certain types of economic re-distribution and provision of social services: In order to acknowledge the "rationality" of such contractors' decision to prefer a state-enforced common good to the unhindered self-determination of non-mainstream moral minorities, one must already have decided concerning the superior "rationality" of behavior that adjusts to mainstream preferences. Original contractors must thus be supposed to be constitutionally unable to decide impartially between the moral weight exercised by the quantity of confessors of a moral creed and the quality of their commitment. These contractors, in other words, must already be either committed to a dominant-culture middle-aged-male mindset, or constitutionally unable to fathom the fires of hell.

33. The most important texts are his *Grundlegung zur Metaphysik der Sitten* (1786), his *Kritik der Praktischen Vernunft* (1788), and his *Metaphysik der Sitten* (1797).

Kant himself does not use the term "ethics expertise". His practical philosophy however addresses the very problem of the legitimacy of political power, which frames this essay's concern with ethics expertise. This philosophy offers a moral account of the state, which endeavors to render respect for humans' autonomy rationally compelling. This account thus provides a basis for modern polities' normatively liberal commitments. Accordingly, Kant's theory of the state is generally acknowledged – at least as far as claims to rationality are concerned – to offer the most stringent foundation for the ethical acceptability of modern liberal democracy. At the same time, that theory links humans' entitlement to respect with their moral accountability, or "dignity", thus rendering a rationally compelling restriction of autonomy rights at least conceivable.

34. It is this basic claim by Kant which places his moral and political theory in the category of historical, or bottom-level rationalist endeavors (as distinguished from second-order rationalist theories which may appeal to the sensations of pleasure and pain, cf. note 15).

35. Since in the course of thinking through his moral theory Kant rightly decided that the assumption of an immortal soul and a justly recompensing God are morally unavoidable, that embodiment-supported empirical self cannot even derive normative significance from its indispensability for sustaining its rational counterpart: Once the body has been shed, the soul is supposed to take over on its own.

36. There are, of course, mediate senses in which beneficence is established as morally relevant in the theory of the virtues. But since these must be understood (among other things, like furthering humanity's progress) as functioning as ascetical exercises for training one's ability to obey the moral imperative (see Delkeskamp-Hayes, 2005), these virtues can in particular not require anyone to act in such a way as to oppose what is morally required. *A fortiori* they cannot require politically implementing beneficence by invading persons' property, thus disrespecting their autonomy.

37. It is usually claimed today that the liberal principle itself requires the state's involvement in procuring a common good which consists in publicly granted social services and goods: One cannot speak of autonomy, so it is argued, while the conditions of realizing it through real-world action are absent. Just as Kant included property in what persons are entitled to, so one must include health, shelter, and education. But this argument disregards a crucial difference. To argue that to respect a person's autonomy means to grant him all means of realizing that autonomy disregards the difference between acknowledging the (as Kant terms it) hypothetical property claims which already exist in a pre-political state "of nature" on the one hand, and, on the other hand, positing the state's authority to supplement its function for ratifying and protecting these pre-political givens by additional tasks (of providing health care, shelter, and education, which, since provided by families and communities even in the state of nature, did not establish hypothetical claims of the same universality). In other words, autonomy respect as conceived by Kant involves taking care to leave untouched entitlements which are already societally acknowledged as

such, not however providing people with goods or services to which they may have at most a moral claim.

38. Mellon, 2003, p. 62. But then he is a theologian, not a philosopher. It is interesting to see how another theologian, when discussing a very similar case, classifies such a separation of hospital and patient as evidence for a lack in the hospital's – as he sees it – overriding commitment to the patient (Illhardt, 1999, p. 263). The irresolvable conflict over what that patient's "good" involves is thereby disregarded. Only if from the very start the hospital's refusal to accede to the patient's wish (to be let die) is interpreted as violating the latter's "good" (construed in terms of the patient's preference exclusively), and if the hospital's own moral commitment (to save the patient's life) is discounted as institutional self-service, does such a classification appear compelling. But of course such interpreting and discounting presupposes, again, that the seriousness of some hospitals' religious commitments has been discounted, or that normative diversity is not taken seriously.

39. Such an option will appear unacceptable only as long as one construes the problem of ethics expertise in health care (at large) on the model of counseling in clinical settings, where immediate decisions are inescapable. It is only in terms of this model that one can understand Benjamin arguing in one of his examples (1990, p. 384), that "a policy was needed soon" and that therefore it would have meant "to admit failure" had the committee settled for the impossibility of coming to an agreement in the face of members' irreconcilable diversity. But then, even for the clinical setting, as Engelhardt has shown (1996, pp. 175 ff, 2000, pp. 382 ff), different professional codes could be applied to patients from different backgrounds, or to patients who had declared their consent to the relevant policies when opting for a particular insurance, a particular hospital, or a particular ward. Just as different denominational hospital chaplains have accustomed themselves to sort their clients according to the latter's spiritual commitments or personal preferences even within given hospitals or wards, so could health care teams.

The desirability of such a diversification, even with regard to legal codes, is evidenced by Siep's confession: "In clinical ethics committees one must sometimes counsel physicians who recognize an obligation to act in a way that is legally not permitted" (2003, p. 125; cf. 2002, p. 89). (For an informal way of permitting an existing legal diversity to persist "underneath" agreed-upon ethical consensus statements see the strategic use of indeterminate language in the Oviedo Bioethics convention, as described by Honnefelder, 1998, p. 128).

REFERENCES

Agich, G. J. and Spielman, B. J. (1997). 'Ethics Expert Testimony: Against the Skeptics.' *The Journal of Medicine and Philosophy* 22(4), 381-403.

Bahners, P. (2001). 'Wir sehen weiter.' *Frankfurter Allgemeine Zeitung,* 1.6.

Bayertz, K. (1991). 'Modelle der Applikation. Ethische Prinzipien und moralische Klugheit.' In: Seebohm, T. (Ed.), *Prinzip und Applikation* (pp. 223ff). Stuttgart: Steiner Verlag.

Bayertz, K., and Runtenberg, C. (1997). 'Gen und Ethik: Zur Struktur des moralischen Diskurses über die Gentechnologie.' In Elstner, M. (Ed.), *Gentechnik, Ethik und Gesellschaft*, (pp. 107 ff.). Berlin: Springer Verlag.

Benjamin, M. (1990). 'Philosophical Integrity and Policy Development in Bioethics.' *Journal of Medicine and Philosophy* 15, 375ff.

Birnbacher, D. (2002). 'Wofür ist der "Ethik Experte" Experte?' In: Gesang, B. (Ed.), *Biomedizinische Ethik. Aufgaben, Methoden, Selbstverständnis* (pp. 97 ff.). Paderborn: Mentis Verlag.

Braun, K. (2000). Grenzen des Diskurses. Biomedizin, Bioethik und demokratischer Diskurs. In: Barben, D., & Abels, G. (Eds.), *Biotechnologie – Globaliserung – Demokratie. Politische Gestaltung transnationaler Technologieentwicklung* (pp 409 ff). Bonn: Rainer Bohn Verlag.

Campagna, N. (2000). 'Von der Bioethik zum Biorecht: Demokratietheoretische Übersetzungsprobleme.' In: Kettner, M. (Ed.), *Angewandte Ethik als Politikum* (pp. 280 ff). Frankfurt: Suhrkamp Verlag.

Childress, J.D. (1996). 'Konsens in Ethik und Poliktik. Am Beispiel der Forschung an fötalem Gewebe.' In: Bayertz, K. (Ed.), *Moralischer Konsens* (pp. 203 ff). Frankfurt: Suhrkamp Verlag.

Council of Europe. (1996). *Convention for the protection of human rights and dignity of the human being with regard to the application of biology and medicine: Convention on human rights and biomedicine,* Adopted by the Committee of Ministers on 19 November 1996, Strasbourg: Directorate of Legal Affairs.

Deckers, D. (2001). 'Vom Nutzen der Ethik.' *Frankfurter Allgemeine Zeitung,* 9.4.

Declaration of the rights of man and of the citizen: 1789, August 26. Available: http://www.yale.edu/lawweb/avalon/rightsof.htm.

Delkeskamp-Hayes, Corinna (2005). 'Cultivating health care rights as cultivating health care cultures: Can limits to tolerance be defended?' In: Engelhardt, H.T. and F. Jotterand (Eds.), *Global Bioethics.* In press.

Deutscher Bundestag: Referat Öffentlichkeitsarbeit (Ed.) (2002). *Enquete Kommission: Recht und Ethik in der modernen Medizin. Schlußbericht.* Berlin: Deutscher Bundesetag.

Engelhardt, H.T. (2000). *The Foundations of Christian Bioethics.* Lisse: Swets & Zeitlinger.

Engelhardt, H.T. (1996). *The Foundations of Bioethics* (2nd ed.). Cambridge: Oxford University Press.

Fletcher, J.C. (1997). 'Bioethics in a Legal Forum: Confessions of an "Expert" Witness.' *The Journal of Medicine and Philosophy* 22(4), 297-324.

Fuchs, M. (1997). *Internationaler Überblick zu Verfahren der Entscheidungsfindung bei ethischem Dissens. Gutachten für die Enquête-Kommission "Recht und Ethik der modernen Medizin" des Deutschen Bundestags.* Berlin: Deutscher Bundestag

Gottschalk, N. & Elstner, M. (1997). 'Technik und Politik: Überlegungen zu einer innovativen Technikgestaltung in partizipativen Verfahren.' In: Elstner, M.. (Ed.), *Gentechnik, Ethik und Gesellschaft*, (pp. 143 ff.). Berlin: Springer Verlag.

Hardwig, J. (1994). 'Toward an Ethics of Expertise.' In: Wueste, D.E. (Ed.), *Professional Ethics and Social Responsibility* (pp. 82-101). Lanham, MD: Roman and Littlefield Publishers, Inc.

Höfling, W. (1998). 'Das Menschenwürdeübereinkommen zur Bio-Medizin und die Grund- und Menschenrechte.' In: Neuer-Miebach, T., & Wunder, M. (Eds.), *Bio-Ethik und die Zukunft der Medizin.* (pp. 72 ff). Bonn: Psychiatrie-Verlag.

Honnefelder, L. (1998). 'Der Entwurf einer Menschenrechtskonvention zur Biomedizin des Europarats als Beispiel eines europaweiten Konsensbildungsprozesses.' In: Ammon, U., & Behrens, M. (Eds.), *Dialogische Technikfolgenabschätzung in der Gentechnik: Bewertung von ausgewählten Diskurs- und Beteiligungsverfahren.* (pp. 123ff). Münster: LIT Verlag.

Illhardt, F.-J. (1999). 'Der Kompromiss: Ethik-Beratung gegen moralischen Rigorismus,' *Ethik in der Medizin, ZZZ,* 262 ff.

Jonsen, A.R., & Toulmin, S. T. (1988). *The Abuse of Casuistry. A History of Moral Reasoning.* Berkeley: University of California Press.

Jonsen, A. R. (1980). 'Can an Ethicist be a Consultant?' In: V. Abernathy (Ed.), *Frontiers in Medical Ethics: Applications in a Medical Ethics* (157-171). Cambridge, MA: Ballinger.

Kamm, F. M. (1990). 'The Philosopher as Insider and Outsider.' *Journal of Philosophy and Medicine* 15, 347ff.

Kant, I. (1786). *Grundlegung zur Metaphysik der Sitten*, 2nd edition, Riga: Johann Friedrich Hartknoch Verlag.

Kant, I. (1788). *Critik der praktischen Vernunft.* Riga: Johann Friedrich Hartknoch Verlag.

Kant, I. (1793). 'Über den Gemeinspruch: Das mag in der Theorie richtig sein, taugt aber nicht für die Praxis,' *Berlinische Monatsschrift,* September, pp. 201 ff.

Kant, I. (1797a). *Die Metaphysik der Sitten, Erster Theil, Rechtslehre*, Königsberg: Friedrich Nicolovius Verlag.

Kant, I. (1797b). *Die Metaphysik der Sitten, Zweiter Theil, Tugendlehre*, Könnigsberg: Friedrich Nicolovius Verlag

Kipnis, K. (1997). 'Confessions of an Expert Ethics Witness.' *The Journal of Medicine and Philosophy* 22(4), 325-343.

McConnell, T. C. (1984). 'Objectivity and Moral Expertise.' *Canadian Journal of Philosophy* XIV, 193ff.

Mellon, B.F. (2003). 'Faith-to-faith at the Bedside: Theological and Ethical Issues in Ecumenical Clinical Chaplaincy.' *Christian Bioethics* 9, 57-67.

Menzel, P. T. (1990). 'Public Philosophy: Distinction Without Authority.' *Journal of Philosophy and Medicine* 15, 411 ff.

Mieth, D. (1997). 'Gentechnik im öffentlichen Diskurs: Die Rolle der Ethikzentren und Beratergruppen.' In: Elstner, M. (Ed.), *Gentechnik, Ethik und Gesellschaft* (pp. 211 ff.). Berlin: Springer Verlag.

Momeyer, R.W. (1990). 'Philosophers and the Public Policy Process: Inside, Outside, or Nowhere at All?' *Journal of Philosophy and Medicine* 15, 391 ff.

Moreno, J. D. (1991). 'Ethics Consultation as Moral Engagement.' *Bioethics* 5, 44-56.

Moreno, J. D. (1996). 'Konsens durch Kommissionen: Philosophische und soziale Aspekte von Ethik-Kommissionen.' In: Bayertz, K. (Ed.), *Moralischer Konsens* (pp. 179ff). Frankfurt: Suhrkamp Verlag.

Paul, J. (1998). 'Das bio-ethische Netzwerk.' In: Neuer-Miebach, T., & Wunder, M. (Eds.), *Bio-Ethik und die Zukunft der Medizin.* (pp. 61ff). Bonn: Psychiatrie-Verlag.

Rehmann, C. (1990). 'Ethik, Politik und Gentechnologie.' *Öko-Mitteilungen*, 3, 30 ff.

Rippe, K.P. (2000). 'Ethikkommissionen in der deliberativen Demokratie.' In: Kettner, M. (Ed.), *Angewandte Ethik als Politikum* (pp. 140 ff). Frankfurt: Suhrkamp Verlag.

Saretzki, T.(1998). 'Das Diskursprojekt "Gentechnologie in Niedersachsen": Ein dezentralisiertes Forum für die Kontroverse zwischen "Machern" und "Mahnern" und seine Vermittlungsprobleme.' In: Ammon, U. & Behrens, M. (Eds.), *Dialogische Technikfolgenabschätzung in der Gentechnik: Bewertung von ausgewählten Diskurs-und Beteiligungsverfahren* (pp. 79ff). Münster: LIT Verlag.

Sharpe, V. A. & Pellegrino, E. (1997). 'Medical Ethics in the Courtroom: A Reappraisal.' *The Journal of Medicine and Philosophy* 22(4), 373-380.

Siep, L. (2002). 'Ethik in Anwendung: Der Philosoph in Ethik-Kommissionen.' In: Gesang, B. (Ed.), *Biomedizinische Ethik. Aufgaben, Methoden, Selbstverständnis* (pp. 86 ff.). Paderborn: Mentis Verlag.

Siep, L. (2003). 'Probleme der Ethik-Kommissionen aus der Sicht des Philosophen.' In: Wiesing, U. (Ed.), *Die Ethik-Kommissionen. Neuere Entwicklungen und Richtlinien.* (pp. 124 ff.). Köln: Deutscher Ärzte Verlag.

Singer, P. (1972). 'Moral Experts.' *Analysis* 32, 115-117.

Weinstein, B. D. (1994). 'The Possibility of Ethical Expertise.' *Theoretical Medicine* 15, 61-75.

Wildes, K. W. (1997). 'Healthy Skepticism: The Emperor Has Very Few Clothes.' *The Journal of Medicine and Philosophy* 22(4), 365-371.

Yarborough, M. (1997). 'The Reluctant Retained Witness: Alleged Sexual Misconduct in the Doctor/patient Relationship.' *The Journal of Medicine and Philosophy* 22(4), 345-364.

CHAPTER 10

LISA S. PARKER

ETHICAL EXPERTISE, MATERNAL THINKING, AND THE WORK OF CLINICAL ETHICISTS

This essay results from thinking about ethical expertise in relation to the goals of clinical ethics consultation and the challenges of teaching ethics. My thoughts have centered on two sets of issues. The first emerges from articles, published in philosophy journals in the mid-1970s to early 1990s, that debated the possibility and nature of ethical expertise, as well as the proper role of so-called moral experts in the public arena and institutional settings.[1] The debate was timely because of what Annette Baier then described as the "recent expansion of the socially tolerated role of salaried moral philosophers beyond colleges into hospitals and into the work places of other professions" (Baier, 1984, p. 491). In the first section of this essay, I outline major threads of this debate and suggest various puzzles and inadequacies in the accounts of ethical expertise. I suggest that it may be fruitful to "begin at the end," with the actual activities of putative ethics experts, in order to understand what traits, training, activities, and goals are appropriate for would-be ethics experts.

In the essay's second section, I discuss a set of issues emerging from my experience as one of those philosophers who, as Baier puts it, "do" applied ethics in clinical contexts and from discussions with colleagues similarly engaged. I show how the practical challenges and apparent paradoxes that we face as clinical ethics consultants and teachers relate to philosophical concerns raised in section one, and I propose that examination of a different sort of work — that of mothers — may shed light on this second set of issues. I argue that what Sara Ruddick terms "maternal thinking" helps to characterize aspects of the work of ethics experts in clinical settings and that the sort of ethical caring that must be undertaken in health care institutions suggests justifications for the roles of "professional ethicists" in them. In the essay's final section, I employ analysis of maternal thinking to address additional concerns about ethical expertise raised in the philosophical and bioethics literature, particularly concerns centered on the appropriate role for ethics experts in a pluralistic society with regard to advising institutional and public policy making bodies and influencing public discourse.

L. M. Rasmussen (Ed.), Ethics Expertise: History, Contemporary Perspectives, and Applications, 165-207.

1. ON THE POSSIBILITY OF (MORAL PHILOSOPHERS AS) MORAL EXPERTS

The contour of debate surrounding the possibility and nature of ethical expertise during the second half of the twentieth century is almost as well known as most of its participants. On the question of whether moral philosophers are moral experts, A. J. Ayer and C. D. Broad succinctly state the con position. Taking a broader view of moral philosophy, Peter Singer argues for the relevance of philosophers' skills for developing expert moral opinion. Robert Holmes subsequently disputes this relevance when he, like R. G. Frey, takes the position that what we want from moral experts is not clarity in service of sound argument, but normative opinions that are decisively right, and that such opinions are not likely to be produced by exercise of analytical skill. Frey and Holmes believe that the failure of our normative theorizing to yield consensus on an account of the good undermines the possibility of expertise in making moral judgments. Holmes, in company with Iris Murdoch and Annette Baier, suggests that development of character, wisdom, or the capacity for moral discernment may carry would-be moral experts further than analytical skill *per se*. Baier explores the relevance of philosophers' traditional social roles and intellectual methods for their work with other professionals in various institutional settings. Also wanting to focus on social roles played by ethics experts, Bruce Weinstein analyzes the concept of expertise, argues that it is possible in the domain of ethics, and proposes a taxonomy of different types of ethical expertise. Despite its flaws, Weinstein's account suggests that examination of what moral experts actually do and the social practices that involve them may be the most fruitful way to ground an account of their expertise. His arguments "play past" the conceptual problems raised by Frey and Holmes, but raise a different set of considerations about what we (should) want in our ethics experts. During these same decades, Dan Brock, Jan Crosthwaite, Rosemarie Tong, Robert Veatch, Alan Weisbard, and Dan Wikler reflected on the role of philosophers and bioethicists — and their allegedly expert opinions — in public policy making.[2] Finally, in bioethics, as if emerging from all this published debate and yet actually responding to the fact that ethics experts were functioning in hospitals — whatever it may mean to function as an expert — a Task Force of the prominent bioethics professional societies served as a consensus panel to consider the nature and goals of health care ethics consultation and published a set of core competencies for engaging in such consultative work (Task Force, 1998).

Complicating the discussion of ethical expertise are the different points of entry and foci of the authors engaging the debate during the past fifty years. Some are concerned primarily with what the possibility of moral expertise means for philosophy, and what philosophy can contribute to situations that demand moral expertise. What seems to be at stake is partly a matter of disciplinary territoriality, where the question is not whether there are moral experts but whether philosophers are best (or solely) suited to be such experts. Others are more concerned with what ethical expertise would mean for other social practices, from living well to practicing medicine, or for both the good life and public life in a pluralistic democracy. Parties to the debate take different stands or employ different assumptions about how moral expertise is manifested, for example, in character, conduct, or cognitive skills. The question of whether moral philosophers are well or uniquely suited to be moral experts — or perhaps quite ill-suited — depends in part on whether expertise is a matter of knowing, thinking, doing, living, or being. Different answers to this question invite (or depend on) different conceptions of the nature and goals of moral philosophy. For this reason, debate about so-called moral experts may be as important to discussions in academe's seminar rooms as it is "in the trenches" of health care institutions and other "applied" settings.

Indeed, in what may turn out to be the most insightful contribution to the moral expertise debate, Annette Baier explores parallels between the traditional roles of philosophers and their newer ones in institutional settings.

> Our own professional activities, until recently, have been restricted to teaching and writing. Just as most of us teach without having been trained in the art of teaching, learning, if at all, from sorry experience, so those who consult and advise other professionals are learning by trial and error, not usually by any study or expert training in the new areas. I suppose that it is no worse for philosophers to think that, simply by being philosophers, they can make a contribution in a hospital, than to assume, as most of us do, that merely by philosophizing aloud in front of our students we thereby contribute constructively to their education (1984, p. 491).

Taking us back to the central issues of the *Apology* and *Crito*, Baier suggests that philosophers would do well to turn "self-conscious attention" not only to philosophy's intellectual methods, but also to its social roles. She points to philosophers' reluctance to plumb philosophy's educational or cultural purpose: "our sacred ability to be critical and questioning seems to have as its limit a refusal to question the value of becoming, in our sense, philosophical and critical" (1984, p. 491).

1.1. Moral Philosophy and Moral Expertise

In fact, the initial thread of twentieth-century debate over ethical expertise simply assumes the value of philosophy's critical intellectual methods. The question is then whether superior facility with those methods or the development of particular cognitive capacities amounts to moral expertise. This question of whether moral philosophers are moral experts rests on whether having moral expertise is a matter of *acting rightly* and living a morally good life, or a matter of *thinking clearly* about a body of knowledge and particular moral questions, perhaps even *thinking correctly*, in a moral sense, about moral questions. A. J. Ayer and C. D. Broad take moral expertise to reside outside the purview of philosophy. Ayer comments:

> It is silly, as well as presumptuous, for any one type of philosopher to pose as the champion of virtue. And it is also one reason why many people find moral philosophy an unsatisfying subject. For they mistakenly look to the moral philosopher for guidance (Ayer, 1954a, p. 246).
>
> There is a distinction ... between the activity of a moralist, who sets out to elaborate a moral code, or to encourage its observance, and that of a moral philosopher, whose concern is not primarily to make moral judgments but to analyze their nature (Ayer, 1954b, p. 7).

Broad opines:

> [I]t is no part of the professional business of moral philosophers to tell people what they ought or ought not to do or to exhort them to do their duty. Moral philosophers, as such, have no special information not available to the general public, about what is right and what is wrong; nor have they any call to undertake those hortatory functions which are so adequately performed by clergy-men, politicians, leader-writers, and wireless loudspeakers (Broad, 1952, p. 244).

Arguing against such assertions, Peter Singer concludes that moral philosophers deserve the appellation of moral expert because of the relevance of thinking clearly for being able to decide to act rightly (1972). He takes up a "plausible argument" made by Gilbert Ryle to the effect that in order to know the difference between right and wrong, one must care about it, and that therefore, knowing the difference is not really a proper instance of knowing (Singer, 1972, pp. 115-116; Ryle in Melden, 1958, pp. 157-158). The connection between knowing and caring is a point to which I return below. Singer, however, pursues the connection by asserting that while "caring about doing what is right is, of course, essential, ... it is not enough, as the numerous historical examples of well-meaning but misguided men indicate" (1972, p. 116). All of these points are in service of Singer's claim that what is necessary to be able to decide what is right involves gathering information, discerning its relevance, combining it "with a basic moral position," and eliminating bias (1972, p. 116). For these tasks, he argues, the moral philosopher is well-equipped.

> First, his general training as a philosopher should make him more than ordinarily competent in argument and in the detection of invalid inferences. Next, his specific experience in moral philosophy gives him an understanding of moral concepts and the logic of moral argument. ... Clarity is not an end in itself, but it is an aid to sound argument, and the need for clarity is something which moral philosophers have recognized (Singer, 1972, p. 117).

Finally, Singer opines that if moral judgments are to be based on something other than "unreflective intuitions," time is needed for "collecting facts and for thinking about them," time which philosophers are presumed to have in the context of their professional work. With Ayer and Broad, Singer shares a view of philosophy as concerned with conceptual analysis and logical argument, but his view differs from theirs in thinking of such activity and skill as being central to moral judgment and thus moral expertise. Moral expertise is needed in situations in which one must decide right from wrong, and "given a readiness to tackle normative issues," philosophical skills (and the time, or perhaps leisure, afforded or required by the philosophical life) can prove helpful, perhaps essential. "Only if the moral code of one's society were perfect and undisputed, both in general principles and in their application to particular cases, would there be no need for the morally good man to be a thinking man" (Singer, 1972, p. 116).

Most people unfamiliar with mid-twentieth century ethical theory would be surprised that a moral philosopher's "readiness to tackle normative issues" could ever be in question. Indeed the following passage from C. D. Broad's Preface to *Five Types of Ethical Theory* seems a bit tongue-in-cheek, quaint, and yet somewhat disturbing in light of the social roles actually played by moral philosophers. Broad writes:

> It is perhaps fair to warn the reader that my range of experience, both practical and emotional, is rather exceptionally narrow even for a don. Fellows of Colleges, in Cambridge at any rate, have few temptations to heroic virtue or spectacular vice; and I could wish that the rest of mankind were as fortunately situated. Moreover, I find it difficult to excite myself very much over right and wrong in practice. I have, *e.g.*, no clear idea of what people have in mind when they say that they labour under a sense of sin; yet I do not doubt that, in some cases, this is a genuine experience, which seems vitally important to those who have it, and may really be of profound ethical and metaphysical significance. I recognise that these practical and emotional limitations may make me blind to certain important aspects of moral experience. Still, people who feel very strongly about any subject are liable to over-estimate its importance in the scheme of things. A healthy appetite for righteousness, kept in due control by good manners, is an excellent thing; but to "hunger and thirst after" it is often merely a symptom of spiritual diabetes. And a white-heat of moral enthusiasm is not perhaps the most favourable condition in which to conduct the analysis of ethical concepts or the criticism of ethical theories (1930, pp. xxiv-xxv).

Nevertheless, expanding the purview of moral philosophy and the content of moral expertise beyond skills of critical thinking to include the normative products of such thought is precisely the goal pursued by some philosophers with measured arguments, if not a white-heat of enthusiasm. R. G. Frey recounts:

> The strength of Singer's position lies in its plausible — I myself am inclined to say accurate — portrayal both of the good man as a thinking man and of the moral philosopher as a skilled critical thinker; and this portrayal explicitly allows there to be moral expertise, in that it allows for there to be increasing skill on the part of individuals in the critical assessment of moral arguments and issues. Nevertheless, I think Singer's position is unsatisfactory so far as providing support for a claim to moral expertise on behalf of moral philosophers is concerned; for it fails to come to grips with the crucial difficulty surrounding any and all claims to moral expertise.
>
> The difficulty in question is quickly sketched. ... [N]o one doubted that moral philosophers could go much further and deeper than ordinary people in unravelling the logic of moral argument; ... what was at issue was whether their acknowledged skills in the endeavours of meta-ethics ... made their specific moral judgments any more likely to be correct or true or in the right direction than the moral judgments of ordinary people. ...
>
> [Q]uestions of moral expertise are not concerned with the skills that go to comprise the critical examination of moral issues but with the outcome of the use of those skills in terms of particular moral judgments (1978, pp.48-49).

Frey argues that the contents of one's moral judgments, at least insofar as one would undertake to defend one's judgments, depend upon one's normative ethical theory. Being able reliably to determine which acts are right depends on being able to answer the question — "what makes right acts right?" — whose answer is supplied by a normative ethical theory. This understanding of what is requisite for the possibility of defensible moral judgment thus undermines the (at least current) possibility of ethical expertise.

> [T]he question of whether I am a moral expert is linked in its normative aspect with the adequacy of my normative ethic, and until I can demonstrate that my theory is adequate, for which purpose I first need to specify and then go on to justify some test of adequacy, I cannot even begin to satisfy the normative aspect to claims of moral expertise (1978, p. 51).

This test of adequacy, an Archimedean point outside of all competing normative theories, Frey correctly notes, "so far has eluded us," and he argues that development of such a test is a prerequisite for establishing an adequate normative ethical theory possession of which would be a necessary, though not sufficient condition, for moral expertise (1978, p. 51).

Robert Holmes agrees and applies the argument to bioethics. He contrasts the three components of philosophical ethics — metaethics, normative theory, and applied ethics, which is comprised of "*analyses* of specific moral problems like abortion or animal rights" — with "substantive morality, the ongoing process of *making moral judgments* that all of us engage in during the course of living" (1990, p. 145, italics added), and situates bioethics as both a branch of applied ethics and a matter of substantive morality, because it is concerned with resolving specific moral problems. He concludes for reasons similar to Frey's, that expertise in any of the three areas of philosophical ethics "provides one with no particular warrant for presuming to be able to make better moral judgments in particular situations than anyone else" (1990, p. 147). Whether applied ethics is conceived as a matter of "applying" theory to specific problems, as Holmes suggests bioethics textbooks purport, or whether it proceeds by allegedly neutral ethical analysis, which he claims philosophers like Jonathan Bennett and Judith Jarvis Thomson attempt, the adequacy of particular moral judgments resulting from applied ethical analyses rests upon the adequacy of particular normative theories. He claims that "the attempt to engage in neutral ethical *analysis* only obscures, rather than avoids, the problems in trying to apply ethical *theory* to practice" (1990, p. 157). Thus, he worries, "personal moral convictions, as well as class, gender, racial and other biases, may govern the philosophical treatment of problems in bioethics more than an unvarnished dedication to the truth" (p. 156), just as such "predispositions can enter significantly into one's choice of a normative ethical theory when it comes to applying normative ethics to practical problems" (p. 157). He is concerned that, "not being a science, ethics aspires to the next best thing, and that is [to be] scientific. But just as science cannot by itself yield answers to moral problems, ethical analysis that looks to science for its model cannot do so either" (1990, p. 157). Echoing, albeit less emphatically, charges lodged by Cheryl Noble (1982) against philosophers-as-moral-experts, Holmes suggests that as a consequence of its attempt to be scientific, philosophical ethics abstracts moral problems from the context of their development and analyzes them in detached neutral terms that reflect the biases of those philosophizing. He concludes that, in the absence of consensus on a normative theory or, to use Frey's terms, absent a test for the adequacy of such theories,

> [philosophical ethics] can provide conceptual clarity ...; it can provide the categories by which to discuss the problems theoretically; it can sometimes show that, *if* a certain theory or analysis is accepted, that has certain implications for substantive morality, and it can produce valid arguments .... But, while these are useful undertakings, they are insufficient to resolve the problems of bioethics or of applied ethics in general. They are not, I suspect, even necessary conditions of resolving those problems. The cultivation of a morally sensitive, caring, and compassionate character probably counts for more in the end than these analytical skills (1990, p. 157).

Thus, Holmes recommends, philosophical ethicists should view their training with greater humility as it is "no guarantee of — and a poor substitute for — moral wisdom" (1990, p. 158).

1.2. The Concept of Expertise and the Possibility of Four Types of Ethics Expert

Given the deep disagreements about the nature of both moral philosophy and moral expertise that underlie this debate, Bruce Weinstein's introduction of an explicit analysis of expertise should be most welcome. In fact, it both complicates and advances the discussion. He analyzes expertise "as a capacity either to provide strong justifications for claims in a domain," epistemic expertise, "or to exercise a skill well," performative expertise (1993, p. 61). He claims that the two types are "conceptually and logically distinct" and that there may be ethics experts of both types, with epistemic expertise in ethics corresponding to the subdomains of descriptive ethics, metaethics, and normative ethics, while a "fourth sense of ethical expertise corresponds to the performative subdomain of ethics: ... expertise in *living a good life*" (1994, p. 62). Examination of Weinstein's analysis reveals threads of the debate that his account fails to resolve, largely because his account of being an expert relies heavily on the socially accepted notion of an expert as someone who is better than others at a skill or at justifying his opinions — a socially defined authority — rather than on a conception of expertise as authoritative in some absolute sense. It is a specific strength of his account that his analysis begins to direct attention away from ethical expertise in an idealized, absolute sense and to develop an analysis grounded in actual social practices surrounding ethical expertise. On the other hand, Weinstein's account may permit the threshold for being an epistemic ethics expert to be set fairly low because he fails to note the interconnectedness of the subdomains of epistemic ethical expertise. Moreover, his somewhat confusing analysis of performative ethical expertise both neglects basic features of what it means to live an ethically good life and fails to give an account of what putative ethics experts actually do when they perform in that social role, an admittedly different sense of performative expertise than Weinstein intends. (This essay's second section begins to supply such an account.)

Weinstein considers it uncontroversial that there can be epistemic expertise in metaethics and descriptive ethics (1994, p. 63). Metaethical expertise involves analysis of the language and logic of moral arguments, while expertise in descriptive ethics involves being able to give an account of the moral beliefs of a particular culture or tradition, e.g., the moral code of the Ik or the Roman Catholic position on abortion. Expert descriptive accounts can become out-of-date, and therefore nonexpert; a pre-1995 account of Roman Catholic positions concerning abortion that does not reflect the 1995 Encyclical on the Value and Inviolability of Human Life could no longer be considered an expert account. Moreover, at any given time, expert descriptive or metaethical accounts might differ, although the assumption is that one account is more accurate than another even if we lack the capacity or criteria to determine which. In both descriptive ethics and metaethics, some accounts are more strongly justified than others, and therefore, Weinstein says, we may uncontroversially speak of expertise in each.

Thus we see that Weinstein understands 'expert' as a threshold concept and as a concept relative to time, domain, and a population or a comparison class of laypeople or nonexperts (1993, p. 70). Epistemic experts provide stronger justifications for their opinions in the relevant domain than nonexperts do. Status as an expert thus depends in part on one's own capacity to justify one's claims and in part on the level of that capacity of others. The authority of an expert is thus nonabsolute. Moreover, to be an expert, one need not offer absolutely authoritative opinions, but merely be capable of providing strong justification for one's claims, justification that meets some minimal threshold and is stronger than what most could offer. We defer to experts, when we do so justifiably, because we recognize their *capacity* for giving stronger justifications than most; for practical reasons involving conservation of time and effort, we do not require that they offer such justifications for every opinion they offer in their domain of expertise (1994, p. 63). It is at least in part for similar reasons of conserving time and effort that we defer to experts at all, rather than developing the requisite skill or knowledge base ourselves, if indeed we have the capacity to do so. Moreover, we recognize that an expert's expertise is relative to a particular domain; we do not (or ought not) generalize her expertise beyond that domain. Therefore, a person's status as expert rests on her capacity to strongly justify her opinions in her domain of expertise, not, for example, on a cult of personality surrounding her (1993, p. 70; 1994, p. 63; Veatch, 1973). Justifiable claims of, and deferrals to, expertise involve at least implicit appeals to reason, or attempts to persuade, rather than resort to manipulation or reliance on other sources of power.

When Weinstein turns to claims of expertise in normative ethics, which he acknowledges to be controversial, the importance of his understanding of 'expert' as a nonabsolute threshold concept relative to a population emerges more clearly (1994, pp. 64-66). Rejecting the noncognitivist position, he argues that because epistemic expertise is "a capacity to provide strong justifications for claims in a domain," "it is legitimate to speak of expertise in normative ethics, even if there is disagreement among normative ethicists," because strong justification of particular moral judgments is possible (1994, p. 67). To illustrate, he analyzes the differing positions of Paul Ramsey and Richard McCormick regarding the permissibility of nontherapeutic pediatric research. He concludes that "even though they reach contrary conclusions," both may be considered experts on the ethics of such research, because

> both moral positions are strongly justified, because both are supported by appeals to ethical rules, principles, and theories. Neither position can be mistaken for moralizing, or the mere assertion of moral beliefs. Both ... give well-supported reasons for their beliefs, and the conclusions of their arguments logically follow from the premises (1994, p. 68).

Here, Weinstein explicitly treats being an expert as a threshold concept, saying that one might wish to assess "on the basis of their justifications, which account is *more* expert, and thus more likely to be true, but this is a different task from considering whether Ramsey and McCormick are experts on the issue ... or whether their judgments are expert ones" (1994, p. 68). Recognizing that being an expert, in part, means achieving a threshold of justification-giving, but that expertise may still admit of degrees (beyond that threshold) is a significant point; however, Weinstein does not sufficiently explore the ramifications of this view. Doing so, it becomes clear that the subdomains of ethical expertise are more interrelated than Weinstein acknowledges.

If, for example, we sought to determine whether Ramsey or McCormick is more expert, we could not assess the greater likelihood of the accuracy of one or the other's proffered account *on the basis of their justifications* alone, but would need to judge their justifications on some other basis such as the adequacy of the normative theory to which each appeals. If we were only to assess their degree of expertise within the framework formed by their justifications, then we would merely assess how well they utilize those concepts or manipulate theory to support arguments. In fact, Weinstein points out, Ramsey and McCormick differ over the relevance of clinical information to determination of the permissibility of the research in question. This difference of opinion reflects different normative commitments of each. To determine that one position is superior to another, which we presumably would want to do if we care about enrolling children in nontherapeutic research, we would need to judge the relative superiority of their normative commitments, which, as Frey argues, we thus far lack the criteria to judge.

If Weinstein's view is much more modest, and if as the mark of normative expertise he means merely that someone with the capacity to provide strong justifications (based on a particular normative theory) is more likely than others to provide (again based on that theory) strong justifications for specific moral judgments, then his account amounts to Holmes's concession that training in moral philosophy enhances one's ability to claim that *if* a certain theory is accepted (or true), then it has certain implications for substantive moral judgments. Weinstein's conception of ethical expertise would then look much like that which Singer described — namely, facility in gathering and discerning relevant information and constructing logical arguments employing both that information and relevant moral concepts. What is at issue is what we want from our normative ethics experts, or what we take them to be providing when we defer to them or utilize what they have to offer. Singer, Frey, and Weinstein might all agree that what we want is an increased *likelihood* that expert judgments are reliably right in an absolute, not a theory-relative, sense. Of course, all would agree that the expert's right judgments will depend on some theory or background set of beliefs, but according to Frey a *real expert* must depend on the *right* theory. For Singer, analytical skills, and for Weinstein, strong justification, can carry one across the threshold of a *sufficiently increased likelihood* of correct judgment to constitute moral expertise. Frey wants more of a guarantee that barring the occasional human error, experts will make morally correct judgments. Since we do not have consensus on the correct moral theory, we might seek a theory-neutral account of ethical expertise, but none is possible beyond the circular account claiming that experts are especially good (i.e., expert) at justifying their positions. Weinstein's advocacy of a nonabsolute population-relative account of normative ethical expertise is a useful move, but he fails to recognize that where we set the threshold for being said to possess such expertise is dependent not only on how well the relevant nonexpert population justifies claims in that domain, but also on the strong justification of accounts in other subdomains of ethics.[3]

Even Weinstein's notion of expertise in descriptive ethics may be less independent of metaethical and normative commitments than he anticipates. Like a social scientist who, for example, observes and reports the drug-seeking behaviors of addicts or the professional conference behaviors of Cambridge dons, a descriptive ethics expert may be able to observe and report reliably what Roman Catholic Church officials have to say about abortion, for example, or what members of a particular culture say about lying or how they respond to those they label as liars.[4] It is not clear, however, that deeper descriptions are possible without developing some greater understanding of the group's normative and metaethical views, the beliefs and evaluative commitments that ground their pronouncements and prompt or justify their actions. Bernard Williams, for example, comments that

> how we 'go on' from one application of a concept to another is a function of the kind of interest that the concept represents, and we should not assume that we could see how people 'go on' if we did not share the evaluative perspective in which this kind of concept has its point. An insightful observer can indeed come to understand and anticipate the use of ... [a] concept without actually sharing the values of the people who use it .... But in imaginatively anticipating the use of the concept, the observer also has to grasp imaginatively its evaluative point. He cannot stand quite outside the evaluative interests of the community he is observing, and pick up the concept simply as a device for dividing up in a rather strange way certain neutral features of the world (1985, pp. 141-142).

In order to offer a descriptive ethics account, the expert must either report the actions and statements that the group itself labels as ethical, or report actions and statements that she interprets as relevantly ethical in nature. If she does not evaluate whether the observed group is correct (according to its own code) in labeling specific actions and statements as ethical, then she is not reporting (their) ethical knowledge *per se*, but just gross behaviors. She would be wrong or at least overreaching to call this descriptive account a descriptive *ethics* for she really has no strong justification to attribute ethical commitments to them.

If on the other hand she does evaluate sufficiently what they take themselves to be doing in their actions and statements in order to be able to give a rich ethical account, then she must employ some set of metaethical and normative commitments to which to compare what she observes. She could thus distinguish the group's attributions of agency and responsibility (e.g., holding children responsible for being late to school and punishing them by sending them to the hall) from mere causal explanations and attempts to fix something (e.g., asking a repairman to fix a toaster that takes too long in toasting a slice of bread). If they paused to reflect on their different treatment of tardy children and slow toasters, members of the observed group would, she thinks, offer an explanation that reflected the normative and metaethical commitments that she attributes to the group in her description of their moral code. In order for someone to be an expert in descriptive *ethics*, she must have sufficient expertise in normative ethics and metaethics to justifiably attribute some set of such commitments to the group whose actions and statements she is describing in ethical terms.[5]

The deep interrelatedness of epistemic expertise across the subdomains of ethics presents a problem for Weinstein's account primarily because he does not consider it. This flaw is not fatal if we are willing to accept that ethical expertise is not absolute, but is instead a relative concept grounded in social practices. But then we need to appreciate the interconnectedness of strong justifications offered in the ethical subdomains and the relevance of justifications offered in one subdomain for evaluating the strength of justifications offered in another. Someone can supply a more strongly justified account of the moral code of the Ik than other people without being able to understand how the Ik might "go on" to novel cases, or one might have superior capacity to argue against emotivism without being able to justify a stand on abortion or even without being able to give an account of what makes right acts right. We are justified in calling these people ethics experts in virtue of the relative superior strength of their justifications. But when we have competing expert accounts of the Ik's code or the permissibility of abortion, then it is appropriate to examine the strength of justifications from other ethical subdomains that support those accounts.

While Frey's and Holmes's demand for an (absolutely) authoritative account, rather than a (merely) strongly justified one leads to a regress that may be stopped only by a (nonexistent) Archimedean point outside the realm of normative ethical debate, Weinstein uses our actual social practice of relying on ethical expertise to ground an alternate account of ethical expertise. Weinstein suggests that we ought not trouble ourselves too much that we thus far lack consensus or criteria of adequacy. Given the usefulness of ethics experts, we should reject Frey's overly stringent demands, embrace Singer's claims of the usefulness of analytical skills, and proceed to rely on those whose justifications are stronger than ours, so far as we can currently discern. If we need to decide this week whether to approve a proposal of nontherapeutic pediatric research or need to answer an exam question about the Ik, it seems correct to rely upon the strongly justified accounts and opinions of our recognized ethics experts; in some cases, we may have an obligation to do so. On the other hand, I would suggest that if we are considering the question of what constitutes ethical expertise we need also to consider how we may best move beyond what we can currently discern.[6]

Given Weinstein's commitment to grounding his account of ethical expertise in our social practices of recognizing and justifiably deferring to experts, his account of performative expertise is perhaps most troubling. It is disappointing in two ways. First, in the area of ethics it seems mistaken to claim that epistemic and performative expertise are logically and conceptually distinct and thus to sever performance of the skill(s) of ethics from the capacity to supply justifications. Weinstein states that "expertise in the performative sense refers to the ability to perform a skill well, so performative expertise in ethics suggests an ability to live well, or to live a good life" (1994, p. 70). He acknowledges that "there is valid disagreement concerning how the good life is to be understood," but we need not resolve that disagreement "in order to recognize that certain people live better than others according to the rules and virtues of a particular moral tradition" (1994, p. 71).[7] But does merely following the rules and exhibiting the relevant virtues constitute ethical expertise? Weinstein suggests not when he writes that "a performative expert is not merely a rule-follower, but one who realizes the virtues of a particular practice" or skill (1993, p. 61). It is to be hoped that by 'realizes the virtues' he intends to invoke an epistemic state of appreciation or even understanding. Yet because he divides the epistemic from the performative, he may intend the more modest interpretation that the performative expert realizes *or displays* the virtues of or excellence at a skill.

One may be an expert at juggling, ballet, or shooting pool while utterly lacking the capacity to explain how one does what one does. Giving an account or having the capacity to justify one's "doings" is in no sense integral to what a juggler, dancer, or pool shark does. Only if we learned that the dancer was externally (or perhaps illicitly) controlled might we question her expertise.[8] We at least want our performative experts to be the "authors" of, or the agents relevant to, their performance. It is, therefore, troubling that Weinstein proposes an autistic savant as a paradigm case of performative expertise with respect to counting (1993, p. 58). I would argue that the autistic savant is not expert because he is not responsible for his capacity to count or for his exercise of that skill. Indeed it is the fact that he is not responsible for his counting skill in any usual sense that marks him as an autistic savant. The sum "comes to him," and he cannot account for his counting ability. Recent research indicating that it is possible to use electromagnetic pulses to induce in nonautistic adults superior drawing ability similar to that of autistic children suggests that we may need to decide whether one's electrode-induced capacity to draw realistic scenes better than others would render one an expert (Snyder and Mitchell, 1999; Osborne, 2003). It seems unlikely, however, that we would judge that an autistic person's condition-based inability to deceive makes them morally good in a sense that is on a spectrum with performative ethical expertise at one end. They may be morally good insofar as they do not lie; even though they simply lack the ability to do so, they are compliant with the ethical rule prohibiting lying. But their compliance is not meritorious and does not put them on the road to being performative ethics experts. In contrast to dancing or counting, it is the mark of moral life that one act or develop and display character traits for the "right" reasons. Even if we continue to disagree about the content of the right reasons, we agree that in ethics reasons matter.[9] It seems odd that Weinstein, whose account of epistemic expertise (in ethics and other domains) depends on the capacity to supply strong justifications for one's claims, divorces this dimension from judging and acting ethically and living an ethically good life. He notes that for practical reasons, an epistemic expert need not always supply her strong justifications; it is sufficient that she have the capacity to do so. Similarly, a morally good performer need not always supply her reasons. However, if asked for them, she cannot, like an autistic savant, claim to have no idea why she acts as she does and still be considered a performative ethics expert (or even a responsible moral agent).[10]

Weinstein's examples of performative ethics experts are confusing. He cites an observant orthodox Jew and a secularist pursuing G. E. Moore's conception of the good life and notes that each "has a different *understanding* of performative ethical expertise" thereby reconnecting *knowing that* God or Moore commends a particular path with *knowing how* to pursue the good life (1994, p. 70, italics added). Yet, while both of his exemplars may be morally good people, it seems at best hyperbole to term them performative experts unless they can give an account of why they act as the Talmud or *Principia Ethica* advise, indeed an account that provides strong justification for why they act as they do. Weinstein would likely respond that supplying such an account is a matter of one of the epistemic subdomains of ethics. I would argue that in ethics Weinstein's notion of performative expertise dissolves either into epistemic expertise or into living well without any claims to expertise. Weinstein is again confusing when he proposes a physician who refuses to discriminate against HIV infected patients as an exemplar performative ethics expert, because he states that even though she or he "may not be capable of providing a strong moral justification for what we ought to do, she or he can, through example, inspire others to live a morally good life" (1994, p. 71). I would claim that those of us who are thereby inspired to live or act in a morally good way must supply the reason that she acts as she does; only thus can her action be inspiring. Why else would we (justifiably) seek to emulate her? Such a physician's conduct is right, and perhaps she may be said to be morally good in this regard; however, she is not an expert.

At the same time that he offers this muddled account, Weinstein disappoints in a second way by neglecting another possible sense of performative ethical expertise. We might think of performative ethical expertise as a "doing" of ethics in the sense that Annette Baier describes when she refers to philosophers "doing" applied and professional ethics. If we think of the performative aspect of normative ethics — the making of judgments, providing of justifications, and deciding about particular questions — it is clear that people engage in these activities when they are simply living a good life or acting well, or engaging in substantive morality to use Holmes's term. When some engage in these activities, however, they may be performing as ethics experts, occupying a relatively public social role, as Baier describes when she asks what social role or social good professional philosophers perform. In thinking about what performative ethical expertise might mean, it may be worth exploring these performances, the publicly observable activities of moral philosophers and others currently engaged as ethics experts. In the next section I examine those activities and the intellectual work that informs them, and compare such work to the work of mothering.

2. MATERNAL THINKING AND THE WORK OF CLINICAL ETHICISTS

The previous section attempted to respond to the conceptually oriented question: "what can these philosophers justifiably think they are doing in hospitals on ethics committees or as part of the health care team?" This section attempts to address the clinical ethicist's own anxious question: "what *are* we doing here?" Colleagues and I have returned from ethics consultations and confided, "I'm not sure they really needed me. The ethical issue wasn't the problem; it was really more a matter of hand holding. They needed a social worker or a counselor to help them communicate and clarify their values and positions; I wasn't trained to do that as a philosopher or ethicist." Or, we have returned from clinical ethics teaching sessions bemoaning that "I didn't really teach ethics so much as engage in consciousness raising. Sure the students need that, but do they need a philosopher to do it? Moreover, it would probably be more effective if one of their own modeled moral sensitivity and discernment of ethical questions for them." Finally, we sometimes complain of being victims of our own success: "It has been over a year since we were invited to do an in-service training session in XYZ clinical service. We used to be invited there all the time, but they seem to have outgrown the need for our services. They don't even call for many ethics consultations anymore; Dr. _______ seems to do what we used to." These concerns may be labeled as, respectively, concern about hand holding, consciousness raising, and planned dispensability, and they provide the impetus for the thoughts presented in the balance of this essay.

Each of these expressions of discomfort or concern is also mixed with a belief that we are being helpful, making clinical care better for patients, increasing awareness of potential for conflict or moral wrong, and helping to avoid or resolve conflicts and other problems. We believe in what we are doing, yet we have enough awareness of ourselves and our limitations to have some doubts. At the same time, we may be reluctant to explore our doubts in climates — both institutional and more broadly social — that seem somewhat hostile to the humanities and to aspects of medicine and other social practices that are not clearly "evidenced-based" with statistically significant results to support claims of benefit. In day-to-day clinical dealings, few seem to share Iris Murdoch's view that

> [i]t is totally misleading to speak, for instance, of 'two cultures', one literary-humane and the other scientific, as if these were of equal status. There is only one culture, of which science, so interesting and so dangerous, is now an important part. But the most essential and fundamental aspect of culture is the study of literature, since this is an education in how to picture and understand human situations. We are men and we are moral agents before we are scientists, and the place of science in human life must be discussed in *words*. That is why it is and always will be more important to know about Shakespeare than to know about any scientists ... (1971, p. 34).

In part for reasons alluded to in the previous section,[11] success in ethical reasoning — let alone success in teaching others to engage in ethical reasoning — is decidedly difficult to measure (Noddings, 2002). As with raising children, for example, instances of clear success and clear failure seem obvious, but there is a vast and contentious area in between. Philosophers, however, have generally rejected an approach to applying standards reflected in Justice Potter Stewart's comment that although he could not define pornography, "I know it when I see it." We should welcome applicable criteria by which to judge and justify our success as teachers and consultants, yet demarcations between "good enough" and borderline inadequate remain difficult to draw, in part, because they are grounded in contested normative accounts of the good.

In addition, causal responsibility even in the so-called clear cases of success and failure may be impossible to establish. If, after implementation of ethics training, a clinical team reliably makes sound ethical decisions, or if a physician respectfully asks all of her geriatric patients about end-of-life treatment preferences, can the ethics experts truly claim credit for the (alleged) success? There are those who would claim that the Patient Self-Determination Act (PSDA) imposes unnecessary burdens on patients, physicians, and their interactions; even among those who endorse the PSDA's ethical goals, many would argue that clinical ethicists have largely failed in helping doctors and patients achieve them. Experience shows what the Frey-Holmes line of argument contends, whether the ethicist is an expert, and whether her teaching or consultation was successful, partly depends on one's view of the ethical issue and one's normative ethical theory grounding that view. Moreover, there are those who would argue that the clinical team members' characters were formed long before they entered medical school, that clinical ethics education sessions do little more than remind them of regulations and give them a chance to air feelings, and that, after all, one person's sound ethical decision is another's mere preference dressed up in socially acceptable reasons, which in turn, is contrary to still another's reasoned moral point of view (Vertosick, 1994).

These concerns about the possibility of ethical expertise were explored in the first section, and I want to put them aside for the moment. I want to recognize that people — usually philosophers, theologians, or those with special training in these disciplines — are currently functioning as (i.e., are being regarded as) ethics experts in clinical contexts, and then try to make sense of some of the discomfort these experts experience in their work. Beyond the social discomfort of being "strangers in a strange place" and the emotional discomfort of witnessing the suffering of patients, their families, and hospital caregivers, clinical ethics experts experience intellectually grounded discomfort, worries grounded in thinking about the goals and methods of their work. I want to proceed by comparing their work to that of mothers and the challenges of maternal thinking.

Initially, the pairing of mothers and ethics experts sounds like the opening of a joke. Indeed, both are typically relied on to supply the right answer or action, are frequently ignored when they do, and are sometimes ridiculed if they purport, or are thought to purport, to have expertise regarding the right and the good. Beyond prompting such seemingly inconsistent, even neurotic responses in others, however, what do the two have in common? I shall argue that like mothers, clinical ethicists teach, nurture, and provide the conditions necessary for the development and exercise of particular moral capacities. Granted, the work of mothers and of clinical ethicists is as different as it is similar; however, there are considerable and informative parallels between the goals, methods, and challenges of the work of ethics experts and those of mothering.[12]

2.1. Prelude to Maternal Thinking

As a prelude to employing Sara Ruddick's analysis of the work that mothers do and the maternal thinking that informs it (1989), I want to embrace some of the preliminary cautions that Ruddick presents. First, those who engage in maternal work and who occupy the social role of mother need not be women and need not have children either in the sense of giving them life or raising them toward adulthood.[13] Like Ruddick, I worry that referring to "mothers and mothering" may mislead or alienate some readers. Referring to the work of parents might seem more inclusive and less stereotypically gender-based. Yet it is to the traditional work of mothers that she and I refer, while recognizing that others, including fathers and a network of formal and informal caregivers, paid and unpaid workers, engage in mothering work. Moreover, I want to suggest that at least some aspects of mothering work and maternal thinking can be directed toward parties other than one's children, including most obviously other people's children, but also one's students, patients, and those who seek one's counsel. I trust that these possibilities will become clearer upon exposition of an account of mothering and maternal thinking.

In contrast to the stereotypical paternal role, a mother's relationships *vis à vis* her children is one of being in a noncontractual, given, or found relationship (Baier, 1986), and this may seem to be a point of contrast between mothers and clinical ethicists. Nevertheless, although those of us who teach and consult in a clinical setting choose to do so, on a day-to-day or class-to-class basis we take our charges as we find them. For the medical students whom we teach, the teacher-learner relationship is usually far from freely chosen. Although both ethics and anatomy may be mandatory parts of the curriculum, the latter is usually recognized as being integral to the field which the students are eager to enter, while ethics is perceived as an "extra", mandated and taught by outsiders who intrude nonmedical perspectives on the system students want to embrace and to be embraced by. The patients and families with whom we consult are brought to us by the most unchosen of circumstances, first illness, and then perceived moral failing on the part of professionals and institutions on whom they are forced by circumstances to rely. The professionals who seek our ethics consultation similarly feel forced by circumstances, and often by desperate hope that we will ratify their choices or extract them from an ethical quagmire in which they find themselves. None of these circumstances are the same as finding oneself "with child," or as being a child dependent on a mother's good will. But like much of moral life, the ethicist's relationship to her students or consultees is far from a quasi-contractually governed relationship of equals, although should the ethicist's interventions prove inadequate, legalistic structures do lurk in the background (e.g., the possibility of guardianship proceedings, law suit, or professional and governmental regulation). This is another reason for focusing on mothers in contrast to fathers whose responsibilities have traditionally been conceived as financial and legal with the state in the background, not a child in the womb or the house, to impose them.[14] Historically the paternal role has been characterized in terms of rights, property, and contract, while the maternal role has focused on responding to the needs, vulnerability, and sheer presence of children (Ruddick, 1989; Locke, (1690) 1980; Pufendorf, (1682) 1991).

Two final cautionary notes are warranted by my employing a stereotypical notion of mothering. First, I am being selective with respect to the aspects of stereotypical mothering on which I focus. By my failure to discuss other aspects, I do not want to reinforce stereotypes, social systems, and cultural practices that disempower women both with and without children.[15] Second, by referring to the stereotype or cultural construct of mothers and mothering, I am referring in part to an ideal. Nevertheless I want to draw my analogy between ethics experts and "good enough" mothers, not Supermoms or ideal mothers, who never make a mistake and never act nonmaternally. If my analogy has value it must be for understanding the work of ethics experts at the bedside and in case conferences, where they will be human and fallible, and yet good enough to fulfill a role of ethics experts.

2.2. The Work of Mothers and Clinical Ethicists

Mothers see a child's physical and social vulnerability as "socially significant and as demanding care" (Ruddick, 1989, p. 18; Noddings, 2002, p. 8). Those who are committed to mothering see the fact of the child's situation as demanding a nonoptional response of protection; however, for those who do not share (or have not yet entered into) the maternal point of view, whether to respond with care in order to protect, preserve, and sustain requires a decision. Once the maternal vision is developed, however, actions that flow from it and seek the preservation of the child may seem to impress themselves upon the mother-er and to leave her no choice.[16] Perhaps for this reason, among others, the work of mothers has generally not been regarded as work, as *ethical* work, or as an endeavor that grows out of moral judgment. Yet, Ruddick's description of mothers' response to vulnerability is reminiscent of Iris Murdoch's description of what follows upon development of the capacity of moral attention and adoption of its point of view.

> One is often compelled almost automatically by what one *can* see. ... [I]f we consider what the work of attention is like, how continuously it goes on, and how imperceptibly it builds up structures of value round about us, we shall not be surprised that at crucial moments of choice most of the business of choosing is already over. This does not imply that we are not free, certainly not. But it implies that the exercise of our freedom is a small piecemeal business which goes on all the time and not a grandiose leaping about unimpeded at important moments. The moral life, on this view, is something that goes on continually, not something that is switched off in between the occurrence of explicit moral choices. What happens in between such choices is indeed what is crucial (Murdoch, 1971, p. 37).

If we ignore the prior work done by attention, then choice, decision, intention, and freedom to choose assume prominence in moral judgment and are the focus of ethical analysis. If, on the other hand, we concur with Murdoch that development of the capacity for moral attentiveness is a worthy and necessary task of moral philosophy, then we can begin to see as ethical in nature and as ethically mandated much of mothering work and some of the less explicitly analytical aspects of ethicists' work.

Murdoch comments, "I have used the word 'attention', which I borrow from Simone Weil, to express the idea of a just and loving gaze directed upon an individual reality. I believe this to be the characteristic and proper mark of the active moral agent" (1971, p. 34). Contrary to Ryle's concern that if caring is required to know something, then the object of one's care cannot be a proper object of knowledge, Murdoch and Ruddick argue that caring may be necessary in order to discern what may indeed be available to be known.

> Moral change and moral achievement are slow; we are not free in the sense of being able suddenly to alter ourselves since we cannot suddenly alter what we can see and ergo what we desire and are compelled by. In a way, explicit choice seems now less important: less decisive (since much of the 'decision' lies elsewhere) and less obviously something to be 'cultivated'. If I attend properly I will have no choices and this is the ultimate condition to be aimed at (Murdoch, 1971, pp. 39-40).

Following the lead of Murdoch, Ruddick, and other moral philosophers,[17] I submit that part of what is involved in teaching the demands of ethics to others – from one's children to one's medical students – is to create in them this vision, this capacity for seeing the world in a way that makes particular demands upon them, including demands for care, justice, and humility.

Ruddick begins her discussion of mothering as all who seek to understand a social practice should, by articulating the goals or demands of the practice. The demands of mothering are, she says, preservation, growth, and social acceptability; "to be a mother is to be committed to meeting these demands by works of preservative love, nurturance, and training" (Ruddick, 1989, p. 17). Maternal thinking is the intellectual work that balances these demands and devises strategies to fulfill them. The protect-nurture-train schema is captured by this scenario:

> A child leans out of a high-rise window to drop a balloon full of water on a passerby. She must be hauled in from the window (preservation) and taught not to endanger innocent people (training), and the method used must not endanger her self-respect or confidence (nurturance) (Ruddick, 1989, p. 23).

Although it is fortunately the case that most who receive clinical ethics consultation and teaching are not as physically vulnerable as a child leaning out of a high window, a clinical ethicist's trainees and consultees are vulnerable.[18] Many patients, in fact, are physically vulnerable due to illness, injury, or incompetence. Clinical ethics consultants, as well as ethicists who help to draft institutional policies and teach clinical ethics, view themselves as helping to preserve patient autonomy and well-being in the face of patients' vulnerability. Moreover, the work of clinical ethicists may afford a measure of physical protection from what in rare instances might otherwise erupt in violent action (Anonymous, 1989). More frequently, clinical ethics consultation and teaching may be thought to preserve trust and provide forums and conditions that help to preserve respectful relationships between members of the health care team, patients, and their families. Resort to the legal system, for example, may result from a failure to facilitate communication and achieve consensus in a particular case (ASBH, 1998, p. 8; Entman et al., 1994; Hickson et al., 1994; Levinson, 1994). In the face of a health care system that can crush patients' spirits and obliterate their individuality in its attempt to use evidence- or statistically-based medicine to treat their individual maladies as efficiently and effectively as possible, clinical ethicists help to reassert the moral significance of the particularities of the individual patient.

In addition, in the face of hierarchical medical training that can be physically and emotionally brutalizing, the clinical ethicist tries to preserve and validate the self-esteem and humanity of trainees, while nurturing their respect for the same in their patients. The vulnerability of medical trainees demands preservative work of clinical ethicists that is less concretely life saving than that of mothers. Clinical ethicists seek to protect the values, autonomy, dignity, and integrity of trainees and professionals, as well as that of their patients, rather than their physical selves. Thus, clinical ethics work may more nearly parallel the nurturing work of mothers who protect the "self-respect or confidence" of their children while helping them to "perfect, advance, further" their emotional and cognitive capacities (Ruddick, 1989, p. 82). With children, this gradual growth may seem to progress "'naturally', provided favorable conditions for growing," yet "each child grows in her or his distinctive, often peculiar way ... [and] their cognitive and emotional development is easily distorted or inhibited. They 'demand' nurturance" (p. 19). Ruddick observes that recognition of this demand for nurturance is "historically and culturally specific to a degree that the demand for preservation is not" and that "some people ... claim that if children are protected and trained, growth takes care of itself" (p. 19). Similarly, in medicine, the assumption is frequently made by clinical instructors that training in the basic sciences and their clinical application, plus a dose of professionalism, is all that medical students need to grow into doctors (Ruddick and Finn, 1985, pp. 42-43). It is assumed that unless the students are somehow morally deficient at the beginning of the process, they do not need any particular nurturance or ethics education — except perhaps a primer on applicable laws — in order to practice medicine in an ethically acceptable and emotionally sustainable manner.

In contrast, some would argue that medical training does not present the most "favorable conditions for growing" and that it instead presents new emotional and psychological challenges, and raises ethical questions, so that students need both nurturance and ethics education (e.g., Charon, 1996, p. 303; Kasman, 2001; Kasman et al., 2003; Ruddick and Finn, 1985, p. 43). Medical training requires what would be violations of ethical codes in any other context: dissecting dead bodies, opening chest cavities, prying into people's private lives and the orifices of naked bodies, and continuing to perform competently in circumstances where others would take time off to mourn or even just to sleep. That these practices peculiar to medicine are not just permitted, but mandated, and are eventually both highly remunerated and held in awe by the "outside world", creates particular vulnerabilities in medical trainees. They must develop the capacities — intellectual, emotional, and ethical — to justify self-confidence without falling into an abyss of insecurity and despair on one side, or falling prey to the temptations of hubris on the other. Such development demands nurture of cognitive and emotional capacities.

Of the three demands of mothering, the third — training for social acceptability — has the most obvious parallels to clinical ethics training. Training, unlike preservation and nurture, is based most directly in society's, not the children's/trainees' needs, although an at least minimal degree of social acceptability contributes indirectly to children's and trainees' thriving and even to their preservation (Ruddick, 1989, p. 21).

> Social groups require that mothers shape their children's growth in 'acceptable' ways. What counts as acceptable varies enormously .... The demand for acceptability, however, does not vary, nor does there seem to be much dissent from the belief that children cannot 'naturally' develop in socially correct ways but must be 'trained' (Ruddick, 1989, p. 21).

Medical students learn what they need to know in order to diagnose, cure, and prevent disorders in society's members. They learn the standard of care and how to provide it. Meeting that standard protects them from liability. In clinical ethics training, they learn the ethical and legal norms governing their practice. It is here that they most directly learn how to employ their medical knowledge in socially acceptable ways (although they implicitly learn from the example of their clinical mentors and more advanced trainees what is socially accepted within the profession). A clinical ethicist tries to impart to trainees at least enough knowledge of the content of clinical ethics so that as health care professionals, they will know when to seek additional information, advice, or consultation. Medical students must, for example, learn about the determination of patient competence and the process of informed consent, just as they must learn about organ systems and diseases. Although students do not learn all that there is to know about every aspect of the human body or about clinical ethics, they are to learn enough to be able to identify a problem — physiological or ethical — and to begin to seek its solution. Indeed, preventive medicine and its moral analog, preventive ethics, require that health care professionals learn enough to anticipate potential problems and to take or recommend steps to prevent them from materializing (Forrow et al., 1993). Adoption of this anticipatory stance is part of the development of the capacity of moral discernment. As Baier suggests, a sense of history and "something of the novelist's imagination of the detailed human consequences" of proposed actions and policies are indispensable for those who engage in moral reflection (1984, p. 494).

Clinical ethics teachers seek to impart at least part of the recognized body of bioethical knowledge. But they do not want primarily to teach and enforce rules. Like mothers, clinical ethicists recognize that they will not always be there to guide their charges and enforce rules. Both mothers and clinical ethicists train for independence, for the sake of their charges' future autonomy. They recognize that achieving at least some degree of their own dispensability is a goal of their training. "To be responsible for children's moral well-being means helping them become people who will be reliably moral when they are alone or among peers. This means turning over moral initiative to the children themselves" (Ruddick, 1989, p. 108). Most importantly, both mothers and ethicists recognize that in moral life more than choices and conduct matter. A mother cares deeply about what sort of person her child is and becomes. A mother cares about a child's belief system, vision of the good, values, and personal commitments. More than caring that her child behaves well, a mother cares that he behaves well for the right reasons, reasons that matter to her, probably reasons that she shares. "The criteria of acceptability consist of the group values that a mother has internalized as well as the values of group members whom she feels she must please" (Ruddick, 1989, p. 21). The demand to train for social acceptability is not merely externally imposed on the mother by society and its constituent groups; it is a demand whose specific content she both adopts (in large measure) and helps to shape (to varying degrees based on her social power). The content of conduct and character that meets this demand for social acceptability is, therefore, not merely conventional in two ways. First, the content is filtered and shaped by the individual mother or ethicist. Second, successful training in the most important arena of social acceptability — morality — cannot be evidenced by mere conventional behavior on the part of the child/trainee; good reasons and character, as well as conduct matter, and good reasons or character may sometimes require unconventional conduct.

Indeed, clinical ethics training may demand unconventional conduct in the sense of actions and patterns of thought that depart from the norms that dominate medical training. Yet, because the ethos of medicine and the hierarchical structure and isolating environment of medical school are very powerful, the desire of students to identify with their medical exemplars is typically very strong. Against this backdrop, ethics training may prove to be an unwelcome reminder of medicine's fallibility, of medical uncertainty and moral ambiguity, and of other values besides health and life. Clinical ethics discussions frequently invite students to (re)identify with the (nonmedical) needs and interests of patients, with outsider perspectives on health care institutions, and with less powerful members of society. Even discussions that focus on ethical concerns within the health care team inherently suggest that there are chinks in the armor of the health care hierarchy that ought to be investigated. The hierarchical structure of medical schools also presents an unwelcome reminder to clinical ethicists of their own relative impotence and may lead some toward self-doubt or a feeling of inauthenticity about their role and about the positions they espouse. The situation parallels that of mothers who simultaneously recognize, on one hand, their power and responsibility with regard to their children and, on the other hand, their relative impotence in the world.

As protectors and teachers of their children, mothers seek to intervene and control; they choose priorities for their children's development and seek to influence not only their children's wills, but also the circumstances of the world, to effectuate developmental lessons. At the same time, however, they are constantly reminded of how much they cannot control, how easily chance can contravene their best efforts, and how much "people, policies, institutions, or natural happenings" contradict their "values and disciplinary strategies" (Ruddick, 1989, p. 110). As a result of this double bind, Ruddick suggests, most mothers experience a loss of confidence, self-disrespect, and susceptibility to the "gaze" of others. They may respond by submitting to authority and acting inauthentically, repudiating their own values and perceptions.

> Inauthentic mothers construct, before the eyes of their children, a world in which maternal values do not count. Although they teach appropriate behavior, the purposes of that behavior is not theirs to determine. ... [T]hey replace ... the idea of conscience with that of submission. ...
>
> Inauthenticity is an attitude that admits of degrees. Mothers experience typical progressions from occasional submission to the habit of submitting to a principled defense of submission that has become habitual. ... A mother may refuse to let her child watch a particular TV program but simply walk away when the Father gives permission. ... Mothers are no more — and no less — blessed with stamina and gifted in courage than other women and men (1989, pp. 112-113).

Ruddick distinguishes this abdication of the mother's authentic self and her maternal authority from the situation "in which a mother is rendered powerless in her children's eyes because she is treated with contempt" (1989, p. 111). "Fathers, landlords, welfare workers ... and numerous other authorities can and do insult mothers, even in front of their children. The abdication I speak of exists when a mother hides from her child her real feelings and the realities of the power situation as she sees them" (1989, pp. 111-112). Social and institutional structures, as well as individuals working within them, help to construct before impressionable eyes a world in which maternal values do not count. Openness, characterized in this case by explicit acknowledgment of both the limits of her power and the value-laden challenges to it, may serve to preserve mothers' self-respect and their children's regard and also to model for them ways of responding to such limitations with integrity.

Clinical ethicists face a parallel double bind as they operate within settings that send implicit messages that their work and values do not count. The subject matter of their work — the right and the good, the stuff that makes life worth living — is acknowledged to be of profound importance. Yet, *qua* ethicists they do not save lives, at least not by employing awesome medical technology. Moreover, ethics centers are not profit centers within health care systems and medical schools. Historically, ethicists have been related to the highly socially-valued enterprise of medical science by invitation or insinuation, by making themselves useful, which is more recently criticized as a form of habitual submission dressed up in principles and pragmatic defenses (Engelhardt, 2002; Evans, 2002; Wolf, 1996).[19] Their work is considered "soft" and aligned with the weaker half of the claim that medicine is both art and science. During lean fiscal years, clinical ethics teaching and consultation are likely to be considered part of the fat that makes medical care and education more palatable, but that is not essential to the meat and potatoes activities of training doctors and curing disease. In short, there are multiple contexts within health care institutions and medical schools in which clinical ethics is treated with relative disregard if not contempt.

It is part of the work of clinical ethics teachers to engage in critical reflection on the priorities of a health care system. All clinical ethicists work at the intersection of competing religious, ethical, capitalist, patriarchal, cultural, and intellectual value systems. In drawing attention to the health care system's uneasy relationship to (and occasional contempt for) the substance and process or their work, clinical ethicists risk being viewed by trainees (and administrators) as self-interested complainers. Yet by drawing attention to the social and institutional priorities that create the double bind in which they find themselves, clinical ethicists can avoid professional inauthenticity and guard against being co-opted into serving as mere handmaids to and apologists for the institutions that employ them. Moreover, by openly discussing how the activities and values which they advocate are regarded — from contempt to neglect to respect — clinical ethicists model such reflection, advocate for ethical values, and support those who hold nondominant values within the sometimes inhospitable environments of health care.

Especially with clinical ethics training in the traditional moral philosophy mold, such teaching may be more disquieting for medical students than the rituals and rigors of medical training. Baier points out that it is the role of philosophers as social gadflies — in the classroom and in society — to disturb old beliefs and to create in others a similar willingness to do so. Those working in clinical ethics also seek to disrupt students' easy acceptance of freshly acquired beliefs in the hegemony of the biomedical model and the undisputed social power of medical science (Ruddick and Finn, 1985). If, however, clinical ethics teaching degenerates into — or is perceived to engage in — "bashing" of the biomedical model, the clinical ethicist both risks creating a hostile class of fearful, defensive medical trainees-cum-moral sceptics, and falls into the false dichotomies of scientific/humane, objective/subjective, and factual/normative that she seeks to challenge.[20] When considering Socrates's charge of corrupting youth, Baier concludes that "initiation into rational criticism" cannot be said to corrupt, but she admits that if successful, it certainly disrupts the status quo and can be distressing; philosophy's intellectual methods "can unsettle and upset, and it can also produce moral skeptics and even cynics. Part of our faith, it seems, is that anything, even a moral sceptic, is better than an unconscious moral dogmatist" (1984, p. 492). Baier affirms that the examined life is worth pursuing, "but only if the mode of examination does not destroy the life." Just as medicine's cures must not kill the patient, ethicists' challenges to the dominant medical ethos (or to priorities in health policy) must neither create despair and insecurity about the world students are entering, nor allow indoctrination into that ethos to kill off future physicians' means of identification with their vulnerable patients. Walking with students through cases drawn from their clinical experience and allowing them to air their feelings as they analyze their ethical options is preservative, nurturing work.

In class and in individual consultations about particular patients, clinical ethicists "hold the hands" of their trainees and consultees as they engage in fact finding, explore options, and reevaluate priorities. With trainees, ethicists challenge their assumptions — e.g., about how responsible individuals are for their health conditions — and urge them to consider other possibilities and perspectives — e.g., perhaps the surgery could be done without transfusions or perhaps there are fates worse than death. When consulting with patients and their families, ethicists challenge old beliefs, though more gently: maybe love demands letting go; maybe you should try to peacefully coexist with cancer rather than go down fighting; maybe allowing toxins to accumulate in the blood of a man whose house and person you've kept immaculately clean for sixty years will allow him a more pleasant death without dialysis machines. In both teaching and consulting, clinical ethicists seek to create openness to different perspectives and new options. In teaching, we use our analytical skills to play devil's advocate and engage in consciousness raising, challenging complacency, creating an openness to difference, and preparing students to deal ethically with those who do not necessarily share their value commitments or professional priorities. In ethics consultations, the openness that we often seek to create is an openness of patients, families, and professionals to the least awful reality when the ideal good outcome is not possible because of the limits of mortal bodies, imperfect medicine, financial capacities, or social will.

When they do their work best, clinical ethicists, like mothers, approach their trainees/consultees with the same capacities they hope to develop in their charges. The training work of mothers and the training and consulting work of clinical ethicists demand that each meets her charge where he is, be fully present to him, and listen empathetically. "[Mothers learn] to ask "What are you going through?" and to wait to hear the answer rather than giving it. She learns to ask again and keep listening even if she cannot make sense of what she hears or can barely tolerate the child she has understood" (Ruddick, 1989, p. 121). Attentive listening or other forms of attentive observation begin with the belief that more of importance is available to the perceiver than what forcibly impresses itself upon the senses. Mothers must believe in their children's "continuous mental life," and in the fact that more than the behavior of their children matters and that more must be understood than their behavior and pronouncements in order to help them to achieve an ethically acceptable outcome. A "mother striving to understand a child cannot assume that his experience is irreducibly episodic, disconnected, and fragmented" and must "make sense of her child's actions" by assuming "that he is moved by interdependent perceptions, feelings, and fantasies" (Ruddick, 1989, p. 92). Similarly, a clinical ethicist must delve into the beliefs, context, feelings, fantasies, and hopes of her trainees and consultees to make sense of their views on the issue at hand. In this way, the "hand-holding" activities of clinical ethicists — attention, being present, recognition of the other — are integral to their exercise of ethical expertise.[21]

> Attention is akin to the capacity for empathy, the ability to suffer or celebrate with another as if in the other's experience you know and find yourself. However, the idea of empathy, as it is popularly understood, underestimates the importance of knowing another *without* finding yourself in her. A mother really looks at her child, tries to see him accurately rather than herself in him (Ruddick, 1989, p.121).

It is at least an ideal of maternal work that a mother observe, listen, and respond to her child without projecting her values and needs into him. It is, most clinical ethicists would agree, an obligation of teachers and consultants to do so with respect to their students and consultees.[22] Commenting on her role as a teacher, Nel Noddings states:

> I do not need to establish a deep, lasting, time-consuming personal relationship with every student. What I must do is to be totally and nonselectively present to the student — to each student — as he addresses me. The time interval may be brief but the encounter is total (1984, p. 180).

By taking their students as they find them and in explicitly acknowledging the importance of doing so, clinical ethics teachers can model for clinicians-in-training respectful and productive ways for them to interact with patients, families, and colleagues.

Yet, in taking others — children, students, patients, and families — as they find them and attending to them fully, mothers and clinical ethicists lay the grounds for more than effective communication. Attentive encounters contain, at the level of perception and feeling, the ingredients of moral life and ethical judgment: recognition of others, assumption of responsibility, awareness of the demands of justice and the scope of one's agency, and the necessity of understanding and giving reasons. As Murdoch suggests, moral attention is necessary for, or is a necessary part of, the exercise of moral judgment; indeed, acute attention may obviate a distinct act of judgment, decision, or choice. Embodied in the act of attending to another are elements essential to the capacity for moral discernment and adoption of the moral point of view. "Attention," Ruddick observes, "lets difference emerge without searching for comforting commonalities, dwells upon the *other*, and lets otherness be" (1989, p. 122). Attending to another as separate from oneself focuses one's attention on difference, on limits of oneself, and on the possibility — and possible desirability — of change (to become more or less different from the other). Recognition of the scope of one's causal responsibility (power) and moral responsibility follows from recognition of difference and a willingness to accept it, at least initially, without judging it as good or bad.[23]

Elaine Scarry argues that attending to beauty — to particular concrete beautiful things — has ethically salient effects (1999). She argues that perceiving something beautiful takes us out of ourselves, directs our attention to others, and presents to our senses a concrete instance of proportion, symmetry, and relationship, essential features of justice. Here I make a similar suggestion that essential ingredients for ethical judgment are embodied in attending to another. In attending to others, one is often forced to discern pain, problems, and needs that are real. These may be the needs and pains of the other, or they may be one's own suffering or need occasioned by recognition of one's own limitations and one's separateness or difference from the other person. In most cases, especially if one has the support of others' caring attention — like students who are cared for by teachers, or children who are cared for by mothers — one will not be destroyed by these threatening realities. Instead, by developing in children an openness to difference and change, mothers help them learn to accept both suffering and limitation. They also help their children develop the ability to accept responsibility for wrongs and harms, when appropriate, without viewing themselves or the world as irredeemably bad.

In this way, mothers both possess and inculcate in their children what Ruddick describes as the capacity for "cheerfulness".

> To be cheerful means to respect chance, limit, and imperfection and still act as if it is possible to keep children safe. Cheerfulness is a matter-of-fact willingness to accept having given birth, to start and start over again, to welcome a future despite conditions of one's self, one's children, one's society, and nature that may be reasons for despair. ...
>
> In circumstances of personal or social disaster that warrant despair, maternal cheerfulness is an extraordinary feat. Many mothers, with resourcefulness and restraint, help their children

> die well; others sustain seriously damaged children in hopefulness. ... In happier circumstances, clear-sighted cheerfulness is not an extraordinary virtue but a common strength. Ordinary mothers school themselves to look realistically at their children and the dangers that confront them, cheerfully controlling as best they can what is never fully controlled, ... so that the day's disasters can give way to the next morning's new beginnings (1989, p. 74).

While a buoyant mood and countenance of good cheer would seem out of place during most ethics consultations or case discussions, the quiet cheerfulness or resoluteness that Ruddick describes may be a goal of both clinical ethics teaching and consultation. It might be thought of as an attitude of hopeful realism. Although one must guard against *demanding* such cheerfulness from one's consultees or students, modeling and showing one's charges a way toward a reserve of hopeful realism can be a great gift to them.

Those who give or receive health care confront medicine's and their own limitations.[24] It is a task of clinical ethicists to inculcate nondespairing acceptance of those limits and to be with those who are mourning loss of their idealized hope that it could be otherwise. Within a system whose priority is to cure and to alleviate suffering, clinical ethicists must suggest that some types of suffering are measures of meaning (Lidz et al., 2003) or are the "costs of relating" (Noddings, 2002, p. 144). These may seem to be modern day tasks of psychologists or chaplains, not philosophers. Certainly some training in the perspectives and practices of psychology would benefit those who serve as clinical ethicists. Yet there is an important ethical dimension to acceptance of the reality of limitation. Those who mother well recognize that they should prepare their children to confront limitation and be resilient in the face of it; they recognize openness to change as an "ethical ideal and a metaphysical attitude" (Ruddick, 1989, pp. 89-93). If children are to be prepared to live some version of the good life, they must be prepared to do so *in reality*; to be allowed to develop false or impossible ideals is to be allowed to develop a vision of the good life that is not fit for living.[25] Indeed it should be considered as much a failing of mothers and of medical training to instill in children and trainees unrealistic expectations (idealizations) of themselves and the world, as it is for them to fail to instill in their children and trainees a vision of the good (ethical ideals). That this goal may appear to conflict with other goals of mothering work is taken up midway through the next section.

3. TRAINING FOR INDEPENDENCE, PLANNED DISPENSABILITY, AND ETHICS EXPERTS IN DEMOCRATIC SOCIETY

The previous section compared the protective, nurturing, and training work of both mothers and clinical ethicists and sought to address some of the questions clinical ethicists raise about their work as ethics experts. This section considers a worry raised about the role of ethics experts in democratic societies.[26] Some argue that the notion of ethical expertise and the social role of ethics experts threaten the premises of democratic decisionmaking in a pluralistic society — specifically, that everyone must, and must be allowed to, decide for himself on matters concerning the good (Scofield, 1994; Shalit, 1997). In showing how reliance on the expertise of ethics experts need not threaten either democratic ideals or the autonomy of individuals, this section suggests further parallels between the work of mothers and ethicists. More detailed examination of how ethics experts ought to respond when asked to give their expert opinion, and comparison of this strategy to that involved in maternal training, completes my examination of how the practical challenges and apparent paradoxes faced by clinical ethicists relate to philosophical concerns about ethical expertise raised here and in section one.

3.1. Ethical Expertise in Democratic Society

That reliance on ethical expertise and thus the social role attributed to ethics experts are incompatible with the demands of democracy is a charge perhaps most outspokenly stated by Giles Scofield. Regarding the role of putative ethics experts as expert witnesses in court proceedings, he comments:

> [T]o believe that someone possesses specialized knowledge in applied ethics presupposes that some individuals are legally regarded as knowing better than others what should, from an ethical perspective, be done in a given situation. The difficulty with this claim is that it is antithetical to the foundational beliefs of a pluralistic democracy. After all, we believe that each individual is the moral equal of every other, and that each is the best arbiter of his or her own moral destiny. While this means that each of us has the right to be wrong, even stupid, it also means that no one — not even the moral philosopher — can be regarded as having a better view of what ought to be done than anyone else does. If that is so, then no one ... can claim expertise in applied ethics, since that is an expertise that each of us is regarded as having (1994, p. 9).

In an article raising questions about the role of ethical expertise in liberal democracy, Scofield is quoted thus: "If you take democracy seriously, then the basic rule is that every philosopher is simply a citizen, and every citizen a philosopher, capable of making decisions that reflect his or her conscientiously held beliefs" (Shalit, 1997, p. 27). Commenting on his experience on a hospital ethics committee, Scofield is again quoted: "We'd all be sitting there, having some sort of conversation, and then, when I spoke up, they would look at me as if I knew what I was talking about, and the rest of them didn't" (Shalit, 1997, p. 27). Despite their overstatement and conceptual confusions, Scofield's comments succinctly sketch the apparent conflict: in matters of social governance that have ethical import, why should each person have one vote and be allowed to decide for himself how to cast it, in accordance with his own vision of the good, when ethics experts could instead be relied upon to lead society on the ethically justified course? Either we must all be philosopher kings, or we must sacrifice democratic ideals and process to appoint ethics experts as our philosopher kings. Responding to this basic concern, as well as correcting the erroneous assumptions informing Scofield's comments, illuminates the valuable role that ethical expertise can serve in liberal democracy.

To begin, we must recognize two distinct levels on which appeals to ethics experts may be made: the level of the individual (e.g., individual decision making and development of one's conception of the good) and the level of social bodies (e.g., development of public and institutional policy or decisions by judges and juries). Scofield correctly recognizes that these two levels interrelate. Individuals make personal decisions regarding which political candidates or issues to support; policy-making bodies — some of them comprised of elected representatives or their appointees — make social decisions. Ethical expertise may be thought to present a somewhat different challenge at the individual and social levels.

At the level of individuals, ethical expertise may be thought to threaten individual autonomy. This is presumably the level at which Scofield claims that "each is the best arbiter of his or her own moral destiny" and that "each of us has the right to be wrong, even stupid." Yet Scofield is mistaken to argue that from each person's right to be wrong and to arbitrate her own moral destiny, it follows that "no one — not even the moral philosopher — can be regarded as having a better view of what ought to be done than anyone else does." Earlier in his paper, in order to suggest that there are no ethics experts, Scofield recounts arguments similar to those discussed in the previous sections of this essay (1994, p. 2). Ethics is not, he argues, a matter of the scientific, technical, specialized, arcane, or esoteric knowledge that may be the subject of expert testimony in US courts, and would-be ethics experts disagree. While his arguments may establish grounds for challenging the medical ethicist as an expert witness in court, they do not challenge the arguments offered by Singer, Weinstein, and this essay in favor of recognizing and utilizing ethical expertise. From the fact that putative ethics experts disagree about the most justifiable course of action — remember the disagreement of Ramsey and McCormick — it does not follow that no one can be more expert than another or that no one can offer a stronger justification than someone else. Thus, while each of us may have the *right* to be right, wrong, or stupid, it seems not only stupid, but perhaps morally wrong, to remain stupid when another option is available. In many contexts outside of the specific realm of ethics, people seek expertise in order to inform their beliefs and decisions. In the context of ethical beliefs and decisions, one often has a *responsibility* to do so.

> If a person knows that someone else is more adept at making difficult moral decisions than he himself is, it seems that there is a duty to seek the help of that other person (assuming, of course, that the person is willing to help and will give sincere advice). If a person cannot discharge his obligation to perform a certain act unless he does some other act first, it is plausible to say that he has a derived obligation to perform this other act. In complex situations one may not even be able to *know* what one ought to do unless one seeks the advice of a moral expert. So, if one knows who the moral experts are, one may have an obligation to seek their advice when a complex moral decision must be made (McConnell, 1984, p. 211).

Far from threatening autonomy, consultation with ethics experts may serve autonomy by informing its exercise in a particular case (Szabados, 1978, p. 125). More generally, consultation with ethics experts, or training by them, may enhance one's exercise of autonomy by developing one's capacity of moral discernment, or by helping one to develop a conception of the good (or set of values). The only sense of autonomy that may be threatened by consultation with an expert is autonomy in the sense of "going it alone," a radical interpretation of the idea that no one can have another make one's moral decisions (McConnell, 1984, p. 210). Such an interpretation must be grounded in an erroneous belief that spontaneously acquired values and preferences are possible (or morally desirable), and that decisions made in accordance with them are autonomous (Sher, 1982).

Most importantly, recognizing that someone else has ethical expertise — i.e., recognizing someone as an ethics expert — does not disrupt foundational ethical and political beliefs in the moral equality of all individuals. To defer to someone with ethical expertise, i.e., someone capable of giving strong justification for opinions within the domain of ethics, does not suggest that one is not the moral equal of that expert, but only that one is not the equal of that expert *in giving strong justifications* for one's ethical opinions. One is no less a rights-bearer, no less an appropriate subject of moral concern, *no less an equal* to the ethics expert *despite the unequal treatment* that the ethics expert may receive (e.g., solicitation of her opinion and deference to her arguments). Scofield seems to conflate two senses of equality (Dworkin, 1985, pp. 190-191) — moral equality or equal moral regard (being treated as equals) with equal treatment (being treated equally) — or else he confuses the moral value of persons with the value of their moral opinions.

Neither moral nor political equality of individuals — summarized by "one person, one vote" — is threatened by ethical expertise. By supplying strongly justified opinions, ethics experts (and indeed others with different domains of expertise) may influence the beliefs, values, and autonomous choices of nonexpert individuals. Insofar as each person must (or at least should) reason about the decisions before him, every citizen is a "philosopher." But ethics experts get only one vote; they are citizens, not philosopher kings.

At the same time, in contrast to most other areas of knowledge and skill, in ethics everyone is required to acquire some degree of ethical expertise, perhaps more properly termed 'ethical competence'. In other words, it is ethically required that everyone strive, to at least some minimal degree, to distinguish right from wrong, to identify ethical issues or discern when moral categories apply, and to develop or embrace a set of ethical values. Although one may permissibly go to one's grave without developing minimal competence in geography, quantum mechanics, or soufflé baking, it is not permissible to remain equally incompetent regarding ethical concerns. Status as a moral agent imposes particular obligations, including the obligation to become trained in socially accepted values and the capacity of moral discernment, as was argued in the previous section. Therefore, one cannot defer to ethics experts in precisely the same way that one may defer to other experts.

> In science an expert is a man who can be appealed to by the ignorant layman. We are wise to take his word even if we do not understand his reasons. In general there is no blame attached to lack of scientific knowledge. On the other hand, moral ignorance (as exhibited in the expression 'He does not know right from wrong') is either blameworthy or a diagnosis of a certain form of mental illness recently referred to as psychopathic personality ... (Szabados, 1978, p. 126).

In part because of the lack of consensus on substantive moral theory or the test of adequacy for normative theory that Frey and Holmes describe, and in part because of the demands of moral autonomy, one cannot choose an ethics expert based upon her reputation or membership in an ethics guild, the way one chooses a certified architect or accountant. When one hires an architect to design an addition to one's house, it is not required that one understand why a closet cannot be carved out of the bearing wall. One need not understand tax law, but may defer, largely in ignorance, to the expertise of one's accountant. Not so with ethical matters. Whether or not credentialing procedures for ethics experts emerge, one must determine whose expert opinion to trust based not merely upon reputation, past performance, or credentials, but also upon one's assessment of the reasons supporting the expert's opinion.[27] Moreover, if one's Great Hall collapses, or if one's taxes are audited, the architect or accountant bears primary responsibility. (Only if one hires negligent professionals and should have known better, does one retain ethical responsibility for wrongdoing discovered.[28]) On the other hand, if one's moral life collapses or one is taken to task for unethical judgments, having relied upon the expertise or arguments of an ethics expert does not relieve one of responsibility; one retains moral and decisional authority and responsibility.

Recognizing that one retains ethical responsibility in relying on an ethics expert and that assessment of an ethics expert's arguments is integral to appropriate reliance may alleviate a consequentialist concern about relying on ethical expertise. Succinctly put, the concern is that

> [i]f agents rely too heavily on experts for advice they will become moral cripples, as it were. If they do not hold their moral principles rationally and autonomously, they will be unable to apply these principles to new situations (or even to ordinary ones) when moral experts are not available to be consulted (McConnell, 1984, p. 211).

Scofield suggests, in effect, that his presence on the ethics committee transformed his fellow members into moral cripples. They deferred to him automatically and almost completely; they ceased to decide issues of ethical import on their own. They viewed his expertise as largely inviolate and indispensable while regarding their own values and reasoning skills as suspect and dispensable. Evidently he and they failed to use the opportunity of consulting him, as an ethics expert, as an opportunity to gain knowledge and develop skills. They viewed consultation as an occasion for diminished, rather than augmented moral and decisional autonomy. Perhaps they sought to shift responsibility to him, rather than to retain decisional authority and moral responsibility for what was decided. Scofield is wrong, however, to indict all reliance on ethical expertise for the problems arising from inappropriate reliance on ethics experts.

So long as the ethics expert supplies her reasons, she provides the opportunity to enhance her consultees' moral and decisional autonomy. This is equally true at the level of individuals making personal decisions and at the level of social bodies making policy decisions. Examining the limits of the authority of experts, Rosemarie Tong argues that

> when it comes to the experts' distinctively ethical advice, decision makers should not defer to it simply because it 'belongs' to the experts. Rather decision makers should think the experts' ethical advice through for themselves, accepting it only if it fits their own reasoned moral point of view. *Second-hand ethics is not substitute for first-hand ethics in a community of equally autonomous moral agents* (1991, p. 415).

She argues for "an ethics of persuasion," that recognizes that supplying reasons for one's ethical opinions is integral to ethics itself and mandatory for ethics experts. Not only does their expertise reside in their ability to supply strong justification for their opinions, but also it is necessary to persuade others by reason-giving in order to acknowledge the moral agency of those advised and to avoid the "de-moralizing" of nonexperts with the result that they become either apathetic or alienated and aggressive

(Tong, 1991, p. 422).[29] Tong argues that "we need to create an atmosphere in which people both seek and heed advise (*sic.*) ... in which experts offer their knowledge to non-experts in order to spare them the pain of unnecessary mistakes" (1991, p. 423). The recommendations of ethics experts should be regarded "as useful reference tools: not the rulings of philosopher-kings, but the attempt of thoughtful people to share their knowledge — albeit imperfect — with other people." In this way, seeking, receiving, and employing such expertise need not threaten the deliberations of policy-making bodies in a democratic society. Indeed in some circumstances, such bodies have an ethical responsibility to consult ethics experts, just as individual decisionmakers do.

At the same time, those who serve as ethics experts have a complementary responsibility: they must offer the most strongly justifiable ethical judgment that they can and also supply the grounds for their judgment. Fulfillment of this responsibility has a seemingly paradoxical result: in sharing the grounds for their expertise and thus the grounds for their status as experts, ethics experts must render themselves dispensable in some sense. In exercising and sharing her expertise, the ethics expert supplies the resources to render herself redundant in a way that architects, accountants, mechanics, and bakers do not.[30] When they do their work well, indeed with maximal success, they do not transfer to others the ability to do-what-they-do in the way that an ethics expert does when she is maximally successful in her work. This paradoxical yet intrinsic feature of their work is one that ethics experts share with those who do maternal work. Both plan for and work toward their own dispensability.

3.2. Training for Independence: Mothers and Ethicists

Once again we return to instructive parallels between the work of ethics experts and mothers. The apparent conflict between reliance on ethical expertise and the demands of democratic decisionmaking has a parallel in the context of maternal thinking. The idea that a mother is supposed to instill in her children a vision of the good life, while simultaneously teaching (and allowing) her child to develop his own, seems to suggest that mothering has contradictory goals.[31] Similarly, bioethics has been charged with espousing a particular set of values while purporting to train students in more neutral ethical analysis and to support patients in pursuit of their chosen paths (Ainslie, 2002; Evans, 2002; Schneider, 1998; Smith, 2002).[32] One way to see that this apparent conflict is not real involves recognizing that development of the capacity for moral discernment, what Ruddick calls 'conscientiousness', is a primary goal of both maternal training and clinical ethics teaching.[33] If creating in one's charges "the ability to identify, reflect on, and respect the demands of conscience" is a primary goal of one's training work (Ruddick, 1989, p. 116), then it is obvious that advocating a set of values, or instilling a way of life, need not prevent children and students from adopting and maintaining their own. Of course, the strategies that one uses to train for social acceptability and to advocate for one's beliefs matter in this regard, as does the content of the beliefs. As Ruddick points out, a mother's "training strategies may be persuasive, manipulative, educative, abusive, seductive, or respectful and are typically a mix of most of these" (1989, p. 21). A respectful, open, and transparent strategy is recommended not only for ethical reasons — respectfulness and nondeceitfulness are virtues and are conditions for appropriate trust — but also because of the demands of training. Here, I argue, there are strong parallels between the strategies that mothers and ethics experts ought to pursue.

Jan Crosthwaite analyzes possible strategies for responding when one is called upon, as a putative moral expert, to advise on an issue about which one has a considered moral opinion. If she were asked, for example, whether fetal tissue should be employed in research, the following questions arise. "What should I do? Should I present my views? Should I survey the positions and arguments? With or without indicating my own assessments?" (1995, p. 367). Crosthwaite identifies three possible strategies that she describes, in turn, as providing moral cartography, expert judgment, or a charted path. She could simply chart the possible positions on the issue in an allegedly neutral manner and "not indicate or argue in favour of the one" she judges best (1995, p. 367); however, this strategy of moral cartography has several pitfalls that she recounts (1995, p. 373). First, it is difficult, perhaps impossible, to present an utterly neutral account of all positions with an equally strong argument in favor of each, when one is sincerely persuaded by the (greater) strength of one of the arguments. Second, in an attempt to be neutral, one is placed either in the ethically improper position of lying to or deceiving one's consultees by claiming to be neutral when one is not, or in the socially awkward and disruptive position of admitting that one finds a particular position persuasive but refuses to disclose which. On the other hand, a strategy that involves only offering her expert judgment — e.g., "In my judgment as a professional ethicist the use of foetal tissue for therapeutic purposes is morally right" — fails to supply sufficient grounds for evaluating her expert opinion. Crosthwaite explains,

> I must at least provide the reasoning by which I arrive at this assessment. In general, expert opinions should be supported by reasons, to allow for peer review at least. In the case of moral consultation though, there are additional bases for this requirement, partly practical considerations to do with clarity and understanding, but more significantly issues connected to how we should conceive of the process of moral deliberation.
>
> The significance of a moral claim may not be clear or fully understood without understanding the reasoning behind it. ...
>
> ... I am assuming that moral judgement is a reason-governed activity. If one were to hold an extreme subjectivism ... then one would see no ground for providing reasons. But neither would one see any grounds for consulting a moral philosopher. Moral decision-making at the social level could be no more than counting votes... (1995, pp. 369-370).

Certainly if one is attempting to train others to make ethical decisions for themselves, then offering not only expert opinion, but also explaining the basis for one's judgment is critical to illustrating how reasons lead to and support moral conclusions.

Therefore, Crosthwaite argues that ethics experts ought to provide a charted path, "a map which marks both the position of the treasure [the ethical conclusion] and the route to it," complete with the directions concerning how and why to avoid the attractive shoals of less appropriate alternatives and the lure of false siren songs. In providing a charted path, Crosthwaite would "present the different moral positions on the issue, outlining their respective rationales, strengths and weaknesses, and indicate the conclusion ... [that she] draw[s] from this about what is right together with the reasoning which supports it" (1995, p. 368). By such a strategy, the putative ethics expert would display the elements that Crosthwaite considers constitutive of ethical expertise. These include, first, skill in clarifying and analyzing relevant concepts and problems, as well as skill in constructing and assessing arguments and viewpoints (1995, p. 363). Second, Crosthwaite contends that ethical expertise requires knowledge of various philosophical questions, positions, and theories, as well as the assumptions, consequences, and criticisms of different philosophical positions. Third, Crosthwaite believes that ethics experts are committed to understanding different issues and views, to challenging the received view and underlying assumptions, and to using reason to evaluate and support beliefs. Reflecting these commitments, skills, and knowledge bases, ethics experts develop "a critically examined moral perspective from which some judgments on particular issues are likely to follow" (1995, p. 364). As evidenced in this essay's first section, ethics experts do not occupy a single critically examined moral perspective, but instead have developed several competing visions of the good from which differing, sometimes opposite, judgments about particular issues are made (recall arguments by Frey and Holmes, as well as the different positions of Ramsey and McCormick on pediatric research).[34] Similarly, families and mothers (and other social groups) embrace different substantive accounts of the good life and seek to inculcate those beliefs and values in their members. Recognition of the diversity of these moral perspectives creates the need for those consulting ethics experts to work through on their own the reasons given by the expert. Indeed the existence of such diversity and respect for pluralism imposes an obligation to do so. While a degree of dogmatism may be permissible within families (and within other groups that share a conception of the good), mothers have an obligation to train their offspring for social acceptability, independence, and pluralistic life outside the family. Similarly, ethics experts who purport to offer substantive ethical guidance, rather than a merely descriptive ethical account from a particular substantive moral view (e.g., that of the Ik or the Roman Catholic church), must provide a charted path that considers multiple moral perspectives while, perhaps, defending a position resting upon one in particular.

Mothers and ethicists best prepare their charges when they train them to give and evaluate reasons for ethical judgments. They must reveal their reasons for acting, advocating, advising, or arguing as they do. Even if mothers do not (or cannot) offer a clearly articulated argument for why honesty is the best policy or for not striking one's little brother, they best serve their charges if they offer some reasons, and indeed some sets of reasons rather than others. Ruddick writes:

> [T]he work of training can become ... a work of conscience: a mother aims to foster and then protect her child's conscientiousness — her ability to identify, reflect on, and respect the demands of conscience. ... [S]he must help them learn that they cannot count on her own or any other person's authority. Or rather, since adults and children seek and trust the authority of others, conscientiousness requires taking responsibility for judgments of trust while maintaining a respectful independence from the authority judged trustworthy (1989, pp. 116-117).

Part of training children for social acceptability in the most profound sense of ethical acceptability, as well as training children for future autonomous decisionmaking, involves teaching the reasons grounding various value judgments. It also involves teaching *that* ethical judgments are grounded in reasons. By relying on reasons — by explicitly entering into the domain of reason-giving — mothers and ethicists demonstrate that they gain their authority from the soundness of their reasons, not from their positions of power — physical or social. Reasons are open to all and are open to criticism. The strategy of relying on reasons as the source of one's authority and supplying those reasons accomplishes three ethical goals. First, this strategy provides an explicit pathway to children's/trainees' independence. Second, it recognizes their (nascent or actual) moral autonomy. Third, relying on reasons to ground one's authority demonstrates the fundamental compatibility between the authoritative role of expertise and moral and political equality.

Imagine a mother admonishing a big sister not to hit her little brother. One belief system and set of underlying values is advocated by explaining that it is unladylike for girls to display power over boys and that it is very important to be ladylike if a girl is going to be liked by others and thus be successful in the world. Another set of beliefs and values is advanced by explaining that it is almost always wrong to hit another person because hitting causes pain, and just as she would not want pain caused to her, she should not cause it to another. By supplying the basis for her admonition not to hit, the mother provides her daughter with resources to begin to judge the issue for herself and to weave and evaluate a web of interrelated values. Of course, the reasons a mother offers are not the only raw materials the daughter needs to reflect on the substance and source of her mother's ethical teaching. In the course of future adult reflection, the daughter may consider the influence of social mores and of familial and psychological dynamics both on her mother's beliefs and values and on her own. However, by offering reasons, mothers explicitly model that instead of being the inscrutable pronouncements of some Delphic Oracle, ethical judgments have a social history and intellectual provenance and that they are interrelated and open to criticism. Remember that Ruddick describes a feature of an appropriate strategy for seeking and relying on "authority" or experts: one who trusts an authority nevertheless retains responsibility for such trust. The authority's provision of reasons in support of her expert opinion, in combination with her credentials including her track record, provide grounds for such trust. In teaching their children how to rely appropriately but not blindly on experts or authorities, mothers prepare their children to make appropriate use of expertise, including ethical expertise, as autonomous adults.[35]

Moreover, as Tong suggests, offering one's own reasons in advocating for one's ethical judgment, treats the other — one's child, trainee, or consultee — as an agent in his own right. At times, of course, a mother must largely ignore the budding agency in her children and shout "No" to preserve them from harm, or respond "Because I said so" to bend her children's wills toward socially acceptable behavior when their immaturity or circumstances render appeals to reason unwarranted.[36] Clinical ethics experts may benefit from similarly recognizing that sometimes their trainees or consultees are not in a position to be engaged in the discourse of reason-giving. Sometimes students are too worried, frustrated, or tired to engage in ethical reasoning. Clinicians may need emotional support before they can benefit from ethical exploration, and at times of crisis family members cannot absorb the charted course through moral terrain that we are trained to provide. For this reason among others, ethics consultation is not like other consultations requested in health care settings; it typically must take place during the course of multiple interactions during which consultees express different needs, only some of which involve assistance in ethical judgment. For several reasons, however, clinical ethicists should be reluctant to offer an opinion backed solely by an implicit "because I said so." First, unlike mothers, we do not see our charges day-after-day over a period of years. We may have only one "teachable moment" on an issue of importance. Second, for pragmatic reasons, it is critical to avoid being cast as ethical imperialists who issue edicts. Third, while mothers' young children are not autonomous moral agents, our medical trainees and those who consult with us in clinical settings are, or at least should be presumed to be, moral agents. Supplying reasons and arguments respects and engages their agency. Fourth, while mothers generally do not occupy the social role of giving strongly justified reasons within the domain of ethical questions — i.e., they are not ethics experts — clinical ethicists do occupy that social role. Demonstrating that ethical judgments rest on reasons is integral to the enterprise of ethics, and offering what Weinstein terms strong justification for what one advances as expert ethical opinion is integral to being an ethics expert.

Despite this final difference between mothers and ethics experts, an important parallel between the goals of their training persists. A mother trains her children in preparation for their future autonomous decisionmaking; she recognizes that they cannot rely upon her forever. She takes steps to achieve her future dispensability. It is a goal of healthy mothering that she will "be with" her children in an emotional and psychological sense through times of physical separation, into adulthood, and throughout their lives in a relationship that is variously described as healthy independence, mature dependence, or mutual dependence. For a mother to continue to "be with" her offspring is for her to have developed in them the values, intellectual skills, and self-concept that enable them to be self-nurturing and protective in the absence of the mother's active nurturance and protection, as well as to seek support from others in whom their trust may be appropriately placed. It is also a goal of healthy mothering that the content of her training — her vision of the good and "social acceptability," or the values that she instilled — will endure. Similarly, clinical ethicists seek to instill values and develop skills in their trainees that may be relied upon when the ethicists are not available, when there is no time to consult them, or when a decision about whether to consult an ethicist must be made. Indeed, it is a primary goal of clinical ethics training to develop trainees' capacity for moral discernment so that they will recognize when moral categories apply, when ethical analysis is demanded, and when reliance on ethical expertise may be necessary. It is a goal of ethics training to ensure that this capacity, as well as the processes of analysis, habits of mind, and ethical values will "be with" trainees when the ethicist is not. It is also a goal of ethics training to create a sense of "professional self" that makes it possible for clinicians to nurture the moral dimensions of their work and to seek support and consultation throughout their careers.[37]

When ethicists achieve their training goals, their training (and some consultation) services may be no longer needed, or at least may be needed less. Like mothers who find their culinary expertise redundant in their offspring's kitchens at Thanksgiving, clinical ethicists may find themselves invited less frequently to provide training in the services where they have trained clinicians most effectively. It is a goal of ethics training that trainees need to rely on ethics experts less frequently over the course of their training and subsequent practice. Precisely because ethical decisionmaking is not something that can be largely "turfed" to experts, it is necessary that clinicians, like all individuals and decisionmaking bodies, develop competence in ethics. Of course, clinical ethicists are not thereby rendered completely redundant or utterly dispensable. Particularly complex or apparently new questions will demand their expertise, just as tenacious lumps in the Thanksgiving gravy or a son's first turkey may occasion a "maternal consultation".

It is tempting to explore an additional parallel between the demands of Thanksgiving day and a typical hectic day in clinical care. In both cases the demands of the multiple tasks at hand, time constraints, and intense emotions fraught with personal and familial meaning make it tempting to call in the expert rather than attempt to address the challenges at hand by oneself. In the face of time pressures, for example, it may be tempting to call for an ethics consultation and leave it to the ethicist to conduct the family meeting, to discern the patient's values and preferences, and to attempt to resolve conflicts between people or between the patient's hopes and medical limitations.[38] Similarly, on Thanksgiving day in the face of multiple generations, too many dishes for the average oven, and conflicting idealizations about the meaning of Thanksgiving, it may be tempting to call in the family matriarch to sort it all out and get dinner on the table. Rather than abdicating responsibility and "dumping" these challenges on the experts, however, a more desirable course of action in both the culinary and clinical contexts is to collaborate. As Tong suggests, it is desirable to seek the advice of experts in order to avoid unnecessary mistakes, but to think things through for oneself in light of the experts' recommendations. Disavowing all responsibility and turfing Thanksgiving to a so-called expert is not an ethical breach; there is no obligation to "do Thanksgiving" oneself. On the other hand, clinicians do have an obligation to fulfill the moral dimensions of their work as well as its scientific and technical aspects. Moreover, ethical concerns do not come packaged with indicators like grocery store turkeys that pop up when the bird is cooked and safe to eat, but remain hidden when more roasting is necessary. Individuals must — for practical and ethical reasons — judge for themselves when a matter of ethics demands additional reflection.

It is by training their children for independence — by developing in their children conscientiousness or the capacity for moral discernment — that mothers avoid contradiction in their goals of instilling a particular vision of the good life while simultaneously encouraging and enabling their children to develop their own vision of the good. Similarly, by providing reasons in support of their expert opinions, ethics experts' work supports rather than undermines both individual autonomy and democratic ideals and processes. In "best case" versions of family and public life, and in clinical contexts, mothers and ethics experts render themselves somewhat redundant, dispensable in virtue of their success in training their charges for independence.

4. CONCLUSION

This essay has suggested that by examining the work that clinical ethicists actually do and by using a framework of maternal thinking to analyze the challenges such ethicists actually face, some of the puzzles plaguing the notion of ethical expertise may be addressed. The argument has taken us no closer to devising the test of adequacy for normative ethical theory that Frey and Holmes seek, but it has suggested that ethics experts may offer more than mere conceptual clarity and facility in constructing arguments as Singer proposed. It has suggested that *expertise* in ethics also demands more than living well; that *being an ethics expert*, in the sense of *performing* in that social role as clinical ethicists do, intrinsically involves giving reasons for one's opinions; and that the importance of giving reasons depends upon the nature of ethics, the nature of expertise in ethics (providing strong justifications), and the goals of the work of ethics experts. These goals and strategies to accomplish them have instructive parallels in the goals of mothers — namely, to protect, nurture, and train individuals for both social acceptability and independence. The intellectual and ethical nature of mothers' work may be obscured by its seemingly natural nature, its responsiveness to need, and its frequent failure to operate through typical ethical capacities of will and decision. In like manner, the intellectual and ethical nature of some of clinical ethicists' work may be obscured by its similarity to other work, especially that of mothers and spiritual or psychological counselors. Nevertheless, the hand-holding and consciousness-raising activities of clinical ethicists are ethical in nature both in virtue of their consequences and because of the circumstances that prompt them. C. D. Broad is wrong to suggest that wireless loudspeakers, or their contemporary equivalent, could guide clinical trainees, patients, and family members as effectively. Whether one must be a philosopher or have formal philosophical training in order to be an ethics expert remains an open question, though I suspect one need not. To be an ethics expert, particularly of the variety relied upon in clinical contexts, however, one needs to have the will and the skill to examine life in the philosopher's sense. Furthermore, one must also exercise the care and skill necessary to nurture life in the sense involved in maternal work.

NOTES

1. I shall generally take 'moral expert' and 'ethics expert' to refer to the same concept, and thus for the purpose of this essay, treat ethics and morality as one and the same. For ease of discussion, I would like to take 'moral philosophers' and 'ethicists' to refer to the same group of people; however, especially in the field(s) of medical ethics, bioethics, or clinical ethics, ethicists are often not trained primarily as philosophers. Therefore, while 'moral philosopher' will be reserved for those with training in philosophy, I intend to use 'ethicist' (and especially 'clinical ethicist') to refer to those whose primary training may be, but is not necessarily, in philosophy. It is perhaps worth noting and debating on another occasion whether to be a philosopher one needs to be trained in philosophy.

2. In a complementary discussion, philosophers of science and those in science studies — including Arthur Caplan (1980), Paul Durbin (1981), Michael Mulkay (1981), Dorothy Nelkin (1981), Kristen Shrader-Frechette (1980), and Marx Wartofsky (1980) — explored the fuzziness of the boundaries between theory and application, and between the normative-evaluative and descriptive-scientific. Richard Evans (2002) argues that participation of such professional bioethicists, for example on national commissions in the second half of the

twentieth century, narrowed public debate about bioethical issues from discussion of appropriate ends to consideration of means.
3. In fact, not only are strongly justified claims in other subdomains relevant, but also of importance is the strength of justification for higher order claims within normative ethics.
4. It is also possible to describe, for example, how Utilitarians would respond to Thomson's Trolley Problem (1986). Insofar as what one provides is an account of how those adhering to a particular normative ethical theory would respond to a particular problem, one is providing a descriptive account, albeit of what we take to be a paradigmatic normative position or theory, and one is displaying descriptive ethical expertise. It is presumably the ability to use, apply, or extend the theories of which one is capable of giving an account that is also a necessary condition of being an expert in normative ethics. This final point raises the question of whether normative expertise is only epistemic or also performative, and thus whether these two types of expertise are always conceptually distinct.
5. Indeed for someone to be an *expert* in descriptive ethics, we might think that she should have sufficient understanding of the moral concepts and norms of the group to be able to describe (predict) how they would or should "go on," to use Williams's notion, in a novel case or in a case substantially unlike those observed by the expert. Given our appreciation for consistency, we would say that in a novel case those being observed should make particular judgments so as to be consistent with judgments made in previously decided cases. Of course if the anticipated case is too different from previous cases then either our observer will be at a loss to predict how they should judge it — an expert may falter sometimes — or she will rely on her understanding of their normative and metaethical commitments to claim how they should judge it. In the latter case, the work of the observer/descriptive ethics expert blurs into constructing the group's normative theory, or into engaging in normative ethics, even if she would withhold her endorsement of the position she justifies using the group's moral commitments and concepts. Again, epistemic expertise in each of the subdomains of ethics seems interrelated.
6. To some degree Weinstein acknowledges this need and advises that, at the level of social practice, we must investigate the influence of differences in power and various biases and differences (e.g., of race, gender, and class) in shaping our views of what counts as strong justification.
7. Objections raised by Frey and Holmes would obviously apply here as well, but I shall not belabor those points. Weinstein understands performative ethical expertise as a relative concept, i.e., one's expert performance is relative to one's metaethical and normative commitments (1994, p. 70).
8. In the science fiction case in which the dancer's movements are controlled by aliens or, in the operatic context, by the mad scientist in Offenbach's *Tales of Hoffmann*, we might hesitate to say that the dancer is expert even if her moves were superior to everyone else's. Moreover, and this point is more controversial, if she had to take steroids or other illicit substances in order to achieve her expert moves, we might call her expertise into question.
9. Even those consequentialists who deny the relevance of motives or obligations to evaluation of the rightness of conduct must nevertheless acknowledge that the consequences of action supply reasons for the judgment of such conduct, and that anticipated consequences supply reasons to act (i.e., supply an ethical prescription).
10. Even more, to count as an ethics expert, there are particular sorts of sorts of reasons that she cannot supply — for example, "voices in the bed pillow tell me what to do," or "I exhibit good character because I hope one day to exploit the trust it inspires in those who observe me."
11. I refer to the lack of consensus regarding the good and our lack of a test for the adequacy of normative theory.
12. While I draw the parallels between mothering and the work of ethicists in clinical contexts, I believe that the comparison largely holds with ethicists and moral philosophers who engage in teaching and consultation in a variety of settings and indeed with a variety of professionals and nonprofessionals wherever what is of concern is analysis of issues of substantive morality in Holmes's sense.
13. Practically speaking, of course, most of those who engage in mothering work are women who are the gestational and social mothers of children. Some children have fathers who engage in mothering work; some children are in systems or settings that seek to approximate the mothering of actual mothers. Some children who have all of these — mother, father, extended family, and formal, paid care arrangements — nevertheless are not the recipients of much mothering.

As an initial parallel it is worth noting that not all who engage in clinical ethics work are clinical ethicists in terms of their professional identity or social role. That, for example, nurses, lawyers, patients, and mothers at various times engage in clinical ethics work does not suggest that such work is indistinct (i.e., that it lacks distinct goals and practices) or, *contra* Vertosick (1994) that there are not (or ought not be) clinical ethicists.

14. The more gender-neutral ideas of parenting and parental roles and responsibilities seem to combine mothering and fathering, to recognize that both sets of responsibilities must be fulfilled for the successful raising of children, and to acknowledge that male or female parents can fulfill either social role, or both. I do not disagree with these claims. However, to discuss the different sorts of thinking that are involved in the different types of work involved in caring for and raising children toward adulthood, it is useful to distinguish some work as mothering.

15. To elaborate, I do not want to reinforce aspects of traditional mothering that make an unrealistic ideal out of a lack of realistic respect, that make a virtue of selfless disregard, or that define a social role in terms that really reflect a failure of social support and the perpetuation of inequitable social arrangements. I do, however, want to acknowledge that mothering is not a mere useful construct but valuable work performed in the real world with a history as well as the possibility of conceptual analysis. Ruddick (1989) offers similar cautions in her second chapter.

16. Thus, we might commend as morally good those who, in the absence of a decision and active willing on their part embrace and rely on their "maternal instincts" to undertake provision of necessities to children or others in need. Because for many people, especially mothers, response to need seems nonvolitional rather than the product of a decision, such responses are attributed to "maternal instinct", hormones, or "hard-wired" responses to the "cuteness" of babies. It is possible to recognize that seeing a child as vulnerable, caring about that vulnerability, and therefore caring for a vulnerable child, is the result of a sort of moral discernment, without denying the possibility of biological contributions to such responses. The caring response is the simultaneous act and product of seeing the situation as one that demands care.

17. This list includes most notably Nel Noddings (1984; 2002) and Joan Tronto (1993). Tronto's analysis is especially instructive. First, Tronto suggests reasons that care work has been culturally devalued. Second, she resists the cultural assumption that care should be provided, in the first instance, privately, and that public, socially supported care should be provided only upon the failure of private caring (e.g., in the domestic setting). She advocates a social and political theory that takes care and responsiveness to need as central. Her theory thus explicates a social and institutional climate in which the work of professional ethics experts might be maximally warranted and beneficial.

18. Depending on institutional structures and mores, those who are properly said "to consult" with clinical ethicists may be patients and their families, or health care professionals. In conducting a consultation, e.g., in trying to resolve a conflict, however, the clinical ethicist will almost always consult with professionals and patients or family members. Note also that nurses or other nonmedical staff members, who may request an ethics consultation, may be considered vulnerable in virtue of their relatively lower status *vis à vis* physicians within the institutional hierarchy. Requesting such a consultation may heighten their vulnerability if in doing so they are implicitly or explicitly questioning the judgment or challenging the authority of physicians.

19. Despite growing attention to research ethics issues and increased availability of funds for ethics education and research, such initiatives continue to relegate ethical research to a position of "piggy-backing" on scientific enterprises, as when a small portion of the Human Genome Project's budget was set aside to fund the Ethical, Legal, and Social Implications Program, or when the Office of Research Integrity or the Office of Human Research Protections (formerly Office of Protection from Research Risks, within the Office of Public Health and Science) were established to provide federal oversight of research to address concerns that self-policing by scientists amounted to allowing the foxes to guard the hen house. Within this veritable barnyard of researchers feeding at the federal trough, not all animals are created equal, and in some cases ethics recommendations may be paid more lip-service than heed.

20. It is an open question worth exploring on another occasion whether the process and structure of some medical training programs "force" clinical ethics teaching into an oppositional strategy because of powerful interests in maintaining belief in false dichotomies and an almost complete disregard for what the humanities and even more qualitative social sciences may offer to medicine. Iris Young's discussion of how the "logic of identity" operates within strongly hierarchical oppressive systems is instructive on this point (1990).

21. The "hand holding" in which clinical ethicists engage is frequently not recognized as ethical in nature and as ethical work; however, responding to vulnerability is a moral response, and modeling such responsiveness is part of ethical training. The hand holding that clinical ethicists offer in the course of "case consultations" does not merely offer comfort; it is not constituted by supportive listening alone. It is first of all an act of ethical recognition of the other. It also enables clinical ethicists to glean ethically relevant information and to use their analytical skills to organize that information and draw connections among various points. Yet they *engage with* their consultees and trainees in this process. They "feel with" the parties to the problem and become invested in the problem's resolution because of the care they feel for those parties, while at the same time recognizing the independence of the other parties and the separateness of their interests. (See Noddings, 1984,

pp. 176-177.) That this psychologically demanding endeavor is a key task of clinical ethics may be overlooked if we focus too exclusively on the analytical methods employed and taught by clinical ethicists. If we focus on skills that are exclusive to philosophically-trained ethicists, we may ignore deeper humane capacities that clinical ethicists possess in common with others, but utilize in the health care context, and also model and attempt to inculcate in their students.

22. Clinical ethicists must guard against projecting onto patients and families the needs of the health care system — e.g., the need for efficient care delivery, improved outcome statistics, and reduced costs — and health professionals — e.g., the need to work miracles, to be highly esteemed, or to work through feelings of impotence or grief (see Wolf, 1996). Similarly, they must guard against projecting their own needs — e.g., for consistency, closure, or a generalizable ethical solution — onto consultees whose emotional and ethical needs they are supposed to serve. Finally, like all teachers, clinical ethicists may best avoid inappropriate projections of their own needs within the teaching context if they receive appropriate respect and support within collegial relationships and within the structure of the school and its curriculum.

23. On the very different result if one cannot accept difference and "let it be," rather than judging it unreflectively as bad-because-other, see Iris Young on the "logic of identity" (1990).

24. Christy Simpson (2002) employs Tronto's ethical framework of care to describe how attending to patients' "hope needs" is an ethical duty of clinicians, particularly nurses, who "have many opportunities to discover what patients hope for and to encourage patients whose hope has been challenged" (p. 90). She proposes that future research explore whether "health care practices destroy patient hope unnecessarily," whether a duty to preserve hope conflicts with other professional (ethical) duties, and how fostering realistic hopefulness in patients may enhance quality of life for those who give and receive health care.

25. Of course I am trading on two senses of 'ideal'. In one sense, as part of a morality of aspiration, for example, ideals inspire ethical conduct and development of moral capacities. Though unattainable in a pure state or all of the time, these ideals may be distinguished from ideals that deflect people from realistic possibilities, as when a person never enters a lasting relationship because she persists in seeking her ideal(ized) partner. In this latter sense, her idealization may be said to hinder rather than promote her pursuit of a realistic vision of the good life. If one holds such ideals regarding oneself, they may be used punitively, rather than inspirationally. See Noddings, 1984, p. 179, for how teachers should reflect to students "the best possible picture [of themselves] consonant with reality," not fantasy and not "expectations," but a confirming view of the students as admirable so that they may gain "the strength to become even more admirable."

26. This section focuses only on the general question of whether reliance on ethical expertise threatens democratic ideals. It does not take up the particular challenges and paradoxes faced by ethics experts advising public policy-making bodies raised, for example, by Dan Brock (1987), Alan Weisbard (1987), and Dan Wikler (1991).

27. If, however, after genuine attempts to understand the ethics expert's reasons, one still cannot, it may be reasonable to defer to the ethics expert's opinion based upon her reputation, her past performance (i.e., success in providing ethically sound opinions), and one's own past performance (e.g., in not reaching ethically appropriate decisions) and current experience (e.g., not comprehending the situation or having developed relevant values). In short, if one is "sufficiently out of one's depth," then it may be more ethically justifiable to defer to a trustworthy expert whose reasons one does not understand, than to muddle through on one's own (McConnell, 1984, 211).

28. Granted, the homeowner and taxpayer are legally liable, but they have legal recourse against the experts they hired, and the fault — the failure of expertise and any ethical fault that attaches to it — truly rests with the experts, so long as the experts were identified and consulted non-negligently. The homeowner and taxpayer retain responsibility for having trusted the experts, but so long as they chose the experts non-negligently, the homeowner and taxpayer appropriately discharged that responsibility (see discussion of Ruddick on trusting authorities below).

29. Evidencing respect for the agency of others by supplying one's reasons echoes an argument offered by Sarah Buss (1999) for the moral importance of manners. Practicing good manners in one's interactions, she argues, directly evidences respect for others and treats others in a way that acknowledges their intrinsic value or dignity. Similarly, I argue, supporting one's claims with appeals to reason implicitly acknowledges that one's interlocutor participates in the realm of reason-giving and is the sort of being appropriately persuaded by reasons, not one properly moved by brute force or manipulation.

30. My claims of dispensability and redundancy are somewhat overstated. The ethics expert may explain her line of reasoning so well that in a particular case she becomes redundant, but there will be new cases that call for her skill, knowledge, and strong justifications. Nevertheless, ethics experts should "transfer" expertise to others, or develop competence in others, in a way that architects, accountants, and other experts need not.

Interestingly, the role of ethics consultant has been compared to that of an architect by Margaret Urban Walker (1993) and by Susan Sherwin and Françoise Baylis (2003). Walker suggests that the ethics consultant designs "moral-reflective spaces in institutional life" (p. 38), what Sherwin and Baylis term the "necessary space and process for appropriately collaborative decision-making" (p. 145). In the clinical context, they point out, these spaces may be relatively private, while in public policy consulting, the ethicist-architect must ensure appropriately public access to the relevant deliberative discourse. Sherwin and Baylis argue that to create an appropriate site for moral deliberation and to design a sound ethical process for such deliberation, the ethics consultant must frequently move beyond conceptual clarification, inquiry facilitation, and mediation to become an advocate, for example to ensure that "people representing diverse interests are involved in the policy-making process, able to express their views and to have them taken seriously" (p. 147). In the clinical context, ethics consultants must reveal the ways that existing power structures constrain the options available and influence individuals' values and preferences, and they must seek to empower those who have difficulty advocating for their own interests (e.g., because of illness or their position in various power structures). The metaphors of architect and advocate, and the tasks that accompany these roles, invite comparison to the tasks of mothering – creating space for their children's diverse forms of flourishing and defending their children's interests when they are too immature to protect, or even recognize, their own. Moreover, avoidance of the pitfalls of advocacy that Sherwin and Baylis explore – the lure of paternalism, the risk of misunderstanding others' needs and interests, the problem of representing others accurately, and the replication of privilege and disempowerment by speaking for others – may be equally challenging for both ethicists and mothers.

31. This apparent paradox also plays a role in the concept of preserving a child's right to an open future, a notion introduced by Joel Feinberg (1980).

32. Reflecting their different disciplinary backgrounds and intellectual concerns, the authors cited on this point present arguments focused on different aspects and interpretations of the claim that bioethics advances its own particular agenda (Evans) or reflects a substantive, comprehensive moral doctrine at the level of policy (Ainslie), rather than seeking to ensure that patients and members of society may pursue their own visions of the good with regard to matters of health and health care. Schneider, for example, is concerned that bioethics may impose a "mandatory autonomy" doctrine on patients who would prefer to delegate medical decisional authority. Ainslie is concerned that bioethics blurs its two tasks of (i) extending comprehensive moral doctrines to questions arising from our biological natures, and (ii) developing policies for biomedicine and health care that accommodate the diversity of reasonable comprehensive doctrines constituting a liberal pluralist society. In the popular press, claims about the specificity of bioethics' agenda are more ideologically grounded (e.g., Shalit, 1997).

33. Indeed, developing the capacity for moral discernment may be *the* primary goal of maternal training; however it is *a primary* goal, but not the sole goal, of clinical ethics training. Within the norm-governed context of medicine and health care, a primary goal of ethics training is also to make trainees aware of — and likely to adopt — the specific ethical values informing the practice of medicine and provision of health care. There is controversy surrounding the content of this ethics training. Ainslie (2002) argues that the values informing health policy and policies governing professional practice cannot justifiably embrace a particular comprehensive moral doctrine. Others bemoan the deprofessionalization of medical ethics and argue for medical ethics as a professional ethics reflecting medicine's specific moral vision (see this thread reflected in an analysis of medical decisionmaking by Whitney et al., 2003, or more generally, Pellegrino and Thomasma, 1988). For insightful discussion of clinicians' resolution of conflicts between their moral commitments and prevailing standards in the profession, see Wicclair (2000) on grounds for conscientious objection.

34. The possibility of such competing conceptions of the good, and the necessity of a liberal society accommodating a plurality of such comprehensive doctrines, is the focus of John Rawls's political theory for contemporary democratic societies such as ours, a theory that Ainslie (2002) employs in his analysis of bioethics. See Rawls, 1993, esp. pp. 3-77.

35. See Ruddick (1989, pp. 118-119) on the conditions and outcomes of inculcating appropriate trust in children. See Baier (1986) on how appropriate trust underwrites the terms of a social contract.

36. Obviously children are sometimes too immature to appreciate the reasons that might be offered in support of an ethical judgment. Sometimes, however, circumstances make providing such reasons unwise, for example, when one is in church and the sermon is about to begin, or when the person who would be hurt by the child's thoughtless comment is within earshot, or when either the child or mother is too fatigued or upset to engage in respectful discussion.

37. It should also be a concomitant goal of clinical ethicists to work to create an institutional environment that nurtures such ethically-charged self-concepts of professionals and supports the ethical dimensions of their work. An example of such an environment would be one that rewarded trainees and professionals for giving

bad news well, telling the truth, and raising political and economic concerns, as much as it rewarded them for scientific discoveries, mastery of medical facts, or diagnostic expertise.

38. Empirical research might be undertaken to ascertain the effect of increased time pressures on the utilization of ethics consultation services. Based on personal institutional experience, I have some reason to believe that although it would seem reasonable for clinicians to rely on ethicists more during especially busy times, clinicians actually request fewer ethics consultations during peak periods or in the light of particular structural changes like understaffing or managed care-based constraints on clinical time. If this impression proved accurate, it would be important to determine whether clinicians lacked time to exercise their capacities of moral discernment to identify ethical needs, or whether they believed that they lacked time for ethics consultation.

REFERENCES

Ainslie, Donald C. (2002). 'Bioethics and the Problem of Pluralism.' *Social Philosophy & Policy* 19(2), 1-28.

Anonymous (1989). 'Could an Ethics Committee have Prevented Linares Tragedy?' *Medical Ethics Advisor* 5(6), 69-73.

Ayer, A. J. (1954a). 'On the Analysis of Moral Judgments.' In *Philosophical Essays*. London: MacMillan; New York: St. Martin's Press.

Ayer, A. J. (1954b). 'Forward.' In: *Ethics*. P. H. Nowell-Smith. London: Penguin Books.

Baier, Annette. (1984). 'Some Thoughts on How We Moral Philosophers Live Now.' *Monist* 67, 490-497.

Baier, Annette (1986). 'Trust and Antitrust.' *Ethics* 96, 231-260.

Broad, C. D. (1952). 'Conscience and Conscientious Action.' In: *Ethics and the History of Philosophy*. London: Routledge & Kegan Paul Ltd.

Broad, C. D. (1930). *Five Types of Ethical Theory*. London: Routledge & Kegan Paul Ltd.

Brock, Dan W. (1987). 'Truth or Consequences: The Role of Philosophers in Policy-making.' *Ethics* 97, 786-791.

Buss, Sarah (1999). 'Appearing Respectful: The Moral Significance of Manners.' *Ethics* 109, 795-826.

Caplan, Arthur L. (1980). 'Ethical Engineers Need Not Apply: The State of Applied Ethics Today.' *Science Technology, & Human Values*, 6(33), 24-32.

Charon, Rita (1996). 'Let Me Take a Listen to Your Heart.' In: *Caregiving: Readings in Knowledge, Practice, Ethics, and Politics*. Eds. Suzanne Gordon, Patricia Benner, Nel Noddings. Philadelphia: University of Pennsylvania Press.

Crosthwaite, Jan (1995). 'Moral Expertise: A Problem in the Professional Ethics of Professional Ethicists.' *Bioethics* 9(5), 361-379.

Durbin, Paul T. (1981). 'A Significant Limit on Applied History, Philosophy, and Sociology of Science and Technology.' *Science, Technology, & Human Values* 6(34), 18-19.

Dworkin, Ronald (1985). *A Matter of Principle*. Cambridge: Harvard University Press.

Engelhardt, H. Tristram Jr. (2002). 'The Ordination of Bioethicists as Secular Moral Experts.' *Social Philosophy & Policy* 19(2), 59-82.

Entman, Stephen S., Glass, Cheryl A., Hickson, Gerald B., Githens, Penny B., Whetten-Goldstein, Kathryn, Sloan, Frank A. (1994). 'The Relationship Between Malpractice Claims History and Subsequent Obstetric Care.' *Journal of the American Medical Association* 272(20), 1588-1591.

Evans, John H. (2002). *Playing God*. Chicago: The University of Chicago Press.

Feinberg, Joel (1980). 'The Child's Right to an Open Future.' In: *Whose Child? Children's Rights, Parental Authority, and State Power*. Will Aiken, Hugh La Fallette (Eds.). Totowa, NJ: Littlefield, Adams, 124-153.

Forrow, Lachlan, Arnold, Robert M., and Parker, Lisa S. (1993). 'Preventive Ethics: Expanding the Horizons of Clinical Ethics.' *The Journal of Clinical Ethics* 4(4), 87-294.

Frey, R. G. (1978). 'Moral Experts.' *The Personalist* 59, 47-52.

Graham, Loren (1981). 'Why Can't History Dance Contemporary Ballet? Or Whig History and the Evils of Contemporary Dance.' *Science, Technology, &Human Values* 6(34), 3-6.

Hickson, Gerald B., Clayton, Ellen Wright, Entman, Stephen S., Miller, Cynthia S., Githens, Penny B., Whetten-Goldstein, Kathryn, Sloan, Frank A. (1994). 'Obstetricians' Prior Malpractice Experience and Patients' Satisfaction with Care.' *Journal of the American Medical Association* 272(20), 1583-1587.

Holmes, Robert L. (1990). 'The Limited Relevance of Analytical Ethics to the Problems of Bioethics.' *Journal of Medicine and Philosophy* 15(2), 143-159.

Kasman, Deborah L. (2001). 'Physicians' Feelings: Toward a Balance in Medical Education.' *Western Journal of Medicine*, 174(1), 58.
Kasman, Deborah L., Fryer-Edwards, Kelly, Braddock, Clarence H. (2003). 'Educating for Professionalism: Trainees' Emotional Experiences on IM and Pediatrics Inpatient Wards.' *Academic Medicine* 78, 730-741.
Levinson, Wendy (1994). 'Physician-patient Communication: A Key to Malpractice Prevention.' *Journal of the American Medical Association* 272(20), 1619-1620.
Lidz, Charles W., Parker, Lisa S. (2003). 'Issues of Ethics and Identity in Diagnosis of Late Life Depression.' *Ethics and Behavior* 13(3), 249-262.
Locke, John (1690, 1980). *Second Treatise of Government.* Indianapolis: Hackett Publishing Company, Inc.
McConnell, Terrance C. (1984). 'Objectivity and Moral Expertise.' *Canadian Journal of Philosophy* 14(2), 193-216.
Mulkay, Michael (1981). 'Applied Philosophy and Philosophers' Practice.' *Science, Technology, & Human Values* 6(34), 7-15.
Murdoch, Iris (1971). *The Sovereignty of Good.* New York: Schocken Books.
Nelkin, Dorothy (1981). 'Wisdom, Expertise, and the Application of Ethics.' *Science, Technology, & Human Values* 6(34), 16-17.
Noble, Cheryl N. (1982). 'Ethics and Experts.' *The Hastings Center Report* 12(3), 7-9.
Noddings, Nel (1984). *Caring: A Feminine Approach to Ethics and Moral Education.* Berkeley: University of California Press.
Noddings, Nel (2002). *Educating Moral People.* New York:Teachers College Press.
Osborne, Lawrence (2003). 'Savant for a Day.' *The New York Times Magazine* 22 June, 38-41.
Pellegrino, Edmund D., Thomasma, David C. (1988). *For the Patient's Good: The Restoration of Beneficence in Health Care.* New York: Oxford University Press.
Pufendorf, Samuel (1682, 1991). *On the Duty of Man and Citizen According to Natural Law.* Cambridge; New York: Cambridge University Press.
Rawls, John (1993). *Political Liberalism.* New York: Columbia University Press.
Ruddick, Sara (1989). *Maternal Thinking: Toward a Politics of Peace.* New York: Ballantine Books.
Ruddick, William, Finn, William. (1985). 'Objections to Hospital Philosophers.' *Journal of Medical Ethics* 11(1), 42-46.
Ryle, Gilbert (1958). 'On Forgetting the Difference Between Right and Wrong.' In: *Essay in Moral Philosophy.* Ed. A. I. Melden. Seattle: University of Washington Press.
Scarry, Elaine (1999). *On Beauty and Being Just.* Princeton: Princeton University Press.
Schneider, Carl (1998). *The Practice of Autonomy: Patients, Doctors, and Medical Decisions.* New York: Oxford University Press.
Scofield, Giles R. (1994). 'Is the Medical Ethicist an "Expert"?' *Bioethics Bulletin* 3(1), 1-2, 9-11.
Shalit, Ruth (1997). 'When We Were Philosopher Kings.' *The New Republic,* 24-28.
Sher, George (1982).'Our Preferences, Ourselves.' *Philosophy and Public Affairs* 12(1), 34-50.
Sherwin, Susan & Baylis, Françoise (2003). 'The Feminist Health Care Ethics Consultant as Architect and Advocate,' *Public Affairs Quarterly*, 17(2), 141-158.
Shrader-Frechette, Kristin S. (1980). 'Technology Assessment as Applied Philosophy of Science.' *Science, Technology, & Human Values* 6(33), 33-50.
Simpson, Christy (2002). 'Hope and Feminist Care Ethics: What is the Connection,' *Canadian Journal of Nursing Research* 34(2), 81-94.
Singer, Peter (1972). 'Moral Experts.' *Analysis* 32(March), 114-117.
Smith, Wesley (2002). *Culture of Death: The Assault on Medical Ethics in America.* San Francisco: Encounter Books.
Snyder, Allan W., Mitchell, D. John (1999). 'Is Integer Arithmetic Fundamental to Mental Processing?: The Mind's Secret Arithmetic.' *Proceedings of the Royal Society of London B* 266, 587-592.
Szabados, Bela (1978). 'On "Moral Expertise".' *Canadian Journal of Philosophy* 8(1), 117-129.
Task Force on Standards for Bioethics Consultation (1998). *Core Competencies for Health Care Ethics Consultation.* Glenview: American Society for Bioethics and Humanities.
Thomson, Judith Jarvis (1986). 'The Trolley Problem.' In: *Rights, Restitution, & Risk: Essays in Moral Theory.* Ed. William Parent. Cambridge: Harvard University Press, 94-116.
Tong, Rosemarie (1991). 'The Epistemology and Ethics of Consensus: Uses and Misuses of "Ethical" expertise.' *The Journal of Medicine and Philosophy* 16, 409-426.
Tronto, Joan C. (1993). *Moral Boundaries: A Political Argument for an Ethic of Care.* New York: Routledge.

Veatch, Robert M. (1991). 'Consensus of Expertise: The Role of Consensus of Experts in Formulating Public Policy and Estimating Facts.' *The Journal of Medicine and Philosophy* 16, 427-445.

Veatch, Robert M. (1973). 'Generalization of Expertise.' *Hastings Center Studies* 1, 29-40.

Vertosick, Frank T. (1994). 'Ethics for (or out to?) Lunch.' *Bulletin* (September), 562-565.

Walker, Margaret Urban (1993). 'Keeping Moral Space Open: New Images of Ethics Consulting,' *Hastings Center Report* 23(2), 33-40.

Wartofsky, Marx W. (1980). 'The Critique of Impure Reason II: Sin, Science, and Society.' *Science, Technology, & Human Values* 6(33), 5-23.

Weinstein, Bruce D. (1993). 'What is an Expert?' *Theoretical Medicine* 14, 57-73.

Weinstein, Bruce D. (1994). 'The Possibility of Ethical Expertise.' *Theoretical Medicine* 15, 61-75.

Weisbard, Alan J. (1987). 'The Role of Philosophers in the Public Policy Process: A View from the President's Commission.' *Ethics* 97, 776-785.

Whitney, Simon N., McGuire, Amy L., and McCullough, Laurence B. (2003). 'A Typology of Shared Decision Making, Informed Consent, and Simple Consent.' *Annals of Internal Medicine* 140, 54-59.

Wicclair, Mark R. (2000). 'Conscientious Objection in Medicine.' *Bioethics* 14(3), 205-27.

Wikler, Daniel (1991). 'What Has Bioethics to Offer Health Policy?' *The Milbank Quarterly* 69(2), 233-251.

Williams, Bernard (1985). *Ethics and the Limits of Philosophy.* Cambridge: Harvard University Press.

Wolf, Susan M. (1996). 'Introduction: Gender and Feminism Bioethics,' in *Feminism & Bioethics: Beyond Reproduction.* Ed. Susan Wolf. New York: Oxford University Press, 3-43.

Young, Iris Marion (1990). *Justice and the Politics of Difference.* Princeton: Princeton University Press.

SECTION III: CONTEMPORARY APPLICATIONS

CHAPTER 11

ROBERT M. VEATCH

THE ROLES OF SCIENTIFIC AND NORMATIVE EXPERTISE IN PUBLIC POLICY FORMATION: THE ANTHRAX VACCINE CASE

1. INTRODUCTION

It is normal to seek out experts in formulating important public policies. They are prominent in government panels and commissions, in government agencies such as the Food and Drug Administration, and in many other groups that formulate public policies. In matters involving medicine and other sciences, the experts are normally selected because of their demonstrated scientific expertise, yet the questions the groups are expected to answer are invariably normative as well. The questions involve ethical and other kinds of evaluative judgments. Hence, one might demand of such groups not only scientific, but also normative expertise.

This chapter analyzes the roles of expertise in some scientific domains and expertise in bioethics and other normative questions in the formulation of public policy. I shall argue that, insofar as either kind of expertise exists, there is no reason to believe that expertise in one implies expertise in the other. That is, if panels of scientists are chosen for their scientific expertise, there is no reason to assume that the members should also have any expertise in bioethical judgment or any other normative expertise. In fact, I shall suggest that any groups of scientists chosen for their scientific expertise will tend to share normative commitments ("biases") that will tend to make their recommendations distorted. Moreover, I shall suggest that even the process of describing what are often taken to be the "scientific facts" will *necessarily* be influenced by the ethical and other normative and conceptual commitments the experts make and that, once again, the experts will tend to share normative and conceptual commitments ("biases") that will tend to make even their scientific conclusions distorted.

L. M. Rasmussen (Ed.), Ethics Expertise: History, Contemporary Perspectives, and Applications, 211-225.

2. THE INSTITUTE OF MEDICINE ANTHRAX VACCINE COMMITTEE

The claims made here are essentially theoretical. They will hold, so it is claimed, for any scientist or group of scientists, working for a government agency or panel or commission. The claims could be illustrated through the use of any example: the scientists employed by the NIH, CDC, or FDA; the members of any NIH Consensus Development panel; or the scientists employed by the Department of Defense or any other government agency. I have chosen to illustrate the analysis by appealing primarily to an important, recent Institute of Medicine group: The Committee to Assess the Safety and Efficacy of the Anthrax Vaccine (Joellenbeck, Zwanziger, Durch, and Strom, 2002).

An anthrax vaccine was first licensed in 1970. It was originally used primarily to protect veterinarians and workers processing animal products such as hair or hides. The Department of Defense vaccinated some military personnel beginning in 1991 and in 1998 made vaccination mandatory for all U.S. service members (Joellenbeck et al., 2002, p. 1). Persistent anecdotal complaints of harmful effects began to emerge and more than 400 military personnel refused to be vaccinated risking either court martial or retiring from the military (Weiss, 2001, cited in Joellenbeck et al., 2002, p. 3). This led Congress to direct the Department of Defense to support an examination of the vaccine by the Institute of Medicine.

The Committee to Assess the Safety and Efficacy of the Anthrax Vaccine convened in October 2000 (a year before the events of September 11, 2001, and the subsequent anthrax-related bioterrorism scare). It was a ten-member committee made up entirely of academic scientists except for one member whose credentials are listed merely as "scientific consultant". According to the report, the committee was selected "for their expertise in microbiology; vaccine research, development, manufacture, and evaluation; post-marketing surveillance of adverse events; regulatory and licensing procedure; epidemiology; biostatistics; immunology; and health surveillance" (Joellenbeck et al., 2002, p. 4). In addition to the ten members, two consultants were engaged, one an emeritus professor of pediatrics at the University of Pennsylvania and the other the Senior Manager of Biological Quality Control, Wyeth-Ayerst Pharmaceuticals. In addition, the Institute of Medicine engaged a six-member staff and twelve independent reviewers of the report, all of whom are employed at academic or industrial facilities or the National Institutes of Health or were retired from such positions. In short, the committee and its associated staff and reviewers were made up of the scientists from prestigious institutions that would be expected from the Institute of Medicine.

I know none of these individuals personally and have never had significant affiliation with any of their organizations. I assume they, as a group, are very competent scientists in the relevantly related fields and, as far as I know, have no personal axes to grind in the assessment of the anthrax vaccine. I provide this analysis on the assumption that they are all excellent scientists attempting to do the best possible scientific work. I shall nevertheless claim that it is the nature of the use of scientific experts in the public policy process that they *must* draw on personally-held normative and conceptual beliefs in order to complete the task that is assigned to them.

My analysis will proceed in two stages. In the first, I will assume, for purposes of discussion, that the committee and its associates collectively came up with the best possible description of the scientific facts related to the anthrax vaccine based on the best possible science. I will at this stage examine how scientific expertise interdigitates with ethical and other normative assessments, claiming that, even if one assumes that the facts are exactly correct, the committee's conclusions require normative judgments about which the committee members can claim no expertise. In fact, I will suggest that the committee's normative judgments should, in principle, be biased in certain ways.

In the second stage of the analysis I shall return to the question of whether the best scientific experts available should be able to at least produce the best possible compendium of the relevant scientific facts independent of their personally held ethical and other normative commitments and their culturally-determined conceptual formulations. I shall, once again, claim that, in principle, it is impossible for this committee (or any panel of scientists advising on matters of public policy) to describe the scientific facts without drawing on normative and conceptual commitments that come from outside the realm of science. I shall further claim that a different group of equally competent scientists who held significantly different normative and conceptual commitments should be expected to describe the scientific reality differently. None of this is meant in any way to be a criticism of this particular committee or any other committee, commission, or panel or public policy analysts; it is meant merely to state the necessary nature of public policy formation in the world of complex science.

3. NORMATIVE EXPERTISE AND SCIENTIFIC EXPERTISE: AN ANALYSIS

For the first stage of this analysis I want to assume, for purposes of discussion, that the Committee got the science exactly right and that somehow they could do so in a way that was value neutral. For a long period in the modern West, people have tended to believe in a fact/value dichotomy. Since the days of David Hume, matters of ethics and other normative judgments were seen as logically independent from matters of scientific fact. By the beginning of the twentieth century, treating matters of ethics as if they were natural fact was seen as committing the "naturalistic fallacy". For purposes of this analysis, the important implication is that ethical, normative, and policy choices could be seen as logically separate from the facts as determined by good science. Even if, in the real world, all scientists were somewhat less than ideal observers and describers of reality, value-free science was considered an ideal. This ideal was even extended to the social and cultural sciences (Weber, 1949). Often that ideal even crept into the related "science" of social policy as if it also could take the ideally value-free scientific facts and convert them into social policy without imposing a set of normative value judgments. In the first stage of the analysis which follows I wish to show that this set of assumptions is still pervasive even in the twenty-first century, but that it is logically doomed to failure.

4. NORMATIVE AND SCIENTIFIC EXPERTISE IN POLICY RECOMMENDATIONS

Appendix A of the Committee's report contains a short statement of the Committee's tasks:

> The issues include the types and severity of adverse reactions, including gender (sex) differences; long-term health implications; inhalational efficacy of the vaccine against all known anthrax strains; correlation of animal models to safety and effectiveness in humans; validation of the manufacturing process focusing on, but not limited to, discrepancies identified by the Food and Drug Administration in February 1998; definition of vaccine components in terms of the protective antigen and other bacterial products and constituents; and identification of gaps in existing research (Joellenbeck et al., 2002, p. 213).

I assume that any normal reader of this charge would view these as quintessentially scientific questions. What the types of adverse reactions are and whether there are gender differences seem like straightforward scientific questions, as do the other tasks enumerated. Even here there are some problems, however. For example, deciding that a reaction is "adverse" is fundamentally not a scientific task. While good biological scientists should be able to determine the incidence rates of a particular reaction in a carefully defined sample, deciding that the reaction is "adverse" takes one beyond the science. Even death may not be an adverse reaction in some people's worlds. (Oregon has recently legalized the use of a class of drugs the sole purpose of which is to kill people.) But aside from labeling certain reactions as good or bad, these are essentially scientific questions.

5. SAFETY AND EFFICACY AS VALUE JUDGMENTS

The "Statement of Task" for the committee goes beyond these issues, however. It asks the Committee to "make specific recommendations on technical aspects regarding safety and efficacy of the licensed anthrax vaccine" (Joellenbeck et al., 2002, p. 213). This may sound like a task still in the realm of science, but that conclusion would be a mistake. Deciding that an agent such as a vaccine (or an automobile or a rifle) is "safe" inevitably must require a range of normative evaluative judgments. Those judgments will be contingent upon judgments of the importance of the intended purpose of the agent as well as how negatively one evaluates the effects that are classified as undesirable.

The Committee made what it undoubtedly thought was a set of "scientific" conclusions and recommendations. It concluded that the product they were assessing was "an effective vaccine to protect humans against anthrax including inhalational anthrax" and that it was "reasonably safe" (Joellenbeck et al., 2002, p. 2).

These conclusions undoubtedly were reached after the Committee reviewed extensive scientific information about the effects of the vaccine. The labels of "effective" and "safe", however, are not themselves scientific. They are judgments imposed on the empirical findings in the light of the Committee's assessment of the importance of the benefits of the vaccine and how negatively the Committee evaluated the effects that were deemed undesirable. They were, most critically, evaluated in the light of the value and importance of the behaviors that warrant protection from anthrax organism exposure.

Consider first the behaviors that warrant protection from anthrax. We have seen that the original use of the vaccine was to protect veterinarians and workers in touch with animal products. Unless one makes some judgments about the value of domesticating animals for food, animal products, and human enjoyment, one cannot determine the importance of the role of veterinarian. Theoretically, one could avoid all risk to veterinarians, by eliminating the role. While I don't for a minute suggest this would be a wise policy, it might prove attractive to some radical opponents of the use of non-human animals for human purposes. Likewise, it is impossible to determine how important it is to protect workers with hides and animal hair without deciding the value of the products derived from these hides and hair.

More controversially, the value of the anthrax vaccine in the military setting is dependent upon the importance of the military objectives that are being pursued. While many people would have sympathy for certain recent military enterprises, the Gulf war, which was the first setting in which the anthrax vaccine was used, was not without its opponents. (The Persian Gulf Resolution authorizing American participation in the war that led to the use of the anthrax vaccine was approved by only three votes in the Senate and generated 183 opposing votes in the House (United States Congress, 1991).) People who believed that the Gulf war was not justified would surely prefer that they not be placed in a position in which they might be exposed to anthrax released by opposing forces. For them, since there is no net benefit in fighting the war, there is no good reason to take any risk with the vaccine. At the policy level, they would prefer avoiding the war altogether rather than trying to decide whether the risks of the vaccine can be tolerated. The point is not that the anthrax vaccine had particularly controversial uses; rather *every* public policy decision involves assessments of potential risks and benefits. No intervention can be labeled "safe" independent of the judgment that it is "safe enough for the objective being pursued." Safety is fundamentally a normative concept, not a scientific one. Any government panel of scientists rendering a judgment about safety is going beyond what their scientific expertise warrants. It converts scientific experts into experts in making ethical and other normative judgments. It involves what I long ago called the "generalization of expertise" (Veatch, 1973).

Extending this analysis to the current bioterrorism crisis, particularly as it relates to postal employees, deciding whether the anthrax vaccine is "safe enough" for an employee to use is dependent on how important it is for the employee to continue working in the post office as well as how acceptable alternatives to the present postal system would be. For the employee whose family is utterly dependent on the postal worker's job when the postal worker has no other likely means of employment, the vaccine may be safe enough. For the worker who has an equally attractive job option away from the post office or for the one who is torn between continued work and retirement, the vaccine may be far too dangerous. Likewise, from a policy point of view, if one thought the postal system replaceable by alternatives such as email, faxes, etc., that were free from anthrax risk, the risks of the vaccine would appear less justified.

Likewise, judgments about effectiveness require going beyond the science. To call the vaccine "effective to protect humans against anthrax" involves value judgments about what failure rates are tolerable morally and practically.

Contraceptives that are 99 percent effective (technically those with a Pearl Pregnancy Rate of 99) will produce one contraceptive failure in every one hundred cycles, that is, a pregnancy every one hundred months (eight years). Since women use these drugs for much more than eight years, a typical woman relying exclusively on oral contraceptives would have at least one unplanned pregnancy. Whether these drugs are "effective" or dismally ineffective, is a function of how bad it is to experience such unplanned pregnancies. While drug advertisers might want consumers to believe that one failure per eight years is "effective," some women might make other value judgments.

In the case of the anthrax vaccine, the Committee recognizes that it cannot conclude based on the evidence that the vaccine will be one hundred percent effective in preventing anthrax. For example, the Committee says it "did not include an evaluation of the threat from biological warfare agents in its purview" (Joellenbeck et al., 2002, p. 4). Presumably that could mean that it did not consider unusual dose levels and unusual routes of exposure that could come with intentional military uses of anthrax spores. Moreover, "consideration of the full range of topics concerning civilian use of the anthrax vaccine was beyond the purview of this report" (Joellenbeck et al., 2002, p. 4). Later the Committee goes on to point out that "efficacy is relative, not absolute" (Joellenbeck et al., 2002, p. 5). It also points out that in some species of animals, responses differ to different strains of anthrax and that "natural mutations or bioengineered alterations of the PA component of anthrax could result in vaccine-resistant strains" (Joellenbeck et al., 2002, p. 7).

None of this is meant to impugn the conclusions of the Committee. They make a case that one should make the value judgment that the vaccine is relatively effective for many of the envisioned uses. My point is that that conclusion is a value judgment, not a fact. The committee could also have concluded that "the vaccine has not been demonstrated to be effective against all military and civilian uses or against all presently existing and future strains of anthrax." Picking from among the long list of factually correct statements the one that properly captures the best value judgment is the essence of making public policy recommendations. It is not science.

The same problem arises in the Committee's recommendations regarding safety. Although the Committee called the vaccine "reasonably safe," it also reported reviewing "120 VAERS [Vaccine Adverse Event Reporting System] reports on serious adverse events associated with AVA [the vaccine in question]" (Joellenbeck et al., 2002, p. 11). It also heard oral testimony regarding adverse reactions. The Committee acknowledged local injection-site reactions "including redness, induration, edema, itching, or tenderness" (Joellenbeck et al., 2002, p. 11), but considered these comparable to those observed with other vaccines. It claims "no evidence that life-threatening or permanently disabling immediate-onset adverse events occur at higher rates in individuals who have received AVA than in the general population" (Joellenbeck et al., 2002, p. 12). It also concludes that "systemic events, such as fever, malaise, and myalgia, are associated with receipt of AVA" (Joellenbeck et al., 2002, p.

175), but notes that similar events occur with other vaccines. It finally observes that there is not enough long-term data to draw conclusions about health effects over longer periods (Joellenbeck et al., 2002, p. 14).[1] Once again, the point is not that the Committee has misrepresented any of the data or reached unreasonable conclusions. Rather the point is that aggregating the data into a summarizing phrase such as "reasonably safe" involves evaluative judgment. Someone else with a different set of values could have taken the same data, emphasized the documented short-term effects and the lack of long-term evidence to conclude "the vaccine has not been demonstrated to be safe especially in the long-run."

6. THE ALTERNATIVE TO SCIENTIFIC EXPERTS

The problem of concern here is that complex questions of public policy that involve obscure and labyrinthine scientific issues must necessarily rely on experts in the science for at least some parts of public policy assessments. Nevertheless, the experts in the scientific issues involved, in this case, anthrax vaccines, cannot be expected also to be experts in the normative questions upon which a public policy must rest. In this case the judgment about the approval for use of anthrax vaccine and its mandatory use in certain employment situations such as the military or (theoretically) the post office should not be decided on the basis of the particular and peculiar normative commitments of the experts in the science. Expertise in anthrax does not an expert in value judgments make. There is no reason to believe that vaccine scientists should be expert in the importance of fighting the Gulf war, whether immediate side-effects are worth it, whether long-term uncertainties are tolerable, or whether a certain percentage of failures to provide immunity is acceptable. These are all value judgments about which, if there is an identifiable expertise at all, it cannot be assumed to rest with those who are expert in the science.

In some cultures it is believed that there are experts in normative judgments. Priests, Talmudic scholars, philosopher-kings, shamans, or generalized wise men are delegated the authority to make the normative judgments upon which public policy rests. They could theoretically be the ones who make the value choices based on the data provided by the scientific experts.

Some people might believe, I suggest mistakenly, that in the modern secular world professional ethicists should be given this role of serving as experts in normative judgment. I am convinced that professional ethicists have an expertise. They are certainly skilled in moral analysis, in the history of ethics, and in what conclusions have been reached by past participants in moral disputes. There also seems to be an empirical basis for holding that those who survive as professional ethicists hold extreme positions less frequently, at least on some issues. There seem to be fewer professional ethicists who are admitted Nazis or blatant racists than exist in the general population. That may, however, simply be that survival as a professional ethicist who can command a paycheck for his or her work eliminates certain extreme views.

Hence, even though the distribution of views of professional ethicists may be somewhat narrower than the distribution in the general public, I know of no evidence that the consensus view of ethicists is superior or more correct on moral and other normative questions than the consensus view in the general public. In fact, I do not even know of any evidence that it *differs* from the consensus of the general public. In a democratic, pluralistic country it appears that we have no generally recognized expertise in moral and other normative judgments (or if there is some willingness to identify tendencies toward such expertise, its expression is suppressed by a commitment to the liberty of the individual and the egalitarian inclinations). As a matter of practice, then, there is no recognized expertise in ethics and other normative judgments even though there is an obvious expertise on matters of the science underlying public policy.

This suggests that on matters such as the anthrax vaccine in which there is a desire for accurate beliefs about the facts and a commitment to having democratic participation in the normative judgments involved, a society will have to use panels of scientific experts in some other fashion than expecting them to make policy recommendations. They might, for example, be relied upon as merely the source of accurate beliefs about the facts on the basis of which ordinary citizens, perhaps through democratically elected representatives, could make policy. The Congress and its subordinates such as the Department of Defense could thus use the Institute of Medicine's Committee not as a source of recommended policy, but rather as a source of objective information upon which Congress or its agents could then make policy. The panel of scientific experts provides "just the facts," not the policy decisions or even the policy recommendations. This would require the committee to avoid phrases such as "reasonably safe" and stick to collections of descriptions—for example, descriptions of the incidence rates of certain effects in certain populations.

7. NORMATIVE AND SCIENTIFIC EXPERTISE IN PROVIDING SCIENTIFIC FACTS

Unfortunately, the fact/value dichotomy and the division of labor implied in it has suffered a serious set-back in post-modern philosophy of science. The scheme in which scientific experts at the Institute of Medicine, the Food and Drug Administration, or the National Institutes of Health will provide the scientific facts upon which policy makers will then, through the democratic process, make normative decisions, is now seen as a naive, unrealistic separation of tasks. We have already seen that many tasks that the public as well as scientists think of as scientific—such as identifying what is a risk and what constitutes effective treatment—really incorporate a heavy overlay of normative judgments. It is impossible to determine that a drug is either safe or effective without making such normative commitments. In pharmacology, textbooks often label drugs as "medically indicated", that is, as "indicated" for certain conditions. However, saying a drug is "indicated" means nothing more than that, in someone's opinion, the expected benefits are greater than the expected harms and that the drug compares favorably to others on the market that could be used for the same purpose. Of course, knowledge about the expected outcome should be based on the best medical scientific evidence available, but that evidence can never tell whether the expected outcome is good or bad.

Since virtually all treatments can be expected at some level of probability to have many different outcomes, someone must make a value judgment about whether the outcomes that person considers benefits outweigh the outcomes that the person considers harms. Evidence-based medicine can, at most, tell us what the expected outcomes are likely to be, never whether those outcomes are good or bad. To say a treatment is "indicated" is to say that the speaker weighs the pluses as greater than the minuses. A treatment can never be "indicated" on the basis of evidence alone.

8. THE IMPOSSIBILITY OF VALUE-FREE SCIENTIFIC EXPERTISE

The latest generation in philosophy of science pushes the questioning of value-free and concept-dependent science much further. Stimulated by the Kuhnian critique that reveals how scientific descriptions are dependent upon world view (Kuhn, 1962) and concept construction (Kuhn, 1977, 1996, 2000) and by postmodernism in history and the social sciences (Kohler, 1966; Berger and Luckmann, 1966; Lyotard, 1984), there has been increasing agreement that scientific description cannot be isolated from normative and conceptual commitments. Even those using the same language are not necessarily saying exactly the same thing. The terminology may be incommensurable (Hoyningen-Huene, 1990; Veatch and Stempsey, 1995). When one realizes that acts of scientific description involve conceptual and evaluative commitments throughout the scientific enterprise, the possibility of a value-free science disappears. Values must first emerge at what I have called the "pre-scientific stage" in which topics are chosen that are believed to be *worth* investigating. They serve as the basis for choosing hypotheses, for choosing methods, for choosing which from an infinite array of data points are *worth* observing, for choosing which observed data are *worth* recording, for choosing which recorded data are *worth* analyzing, and for choosing which analyzed data are *worth* publishing.[2]

Once the data are published, we reach the stage at which the Committee to Assess the Safety and Efficacy of the Anthrax Vaccine finds itself. As a group of scientific experts, its members have an enormous array of data from which they may formulate their views. They have animal data, *in vitro* data, and human trials in many different settings using a wide range of agents. One of the problems facing the Committee is that the manufacturer of the vaccine, BioPort, was forced in 1998 by the FDA to validate its manufacturing process. It stopped manufacture for a while (Joellenbeck et al., 2002, p. 180). When it began again, manufacture was by a new company and under more strict FDA supervision. The FDA did not approve the license supplement for renovations, changes in labeling, or a modified package insert until January 31, 2001. This means that someone must make a value judgment about whether to rely on data generated from the older and now inadequate manufacturing standards or rely only on new data. Likewise, compulsory administration to military personnel provides data on safety and efficacy that may not be strictly comparable to administration to civilians who presumably may not meet comparable physical fitness standards nor receive comparable exposures. In short, the Committee must make judgments about how much weight different kinds of data should be given. In particular, it must decide how much can be deduced about the long-term risks—the risks that really worry many critics of the vaccine—when only short-term data are available from the manufacturing process currently being used.

The only acceptable conclusion is that, even at the level of the activities that almost everyone would consider to be mere reporting of scientific facts, the Committee must constantly make evaluative judgments about the weight to be given to different kinds of evidence as well as the importance of different findings. In the face of at least 120 anecdotal reports of serious side effects, it takes judgment on top of judgment to conclude that the incidence rate is no different from the rate of these effects in the general population. There is no doubt that a different committee of equally expert scientists could have, while viewing the same data, written a very different report that would lead a reader to have a very different impression of the facts. Even if the Committee had ceded to policy makers the task of making policy about whether the vaccine is adequately effective and safe, it would still have left the fingerprints of its values on the data it provided.

Imagine, for example, that all ten of the committee members, all six of the staff, and all twelve of the reviewers were divided into two separate committees, each made up of five members, three staff, and six reviewers. Imagine further that the division were made by constructing a questionnaire that measured all twenty-eight people's ethical and other values relevant to the issues before the committee. The survey might measure basic value orientations (how oriented to the future people were, whether they were "doers" or less activists, and whether they were generally conservative or liberal).[3] It might also measure more specific moral and nonmoral values—attitudes about the use of animals, American participation in the Gulf war, responses to current bioterrorist threats, and so forth. Now suppose one could align the members, staff, and reviewers along a continuum from the most supportive to the most critical of the uses a society might make of the anthrax vaccine. Then imagine that we could bifurcate the members, staff, and reviewers into two separate committees each given the same task, but neither permitted to communicate with the other. The interesting question is whether the two committees would write the same report and reach the same conclusions. The theoretical considerations I put forward here suggest they would not. Insofar as the conceptual and normative commitments of the two groups were different, they would provide different accounts of the scientific facts as well as different recommendations.

The claim is not exclusively (or even primarily) that each group would let its biases corrupt the doing of ideal science. That would, of course, happen assuming the people involved are less than perfect scientists. The more important claim, however, is that in the reviewing of masses of scientific information (as well as in the formulation of policy recommendations), each group is necessarily forced to make evaluative choices about which data are interesting, important, and worth bringing to attention in the report. The incorporation of the group's values into the reports would be a necessity, not a mere corruption of the scientific process.

9. STRATEGIES FOR NEUTRALIZING THE VALUES OF SCIENTIFIC EXPERTS

Thirty years ago I demonstrated precisely this phenomenon by showing that physicians' beliefs about the scientific facts of oral contraception correlated with their moral views as well as with their clinical practices about such matters as initiating conversations about birth control with patients (Veatch, 1976). Physicians were, in effect, saying to patients, "I won't impose my personal values on you, but I can tell you that the birth control pill is dangerous (or safe)." The fact was that those who believed it was safe also believed it was moral while those who believed it was dangerous also believed it was immoral. Further analysis revealed that these physicians were generally not merely concluding that the pill was immoral because it was dangerous (or moral because it was safe). Rather they "recoded" their accounts of the facts. Opponents of birth control would emphasize the carcinogenic cell changes in the breasts of beagles in claiming the pill was dangerous while defenders of the pill would correctly claim that there was no existing evidence that the pill had ever harmed a human being. Both statements were essentially true—at least at the time. Both represented good science. The critical feature was choosing which data to emphasize.

In the setting of clinical medicine, individual patients might have the option of guarding against these necessary value transmissions in the presentation of the data by selecting a physician whose frame matched their own. True, identifying the physician's values could be very difficult for a patient. (Using religious affiliation or ethnic identity of the physician as a predictor of moral views on birth control turned out to be quite unreliable.) Nevertheless, a patient had the option of selecting the slant that her physician would give to the data. Liberal patients could try to pick liberal physicians who would presumably slant the data the same way that the patient herself would slant it, if only the patient could acquire the pharmacological expertise.

In a national panel of scientific experts the citizens cannot individually select the frame they want their experts to bring to the data. Some other strategy will have to be found. I can envision three strategies for protecting against this inevitable impinging of values on the scientific consensus of experts. Public policy formulators might rely on multiple panels, use political criteria for selecting panelists, or adjust (discount) the expert opinions rendered regarding the descriptions of the scientific facts as well as the normative policy recommendations. Each is controversial and problematic.

9.1. Multiple Panels

I have already suggested one strategy: the use of multiple panels. After the panels have been selected in the traditional way, they can be split into two or more sub-committees that could work independently to provide descriptions of the relevant facts. The policy makers can then use the varying estimates of the facts in formulating public policy. If the sub-groups are divided on the basis of relevant normative commitments, the policy makers would then have some estimate of how sensitive the descriptions of the facts are to the normative commitments of the scientists. If the two panels provide essentially the same estimates, the policy makers can be more confident that the descriptions of the facts are those that would be produced by competent scientists possessing a wide range of value persuasions. The problem, of course, is that one can never know whether another panel of equally competent scientists could have been constructed that would have represented still another value perspective. If, for example, the anthrax vaccine committee had been divided based on values related to the importance of the military, animal husbandry, and antiterrorist uses of the vaccine, it might be possible to construct an additional panel of competent scientists who are even more skeptical about the value of the uses of the vaccine or one made up of scientists who believe even more strongly in its value. A group of competent scientists employed by the military, for example, might tend to be more sympathetic to military uses of the vaccine and less worried about abuses of civil rights implied in its compulsory use.

It should be clear that, if the scientists on the various panels are competent, it would be a mistake to assume that the "average" or mean estimates of the panels are the most reliable. In fact, all panels may provide nothing but accurate accounts of the facts based on mutually acceptable, high-quality data. The panels may simply choose to give greater weights to certain data or they may choose concepts and language that incorporate their value perspectives. One panel might, for example, claim there is no evidence the vaccine produces long term harm while another might claim that it cannot rule out that there are long term dangers. Of course, the same data would support both statements. One panel might strongly emphasize side-effects documented in nonhuman animals while another might de-emphasize such data, claiming that there is no proof that the nonhuman animal data can be transferred to humans. One panel might, as the Institute of Medicine panel did, press for more studies on the macaque and rabbit (Joellenbeck et al., 2002, p. 6), claiming that human trials would be unethical, while another might attempt to justify use of human volunteers and a third might consider use of even these nonhuman animals immoral. Multiple panels can disagree on both descriptions of the relevant facts and policy recommendations *without contradicting one another's data*. There can be multiple correct accounts of scientific observations. There can be, as William Stempsey (2000) has argued, a "value-dependent realism" that permits culturally-based differences in accounts without rejecting realist metaphysics.

9.2. Political Criteria for Selecting Experts

Another problem with using multiple panels is that it is certain to increase the cost of obtaining scientific expertise. If a single panel is all that can be afforded (or if a government agency such as the FDA must rely on single scientists performing their public tasks), another option would be to select the scientific experts from among those deemed technically competent, but intentionally impose normative or political criteria on the choice. That, of course, sounds like heresy. The norm of modern science is that good science should be free of political and other evaluative considerations. The conclusion of this analysis and of the growing body of postmodern philosophy of science is that science, even the best, most competent science, cannot be freed from normative and conceptual commitments. The very choice of language and the bestowing of meaning on terms so shapes the accounts of the scientific facts that some culturally-shaped normative and conceptual perspective must be reflected. It seems reasonable that the public policy maker, who is presumed not adequately competent in the science involved, would want a competent estimate of the facts for purposes for formulating policy. What is that person to do, however, if there can be multiple, competent accounts that are dependent upon normative and conceptual commitments of the scientists providing the accounts? One plausible answer is that, while insisting that the scientific descriptions be technically competent, the scientifically ignorant policy maker would want the account that he or she would himself or herself develop if only he or she had the time and resources to become a competent scientist in the relevant fields. That is, the policy maker would reasonably want the competent account that incorporates his or her own cultural norms and concepts. This suggests that, if the account of a panel of scientists must reflect some linguistic and conceptual commitments, the policy maker should arrange that the panel, while maintaining technical competence, should on average reflect the commitments of the society for which policy is to be made. This might mean recruiting some panelists to the left of the societal norms and others to the right, but that the mean view of the panelists ought to reflect the norms of the society. This might not be the same as the average views of the competent group of relevant scientific experts. When the societal normative mean and the scientists' normative mean differ, a strong case can be made that the panel ought to cluster around the societal mean and the policy makers ought to use the estimate of the relevant facts that would come from competent scientists reflecting that mean rather than the estimate of the relevant facts that would come from a panel made up of competent scientists who cluster around the mean norms of the relevant scientific community.

9.3. Discounting of Expert Opinion on the Science

This approach may prove too threatening to both the policy makers and the scientists involved. They may not be able to grasp that it is logical and acceptable for cultural and personal norms and concepts to shape the way that scientific accounts are constructed and that multiple accounts can be consistent with the facts.

If multiple panels are too expensive, and a panel skewed based on political and other normative criteria is too threatening, then the final option must be considered. A society could rely on the community of scientists to select panels in the traditional manner. For instance, the Institute of Medicine, as a representative group of the society's most competent scientists, could select its own panels using its own traditional criteria. We now should understand that any such panel should, on average, reflect the norms and concepts of the relevant scientific community and that those norms and concepts should shape the account the panel will present to the policy maker (Veatch, 1991b). Moreover, those norms and concepts may not match those of the broader society for which policy is to be made. Using the society's norms would predictably lead to different formulations of the accounts of the scientific facts.

It would clearly be irrational for a policy maker, as a representative of a society that, on average, holds norms that differ from those of the community of scientific experts, to use the estimates of the facts delivered by the panel if the panel incorporates the norms and concepts of the scientific community rather than the broader society. Thus, if the panel has not been skewed in its selection to adjust for these normative differences, the only rational course would be for the policymaker to adjust or "discount" the panel's estimate of the facts. He or she should attempt to adjust the estimate of the facts so that it would become the estimate that would have been provided by a panel of competent scientists whose average values match those of the general public. This adjusting or discounting is obviously risky, perhaps more risky than skewing the panel before it undertakes its work.

In the end, perhaps all three of the recommended strategies should be considered and used in varying degrees depending on the resources available and the likelihood that the expert panel conclusions will be sensitive to the normative and conceptual commitments of the panel's members. In a liberal society, there is only one other option. Some public policies may, in fact, unnecessarily ride roughshod over individual freedoms. In the case of the anthrax vaccine, for example, there is no good reason why individuals should not be provided with accounts of the facts, as confusing as they are, and then permitted to decide whether to take advantage of the vaccine based on individually performed risk-benefit assessments. Even in the case of a vaccine for an infectious disease, if the vaccine is reasonably effective, the choice about vaccine use need not be made collectively. It is a mistake to assume that there is a public health risk if some people conscientiously refuse to make use of a vaccine. If the vaccine is reasonably effective, the only people in jeopardy from someone who refuses immunization are the refuser himself and fellow refusers. Those who have used an effective vaccine are not at risk from others who are not immunized. Any externalized costs of the behavior can and should be internalized. Even in the case of the military, it is hard to imagine any justification for compulsory use of a vaccine. Requiring troops to be immunized for anthrax presumably is defended as a public good because it will protect the public from losing military personnel to illness, but it is logically muddled to claim that we know we can prevent losing troops to illness from a vaccine that has not been approved for use by the FDA and has not yet met the approval of a committee of scientific peers such as the IOM committee. Once the FDA and the IOM have endorsed the claim that the vaccine is adequately safe and effective and once those value judgments have been confirmed by policy makers who have adjusted the scientists' estimates of the facts, very few reasonable people should be willing to forgo the vaccine that is deemed safe and effective in preventing a disease that most would consider much worse than participation in their military assignments. Imposing compulsory immunization in the name of fighting for individual rights and liberty is ludicrous. Deciding that the vaccine is adequately safe and effective and giving people the right to use it is a plausible policy. It is not plausible simply because a panel of IOM or FDA scientists say so, but it is plausible if policy makers—civilians in control of the Department of Defense, the Post Office, and the National Institutes of Health—say so, based on estimates of the facts from scientific experts that has been properly adjusted and otherwise controlled for the differences in normative and conceptual values that can be presumed to exist between the community of scientific experts and the general public. In a society that rejects claims for identifiable ethical or normative expertise, there is no alternative.

NOTES

1. I have intentionally highlighted the Committee's citation of effects most people would consider harmful since the Committee's summarizing conclusion was that the vaccine was "reasonably safe". One could also comb the report for more positive observations.
2. For my early analysis of the ways in which values penetrate the doing of science see Veatch, 1976.
3. Such scales have been constructed (Kluckholn and Strodtbeck, 1961) and have been shown to relate to basic values and attitudes (Veatch, 1976).

REFERENCES

Berger, P. L., & Luckmann, T. (1967). *The Social Construction of Reality*. New York: Doubleday.
Hoyningen-Huene, P. (1990). 'Kuhn's Conception of Incommensurability.' *Studies in History and Philosophy of Science 21*, 481-492.
Joellenbeck, L. M., Zwanziger, L. L., Durch, J. S., & Strom, B. L. (Eds.) (2002). *The Anthrax Vaccine: Is it Safe? Does it Work?* Washington, D.C.: Institute of Medicine.
Kluckhohn, F. R., & Strodtbeck, F. L. (1961). *Variations in Value Orientations*. Evanston, Illinois: Row, Peterson and Company.
Kohler, W. (1966). *The Place of Value in a World of Facts*. New York: Mentor Books.
Kuhn, T. S. (1962). *The Structure of Scientific Revolutions*. Chicago: University of Chicago Press.
Kuhn, T. S. (1977). *The Essential Tension: Selected Studies in Scientific Tradition and Change*. Chicago: The University of Chicago Press.
Kuhn, T. S. (1996). *The Structure of Scientific Revolutions, Third Edition*. Chicago: University of Chicago Press.
Kuhn, T. S. (2000). *The Road Since Structure*. Chicago: University of Chicago Press.
Lyotard, J. (1984). *The Postmodern Condition: A Report on Knowledge*. Bennington, G. & Massumi, B. (Trans.). Minneapolis: University of Minnesota Press.
Stempsey, W. E. (2000). *Disease and Diagnosis: Value-dependent Realism*. Dordrecht, The Netherlands: Kluwer Academic Publishers.
United States Congress. 'Authorization for Use of Military Force Against Iraq Resolution' [The "Persian Gulf Resolution"]. Joint Resolution, January 12, 1991.
Veatch, R. M. (1973). 'Generalization of Expertise: Scientific Expertise and Value Judgments.' *Hastings Center Studies* 1, 29-40.
Veatch, R. M. (1976). *Value-Freedom in Science and Technology*. Missoula, MT: Scholars Press.
Veatch, R. M. (1991a). *The Patient-Physician Relation: The Patient as Partner, Part 2*. Bloomington, IN: Indiana University Press.
Veatch, R. M. (1991b). 'Consensus of Expertise: The Role of Consensus of Experts in Formulating Public Policy and Estimating Facts.' *The Journal of Medicine and Philosophy* 16, 427-445.
Veatch, R. M., & Stempsey, W. E. (1995). 'Incommensurability: Its Implications for the Patient/Physician Relation.' *Journal of Medicine and Philosophy* 20, 253-69.
Weber, M. (1949). 'Objectivity in Social Science and Social Policy.' In: *The Methodology of the Social Sciences* (pp. 49-112). New York: The Free Press.

CHAPTER 12

KENNETH CUST

PHILOSOPHERS RETURN TO THE *AGORA*

It was sudden. It was miraculous. Nobody knows why it happened. But on a small rock-bound Mediterranean peninsula 2,500 years ago a handful of people called Greeks roused the human race to a new ambition and sense of purpose and launched it into history.
- *Life*, January 4, 1963, p. 28.

Notwithstanding the third claim above, the reason for this incredible feat can be traced to the earliest Greek philosophers and their contributions to society as idea, knowledge, and conceptual engineers. That philosophers and philosophizing can make significant contributions to human life in general and individual lives in particular is, at least for some, not surprising. Since the turn of the century, owing in some part to the decline of scientific and logical positivism, as well as an increased demand by a better educated public for moral advice, philosophy has taken on some new directions. In particular, some philosophers have increasingly been called upon to focus their philosophical investigations on practical issues that have arisen in the areas of medicine, nursing, reproductive technologies, and other health related sciences, business, law and criminal justice, engineering, the environment, journalism, and other professional disciplines.

Academic interest, as well as public interest in applied philosophy, can be gauged by the increased demand for philosophical consideration of issues across disciplines. This increased demand has led to an increase in the number of academic centers, journals, and societies devoted to concerns in applied philosophy. Since the early 1970s the demand for applied philosophy has led to the creation of at least 40 new centers in applied philosophy, over 20 new journals, and more than a dozen new societies related to applied philosophy. Finally, in 1988 the first Ph.D. program in applied philosophy was established at Bowling Green State University.

In addition to the increased demand for applied philosophers, traditional philosophers experienced a decline in the number of academic job opportunities available to them. Tenure track positions were difficult to find and non-tenure track positions held little promise for the future. Owing to these as well as other

L. M. Rasmussen (Ed.), Ethics Expertise: History, Contemporary Perspectives, and Applications, 227-241.

considerations, the American Philosophy Association (APA) published a small booklet in 1984 entitled *Careers for Philosophers*. The authors of this booklet concluded that

> It seems reasonable to expect that the number of people receiving Ph.D.s in philosophy who eventually choose non-academic careers in philosophy may soon exceed the number who eventually choose teaching careers. Whether the philosophers in the first group come to be recognized as an important philosophical resource, or will choose to downplay their philosophical identifications, remains to be seen. It is quite possible, however, that the kinds of non-academic jobs that philosophers and students of philosophy find – and create for themselves – and the ways in which they make their philosophical training and capacities known will establish new models of what it means to be a philosopher and of what a career in philosophy can be.[1]

The trend towards applied philosophy has continued unabated. In the mid-1990s we find philosophers employed in a wide variety of non-traditional positions. While much of applied philosophy is concerned with ethics, in particular bioethics, business ethics and environmental ethics – for example, in 1994 there were 38 graduate and post-doctoral programs in bioethics and the medical humanities in the US alone – there are also philosophers applying their philosophical knowledge and skills in other areas as well: computer programming and computer science, banking and investment companies, prisons, and as policy writers at the national, state, and local levels. To give two examples of this trend, a colleague of mine has just completing his second Ph.D. in computer science and a former colleague of mine recently accepted a non-academic position at a Center for Clinical Ethics at a well-known medical center.

The number of philosophers who are currently working outside of the academy and the number of those who have seriously thought about doing so seems to be on the rise. The APA, in a recent report entitled "What Philosophy Ph.D.'s are Doing," surveyed 8300 philosophers in 1995. Of those surveyed 2.3% were unemployed, 6.9% worked for a private company, 4.4% were self-employed, 4.6% worked for a non-profit organization, and 4.1% worked for government. This means that in 1995 there were 190 unemployed philosophers, 572 philosophers were working for a private company, 365 were self-employed, 381 worked for a non-profit organization, and 340 philosophers worked for the government. While it would be interesting to know what those 365 self-employed philosophers were doing, unfortunately the APA report does not provide this information.[2] In another report, the APA's "Candidates Per Job Advertised" report – in 1994-95 there were 2.3 candidates per job advertised. This was also true for 1996-97. The only year the number of candidates per job advertised was higher, according to the APA report, was 1995-96; that year there were 2.6 candidates per job advertised. As any philosophy Ph.D. who has recently tried for an academic position in philosophy knows, not only is the competition tough but universities are simply not hiring as many academics into tenure track positions. In 1995 "51 per cent of the full-time professors hired" (Schneider, 1998, p. A14) across all disciplines were not hired into tenure track positions. In addition, there have been numerous reports concerning the continued feasibility of tenure itself. Finally, in another recent APA survey more than 50 percent of philosophy job candidates answered yes to the following question: "Have you ever seriously considered pursuing a career outside the academy?"[3] While this evidence is not conclusive, it does provide some explanation for why some philosophers are leaving the academy and returning to the *agora*.

While this canvassing of empirical evidence about what philosophers are doing outside of the academy is by no means exhaustive, it does represent a broad range of some of the non-traditional activities that some philosophers are engaged in and why they might be seeking nontraditional avenues of philosophical activity. One should not, of course, conclude from this that only non-academic philosophers are applied philosophers. Furthermore, while some of these activities require perhaps less than one's full philosophical knowledge and skills than others, this also should not detract from the main point being made: academically trained philosophers are beginning to apply their philosophical knowledge and skills in non-traditional ways. Philosophical counseling, understood as an area of applied philosophy, is one of the activities that philosophers are exploring outside of the academy.

1. PHILOSOPHICAL COUNSELING

We could understand philosophical counseling to mean something like a particular philosopher who, using his philosophical knowledge and skills, engages in a beneficial interaction with another individual, family, or group in an attempt to assist the latter in resolving some particular philosophically related problem. Understood in this narrow sense, philosophical counseling is similar to what people think of when they think of traditional psychotherapy; that is, an individual psychologist, social worker (or some other appropriately qualified individual) who uses the particular knowledge and skills of his or her discipline in an attempt to help another individual, family, or group resolve a particular non-philosophical problem. However, if we limit our account of what constitutes philosophical counseling to the account suggested above we will have eliminated a significant number of people who thought that what they were doing was philosophical counseling. In order to avoid this problem I stipulate, at least for the purposes of this paper, that the concept 'philosophical counseling' only includes those philosophers who are in fact counseling particular individuals, families, or groups. For those philosophers who are engaged in similar activities but at the institutional or social level I stipulate, again only for the purposes of this paper, that their activities be referred to as, for lack of a better term, 'philosophical consulting'. By stipulating the meaning of these terms I do not mean to suggest that if one is a philosophical counselor that precludes her from being a philosophical consultant, or *vice versa*.

The question (Is there a difference between philosophical counseling and traditional psychotherapy?) first arose, at least for me, at the July 1994 First International Conference on Philosophical Counseling, held at the University of British Columbia.[4] While there was considerable discussion about what philosophical counseling was, there seemed to be no clear consensus on what, if anything, distinguished philosophical counseling from traditional forms of psychotherapy. The discussions that took place at this conference concerning the difference between philosophical counseling and traditional forms of psychotherapy were motivated by two types of concerns, instrumental and intrinsic.

The main instrumental concern expressed during the conference was: what is this new discipline of applied philosophy called philosophical counseling? The motivation behind this concern was pragmatic; potential practitioners of philosophical counseling wanted to know and understand what constituted philosophical counseling so they could decide if this was an area that they might pursue or, at least for some people, whether what they were doing was philosophical counseling. Another instrumental concern centered on possible training methods for philosophical counselors. These kinds of concerns were raised during some of the discussions that took place after the workshops that demonstrated the practice of philosophical counseling. Finally, some people voiced concerns about the potential dangers of philosophical counseling given that philosophers were not that knowledgeable about psychopathologies.

In addition to these instrumental kinds of concerns, conference participants were interested in discovering the theoretical foundations of philosophical counseling for more intrinsic kinds of reasons. How should we understand the relationship between philosophical counseling and traditional forms of psychotherapy? How should we understand the relationship between the different approaches taken by practicing philosophical counselors? What is the difference, if any, between philosophical counselors and philosophers who merely counsel?

More recently, the book *Essays on Philosophical Counseling* was published. While some contributors to this volume undertook to describe their own particular approach to philosophical counseling, only one contributor attempted to explicate what was common among many (although not all) approaches to philosophical counseling. Ran Lahav argued that the principle underlying many current philosophical approaches to counseling is, "loosely speaking, that various aspects of everyday life can be interpreted as expressing views (philosophical, empirical, etc.) about oneself and the world" (Lahav, 1995, p. 5). While Lahav may be correct about the underlying principle that unites much of philosophical counseling this does not, at least by itself, distinguish philosophical counseling from traditional psychotherapy. However, he does go on to suggest that one of the things philosophers have to offer, that traditional psychotherapists do not, is the academic philosopher's philosophical skills: "those related to conceptual analysis, drawing implications, phenomenological descriptions, *et cetera*" (p. 10). We will return to the issue of the philosopher's tool, for the tools one employs as a practitioner of a profession plays an important role in distinguishing that profession from other professions, as well as distinguishing subdisciplines within a profession.

2. OBJECTIONS TO PHILOSOPHICAL COUNSELING

The problem of distinguishing philosophical counseling from traditional psychotherapy is further complicated by the fact that there is a version of psychotherapy called philosophical psychotherapy. William S. Sahakian, an American Professor of Psychology and Philosophy writing in the 1970s, defined philosophical psychotherapy as being "any system that utilizes beliefs, attitudes, convictions, a person's Weltanschauung or philosophy of life, as a vehicle to alter, control or cope with his psychological or emotional problems, thereby changing his behavior pattern, psychological or emotional state..." (Sahakian, 1979).[5] This definition of philosophical psychotherapy, as stated, is far too broad, for not only would most of the traditional psychotherapeutic approaches fall under it, it could also include any activity—such as teaching and/or advertising, to mention only a few possibilities—that used a person's beliefs or attitudes to change a person's behavior, psychological or emotional state. Furthermore, despite a century having elapsed since Freud initiated "talk therapy", there is still no widely accepted definition of psychotherapy.

In 1985, a conference entitled 'The Evolution of Psychotherapy' was held in Phoenix. The question, "What is Psychotherapy?" was discussed by several of the speakers. However by the end of the conference there was no definition of psychotherapy that was acceptable to all the participants. In 1987, the "What is Psychotherapy? project" was established to further investigate the question. Despite contributions from over eighty-one invited respondents, the authors concluded that "it may be impossible to compose an encompassing definition of our field" but nevertheless, they continued, "each of the approaches described in this book has something valuable to offer" (Zeig and Munion, 1990, p. 423).

While still in its infancy, philosophical counseling has generated a lot of publicity[6] and criticism. What is surprising is the response by some philosophers and non-philosophers to the idea that philosophers are practicing their profession outside of the academy and counseling individuals about specific philosophical problems they have in their lives and, in some cases, earning a livelihood from doing so. Roger Scruton, in an article in *The Times* (London, August 11, 1997) entitled "The Return of the Sophist," characterized philosophical counselors, at least as some practitioners have described it, as "sophists". "There are many signs of the sophist," Scruton writes, "but principal among them are these: mumbo-jumbo, condescension and the taking of fees. The philosopher," Scruton continues, "uses plain language, does not talk down to his audience, and never asks for payment" (1997). In a recent article in the *New York Times*

> Dorothy Cantor, a clinical psychologist and former president of the American Psychological Association, dismissed the idea that "philosophical counseling, or whatever the heck they're calling it," has a legitimate claim on dealing with "something as delicate as a person's mental health." Philosophers who consider themselves mental-health therapists, she said, suffer from a "naive assumption" that purely intellectual discourse can address personal problems that are intractably emotional and sometimes severely debilitating (Sharkey, 1998, pp. 4-1, 4-4).

Both of these criticisms are unjustified.

Scruton's criticism is unjustified for the reason that just because philosophical counselors and consultants charge a fee for their services does not make them sophists in the sense that Plato in the *Phaedo* (101e) and Aristotle in *Sophistical Refutations* took them to be. Aristotle said, "the art of the sophist is the semblance of wisdom without the reality, and the sophist is the one who makes money from an apparent but unreal wisdom" (*Sophistical Refutations*, 165 a22). If philosophical counselors are sophists in this Aristotelian sense, then Scruton's criticism has force.

However, philosophers can receive payment for their services without *ipso facto* being sophists. The reasons for this are several.

First, not all the problems a person suffers from over the course of a lifetime are psychological problems. People suffer from a broad range of problems: medical, legal, economic, psychological, and spiritual. But what many do not realize is that there are also philosophical problems. These problems can range from a search for meaning in one's life—in times of crisis or otherwise—to problem solving and decision making, from career-related difficulties to family problems and interpersonal relationships, and from marital problems to end-of-life decisions.

Since "talk therapy" began some one hundred years ago, virtually every kind of personal and interpersonal issue, including those mentioned above, have been psychologized. Psychotherapists now reign supreme in the domain of talk therapy. However, we should remind ourselves of Mark Twain's reported sage piece of advice: "If the only tool you have is a hammer, an awful lot of things will look like nails" (cited in Zeig and Munion, 1990, p. 2). That is, while many problems may look like psychological problems, and thus seem amenable to the tools therapists have at their disposal, not all of them are. Some of life's problems, especially those that wear and tear on the soul, are philosophical problems, and they can only be solved by reflection and critical thinking—that is, by philosophizing.

What philosophical counselors can bring to the metaphorical couch of the client is, in addition to the Socratic requirement of respect for the autonomy of the client, the education of the client on *how* to think, not *what* to think. If the philosophical counselor confines herself with teaching the client what to think, then Scruton's concern about sophistry may be correct. However, by teaching clients how to think, by assisting clients to critically scrutinize their beliefs and convictions with respect to the problems they are concerned about, by identifying assumptions and presuppositions that the clients are unaware of, by exploring with clients their conception of what constitutes living a genuine and worthwhile life, the philosophical counselor remains true to the Socratic notion of being a conceptual midwife (Plato, *Theatetus*, 150c-151d). They are not selling, to quote Aristotle again, "an apparent but unreal wisdom."

The second reason that philosophical counselors who are concerned with teaching their clients how to think, as opposed to what to think, and who charge for their services are not *ipso facto* sophists is that philosophizing can be, and often is, beneficial. William Sahakian reported that he noticed philosophy's therapeutic effect "in classroom lectures and discussions when students reported with amazing regularity that his lectures were more than the usual informative talks; they were psychologically therapeutic experiences" (Sahakian, 1979). In addition, William Lyons argued that one of the by-products of engaging in philosophical study is a gain in "mental hygiene". Lyons wrote

> A good dose of philosophy ought to make one more alert to the dangers of "wooly" thought, imprecise or foggy expression, and fallacious reasoning. It ought to implant in one, at best, a passion to avoid misleading rhetoric, empty crackling cant, and blind unreasoning prejudice. Contrariwise, it ought to give one a desire to be as clear and simple as possible in expressing one's views, as careful as possible in amassing relevant evidence in support of them, and as correct as possible in arguing for them. It should produce a clear head that thinks lucidly and

> expresses those thoughts precisely. Philosophy should not add to the congeries of tangled jargon-ridden nonsense that already clog the world, it should untie knots in our thinking. In short, the study of philosophy ought to be therapeutic (Lyons, 1990, p. 16; see also Segal, 1998).

A philosophical counselor who benefits her clients by educating them with respect to how to think and philosophize about their problems, even if she charges a fee for her services, is not a sophist, at least not in the derogatory sense intended by Scruton.

Dorothy Cantor's criticism is also unjustified and for several reasons. First, Cantor seems to be assuming either that people cannot have philosophical problems or that all personal problems can only be dealt with via psychological therapies. She is wrong on both counts. In addition, philosophical counselors do not consider themselves mental health therapists in a clinical or medical sense. Rather they consider themselves, first and foremost, professional philosophers who have the knowledge and ability to educate others on how to think and/or assist those others in critically thinking about what they believe, the assumptions underlying their beliefs and their justification for holding those particular beliefs. That one of the consequences of engaging in such activities helps others, in the sense of being mentally therapeutic or—as one popular series of books is titled—chicken soup for the soul, does not make the philosophical practitioner a mental health therapist. Furthermore, philosophical counselors do not have patients in the traditional medical/psychological sense; they have clients. And while some of these clients may manifest psychological or pathophysiological symptoms, it does not follow from this that the problems these clients suffer from are necessarily psychological or mental health problems. A client with serious legal problems, for example, may manifest psychological or pathophysiological symptoms but one would not argue that a psychologist, in lieu of a lawyer, would be the best person to assist in dealing with her legal problems.

Having said this one should heed the caution underlying Cantor's worry; that is, that philosophical counselors are not generally knowledgeable about mental health conditions and considerations. Thus private practitioners of philosophy should be concerned not only about the kind of problems their clients come to them with but also about the client's background, and especially their medical and psychological history. The prudent practitioner of philosophy should educate herself not only in psychology and psychopathology, but also in pharmacology. The reasons for this are several. Clients may approach a philosophical practitioner with a prior history of medical, mental, and/or emotional problems. Knowledge of the client's history will not only protect the client and the practitioner, it will also mitigate the chances of an unsuspecting philosopher being held hostage to the psychological manipulations that can occur when a client is indeed mentally ill and the practicing philosopher mistakes the symptoms of her client's illness as autonomous claims made by a mentally healthy person.

Most of the information a practicing philosopher needs to protect her client and herself can be obtained by having each client complete a client history form prior to the practitioner agreeing to accept the client into her practice. A client history form is similar to the kind of forms that physicians and hospitals use for new patients. In addition to asking for name, address, next of kin, medical and mental health history, the form should also contain questions concerning what medications the client is taking—or has taken in the past—whether or not the client is currently under the care of a physician or other clinical specialist, including psychotherapists and, if so, what is their diagnosis.

Since philosophers are not generally knowledgeable about psychology, let alone psychopathological conditions, the responsible practitioner of philosophical counseling should make an effort to learn about basic psychology, personality disorders, and some of the more common psychopathologies. However, simply having knowledge about what constitutes a personality disorder or psychopathological condition is insufficient, since a lack of knowledge about the kinds of symptoms via which such conditions could manifest would leave the unsuspecting philosophical practitioner in the position of not knowing when to refer a client to a more appropriate source for treatment.

Knowledge of pharmacology is also important, for it offers some evidence of the client's history. While some philosophers will have heard of such drugs as Prozac®, Valium, and Diazepam, few will have heard of Chlorpromazine, Haloperidol, or Thioridazine. The former are generally used to treat cases ranging from mild anxiety to mild depression while the latter are powerful antipsychotic drugs used to treat a variety of psychopathological conditions. In addition to knowledge of what conditions particular drugs are used for, the prudent philosophical practitioner should also be cognizant of the potential side effects and contraindications of common pharmacological agents used for the treatment of the mentally ill.

3. THE DIFFERENCE BETWEEN PHILOSOPHICAL COUNSELING AND TRADITIONAL PSYCHOTHERAPY

Philosophical counselors, like psychotherapists, also have something valuable to offer to clients who seek them out, but what exactly it is is as yet unknown. In the remainder of this paper I will attempt to explicate the essential difference between philosophical counseling and traditional psychotherapy and, in the course of doing so, suggest what it is that philosophers have to offer to prospective clients. I want to begin by considering the similarities and differences between five different (in general terms) occupations: carpenters, physicians, lawyers, psychotherapists, and philosophical counselors.

Let us first consider what each of the first three aforementioned occupations share in common. The first thing these occupations have in common is that they involve clients with a problem who approach—at least for the most part—a practitioner who is skilled in a particular area. For example, a homeowner would approach a carpenter to have him fix the hole in his roof; a person with a broken leg would approach the physician to put his leg in a cast; a person charged with a crime would approach the lawyer to defend him at his trial. In each of these cases there is a practitioner, a client, and a problem; of course, for each of these cases the credentials of the practitioner—in terms of training and education—the kind of client, and the kind of problem the client brings to the practitioner are different.

Now consider the tools used by each of the practitioners to resolve the client's problem. In the case of the carpenter and the physician, the tools are—again for the most part—concrete physical things; the carpenter uses a hammer, nails, and wood; the physician also uses things—drugs to control the pain, physical manipulation of the extremity prior to setting it in place, and the materials which compose the cast for setting the leg. In the lawyer's case, however, the tools used are not concrete physical things but more abstract entities, such as ideas, arguments, legal principles and precedents.

If we now consider psychotherapists and philosophical counselors we find that they also share some things in common with the occupations already considered; there are practitioners and clients with problems. The credentials of the practitioners differ and the problems of the clients are more like the problems of the lawyer's client than the problems of the clients of the physician and carpenter in that the problems are less about things that can be easily observed by others. This is not to suggest, for example, that a physician's patient could be ill but not observably ill, for obviously he can. Rather what I am trying to suggest is—again for the most part—that the problems suffered by the clients of each of these practitioners are of a fundamentally (but unspecified) different kind. Finally, the tools used by the practitioners of traditional forms of psychotherapy and philosophical counseling are more like the tools used by the lawyer than the tools used by carpenters and/or physicians in that the tools used by both the philosophical counselor and the psychotherapist are more abstract and conceptual than concrete. The tools employed by traditional psychotherapists extend across a broad range; they include everything from personality types to body types, from Freudian analysis to dream analysis, from Jungian archetypes to Roger's "unconditional positive regard," to name just a few. Furthermore, traditional psychotherapists take a variety of approaches in their clinical practices; some focus on behaviors while others focus on emotions; some focus on the past while others focus on the present and the future; some focus solely on the client while others focus on the relationship between the client and the therapist, while still others focus on the relationships that a client has with others.

In addition, the credentials of psychotherapists are as varied as are the tools employed, the clinical approaches taken, and the theories which underlie these varied approaches. In the "What is Psychotherapy? project" the authors distinguished, using theoretical orientation and treatment format as the defining criteria, nine categories of psychotherapy. Each category was then broken down into some eighty-one subcategories. Given this incredible array of psychotherapeutic approaches, it is no wonder that they were unable to reach an all-encompassing definition of psychotherapy. The lack of a definition of psychotherapy need not, however, deter us from determining whether or not there is a difference between philosophical counseling and traditional psychotherapy. For just as there are differences between the many forms of psychotherapy, there are differences between philosophical counseling and traditional forms of psychotherapy. There are differences in the credentials of the practitioners; there are differences in the types of clients and the kinds of problems such clients have. The goals of both the philosophical counselor and the traditional psychotherapist are for the most part similar; both aim at, generally speaking, bringing about some sort of positive change, via the use of the conceptual tools they have available to them, that is to the betterment of the client. Finally, the tools each brings to the metaphorical couch are different in a fundamental kind of way.

Given that one of the crucial differences between philosophical counselors and psychotherapists are the conceptual tools they use on behalf of their client, I now want to consider what some of those tools are and how a philosophical counselor could use such tools on behalf of her client. The tools a philosophical counselor has are mainly her philosophical knowledge of alternative theories in metaphysics, epistemology, and ethics, and her analytical and reasoning skills. Like traditional psychotherapists, philosophical counselors will employ different conceptual approaches with their clients. For example, Pierre Grimes (1998) of the Academy for Philosophical Midwifery, uses the Socratic dialectic as his principle tool in his philosophical counseling practice while Alex Howard (2000) takes a more eclectic approach by drawing from his knowledge of the history of philosophy and applying it to the problem a particular client brings to his practice. Bauke Zijlstra, on the other hand, employs Rawls' notion of reflective equilibrium (appropriately modified for the circumstances) to assist "in restoring the client's disturbed equilibrium—both his disturbed equilibrium of life and his disturbed equilibrium of thinking about life" (1997, p. 35). Finally, Andrew Gluck has proposed a model for the "relationship between counseling, psychology and philosophy utilizing Karl Jaspers' psychology of meaningful connections and Karl Popper's concept of World 3" (1997). These examples of how philosophical counselors draw upon their conceptual tools illustrates the scope and variety of tools available to the practicing philosopher; they are as numerous and as varied as the history of philosophy, in both the western and non-western tradition, has to offer.

4. TOOLS FOR THE PRACTICING PHILOSOPHICAL COUNSELOR

One approach I have used with some of my clients is based on rational choice contractarianism. While rational choice contractarianism may have had its roots in Hobbesian political theory, and more recently in the political and moral writings of Rawls and Gauthier and the economic analysis of law by Epstein, Posner, and Coleman, such an approach can be beneficial to some clients who seek out a philosophical counselor.

Rational choice theory, for example, is considered by some economists to reflect, at least in part, a theory of human motivation—although as some critics have pointed out, an incomplete and oversimplified theory of human motivation. My suggestion is that philosophical counselors can employ some aspects of rational choice and bargaining theory as one of the tools on the philosophical counselor's belt on behalf of some of their clients. To illustrate how such an approach could be applied in a counseling situation we will consider a non-counseling example followed by an actual example by a practicing counselor. However, prior to considering such examples, we first need to understand the argument that motivates this approach and then we need to briefly explicate some of the concepts necessary for understanding this approach.

For example, the main aim of David Gauthier's recent book *Morals By Agreement* is to argue for a "contractarian rationale for morality" (1986, p. 9). As a contractarian, Gauthier, like Rawls (1971, pp. 4, 13), endorses the idea that society is a "cooperative venture for mutual advantage among persons conceived as not taking an interest in one another's interests" (Gauthier, 1986, p. 10). The rationale for agreeing to enter such a society is quite straightforward: a society, "analyzed as a set of institutions, practices, and relationships" (Gauthier 1986, p. 11), that can guarantee for each of its members that each will benefit from entering such a society, as opposed to what each could expect from remaining in a Hobbesian state of nature, is one that is sure to have the voluntary support of its members. If such a society is possible, then it must be the case that there is a set of conditions under which each person would voluntarily agree to enter into such a society. The details of Gauthier's argument for what these conditions might be need not detain us here, for we are more concerned with the assumptions made with respect to the agents who would be party to such an agreement.

In Gauthier's case the agreement to enter such a society is a hypothetical agreement, not an actual agreement. Furthermore, the people who are a party to this hypothetical agreement do not include everyone. Gauthier explicitly excludes animals and those who cannot contribute to the cooperative enterprise (Gauthier 1986, p. 268). The people who are party to the hypothetical agreement are highly idealized agents; that is, they are conceived to be rational in the sense that they are concerned to maximize their expected utility and they are fully informed with respect to each other's utility function.

Let us now consider a non-counseling example and see how the reasoning of ideally rational and fully informed agents might be employed to the benefit of such agents. Suppose we have two ideally rational and fully informed agents concerned with reaching an agreement concerning some X. In the course of their bargaining over X they specify the terms and conditions of a contract governing the disposition of X, including the rights and responsibilities of each party to the agreement. The purpose of a contract is to allocate risks; that is, who bears what costs in the event that some event S occurs. In the real world it is not possible to fully specify all the terms and conditions of any given contract, for it is not possible to specify all possible contingencies. Furthermore, even if it were possible to specify every possible contingency, it might not be worth the effort, for there are some contingencies whose possibilities are so remote their eventuality is negligible. In any event let us imagine that some contingency, event S, was not specified in the contract and that event S has now occurred. What are the two parties to do since they did not specify what to do in the event that event S occurred? In some cases they may be able to settle the matter privately without the assistance of some external agent. In other cases they may need the assistance of some external agent—in legal cases, a judge—to impose on them *ex post* some conditions that will accommodate the occurrence of event S. Some legal theorists have argued that in situations like this the judge ought to impose upon the contracting parties the following so-called "default or gap-filling rule": "the court should impose on them *ex post* those rights and responsibilities they would have agreed to *ex ante*" (Coleman, 1992, p. 81) had they specified in the contract what to do in the case that event S did occur. The reason for imposing this default rule on the contracting parties is that they would have taken event S into consideration if they had thought of it and they would have reached an agreement about what to do with respect to assigning rights and responsibilities in the case that event S did occur.

One may object at this point to raise the question as to why we, who are neither fully informed nor ideally rational agents, should consider the reasoning of such agents. The answer is that it is rational for us to do so since we know we are neither fully informed nor ideally rational. Given that these two conditions are not satisfied for any actual moral agent, we also know that any actual agreements entered into, such as what actual agents would actually agree to, are less than ideally rational. That is, knowing that our own imperfections lead us to accept agreements that are less than what fully informed and ideally rational agents would agree to, we realize that what we would agree to, if we could, is what fully informed and ideally rational agents would agree to. What fully informed and ideally rational agents would agree to can be determined by rational bargaining theory. By determining what fully informed and ideally rational agents would agree to, we also determine what we (and/or our clients) would agree to, (believe or think) if we (and/or they) could.

5. CASE STUDIES

While the earlier example illustrates what might happen in a non-counseling situation, let us now consider what might happen in a hypothetical counseling situation and then how an actual counseling situation might be analyzed from this perspective. Imagine a situation where a client seeks a philosophical counselor with respect to some problem. In this case the philosophical counselor can view the problem as an interpersonal problem between past, present, and future selves. Imagine that the client made some decision with respect to some attitude, belief, and/or—to use Ran Lahav's term—worldview. The client now finds herself in a situation such that some event has occurred which she did not take into consideration when her past self made the decision with respect to the attitude, belief, and/or worldview in question. The philosophical counselor's role in this example would be similar to the judge's role in the previous example except that the philosophical counselor would not impose upon the client what the default rule would require. Rather the philosophical counselor's role would be simply to help the client to determine what the default rule would entail. In other words the philosophical counselor would help the client determine *ex post* what she would have decided *ex ante* had she considered the event that has now occurred. In addition, since we are viewing this as an interpersonal problem, and not an intrapersonal problem, the philosophical counselor should also help the client as much as possible to determine how her future self will view the decision the present self is about to make.

Let us now consider an actual example from Louis Marinoff's article, "On the Emergence of Ethical Counseling: Considerations and Two Case Studies" (1995). In doing so I am not suggesting that Marinoff employed this approach but rather that Marinoff's handling of this particular example could be understood from within a rational choice/bargaining framework. While Marinoff explained the situation in some detail, we need only consider the following: Students at a school were participating in a fund-raising activity for some charity. As an incentive students were to receive tickets for some prizes that were to be awarded after the fund-raising was completed. Each student who participated in the fund-raising effort was given a ticket for every $10.00 raised for the charity. Students with tickets were expected to deposit their tickets in the box for the prize they hoped to win. However, as it turned out the winning ticket holder of one of the prizes—a bicycle—was a student, Jane, who had not participated in the fund-raising activities. Upon further investigation it was discovered by the Principal of the school that another student, Sally, had given Jane one of her legitimately awarded tickets as an act of friendship. The problem the client presented the philosophical counselor was concerned with who should be awarded the bicycle: Jane, because she held the winning ticket? Or Sally, because she was the one who had legitimately been awarded the ticket that eventually proved to be the winning ticket?

Marinoff actually discusses other possibilities but they need not concern us here, for what actually took place is less important for my purposes than the approach taken to resolve the issues. That is, we can view this example—and Marinoff's analysis—in terms of a hypothetical contract between the Principal's past and present self. The terms and conditions of distributing the prizes were not fully specified, and consequently the Principal found himself in the position of having to decide what to do given that some event took place that he had not considered. In essence Marinoff's role in this case can be viewed like the role of the judge in the previous non-counseling example. That is, Marinoff's analysis of this case helped the Principal discover *ex post* what they—that is, the Principal and other committee members—would have decided *ex ante* had they considered the possibility of students transferring tickets to students who had not participated in the fund-raising activity: they would have inserted, Marinoff argues, a "Tickets are not transferable" rule (1995, p. 190). This solution to the Principal's problem is not necessarily the uniquely rational solution, but the philosophical counselor need not be concerned with discovering the uniquely rational solution, for such a solution, depending on the costs of discovering what it might be, may not be worth the effort it would take to discover it.

Now consider the following actual example. In this case a female student—call her Linda—approached me to discuss her having failed a recent exam in one of my courses. I knew her to be a conscientious student who worked hard to sustain her GPA. She complained of not being able to concentrate on her studies; she was having difficulty falling asleep at night; and she was suffering from a loss of appetite. Upon further investigation she revealed to me that she was concerned about what she ought to do with regard to a situation concerning her best friend—call her Debby. The situation was that she had recently discovered that Debby's boyfriend was cheating on her. She believed that she ought to tell Debby but that doing so would cause Debby to be hurt. She did not want to hurt her best friend and nor did she want to be in the position of not telling her the truth about what she knew about Debby's boyfriend, for she considered truthfulness among friends, especially best friends, essential to remaining friends. She complained that the situation was "driving her crazy."

Like the previous examples, this example can be viewed as an interpersonal problem between Linda and Debby. I asked Linda if she and Debby had ever discussed their views about truth telling in situations where one party might be hurt upon discovering the truth. She replied that while they had discussed the virtue of honesty in passing they had not considered what they would have agreed to in this kind of situation. We then explored what Linda thought they would have agreed to in this kind of situation, and what she would have thought she would have agreed to were the situation reversed. After some discussion about alternative possibilities, the role of intention, the nature of friendship and the issue of "hurting a friend," and how people in general, and Debby in particular, might respond in such situations, Linda decided that what she ought to do was tell her friend the truth about the matter.

The approach taken here, like the ones discussed earlier, is but one of many tools that the philosophical counselor has available that the traditional practitioner of psychotherapy does not. Not only is it efficacious in certain kinds of situations, it also respects the client's autonomy and promotes critical rational thinking on the part of the client. In fact, after a few days had passed Linda returned to see me and tell me the positive outcome of telling her friend about the boyfriend's behavior. In addition she thanked me for providing her with a tool that she would be able to use in future situations if they arose. Not all cases can be viewed as interpersonal problems and thus would not be amenable to this type of approach; however, for those cases that can be viewed this way, certain aspects of rational choice theory may be put to good use by the practicing philosophical counselor.

NOTES

1. *Careers for Philosophers* is a reprint of some material that originally appeared in the *Proceedings and Addresses of the American Philosophical Association* 58:2 1984, pp. 307-308.
2. This report was available on the APA website but it has since been removed. For the most recent APA data on what philosophers are doing both in and outside the academy see http://www.apa.udel.edu/apa/publications/texts/nonaintro.html.
3. *Proceedings and Addresses of the American Philosophical Association* 70:5 (May, 1997). 246 said yes and 223 said no.
4. In addition to my recollection of the 1994 conference, I have also relied on some unpublished notes by Leonard Angel of Douglas College in Vancouver, B.C.
5. While some philosophical counselors claim that philosophical counseling was founded by the German philosopher Gerd B. Achenbach in the 1980s, the fact that Sahakian was writing on the subject in the 1970s seems to undermine this claim. See Lahav and Tillmanns (1995, p. xii), and Schuster (1999, p. 7).
6. See Wexler, L. (1997). Also see "The Philosopher is In: A New Spin on Psychological Counseling" and "Training the Thinkers" in *Self* (June, 1998). p. 56.

REFERENCES

___ (1998). 'The Philosopher is In: A New Spin on Psychological Counseling.' *Self* (June), p. 56.
___(1998). 'Training the Thinkers.' *Self* (June), p. 56.
Coleman, J. (1992). *Risks and Wrongs*. Cambridge: Cambridge University Press.
Cust, K. (Ed.) (1998). *Inquiry: Critical Thinking Across the Disciplines* XVII, 3.
Engel, S.M. (Ed.) (1990). *The Study of Philosophy* (3rd ed.). San Diego: Collegiate Press.
Gauthier, D. (1986). *Morals by Agreement*. Oxford: Clarendon Press.
Gluck, A. (1997). 'Philosophical Practice in Career Counseling and Management Consulting.' In: W. van der Vlist (Ed.), *Perspectives in Philosophical Practice: Collected lectures Held at the Second*

International Conference on Philosophical Practice (pp. 195-213). Groningen, Holland: Vereniging voor Filosofische Praktijk.

Grimes, P., and Uliana, R. L. (1998). *Philosophical Midwifery*. Costa Mesa, CA: Hyparxis Press.

Howard, A. (2000). *Philosophy for Counselling [sic] and Psychotherapy*. London: Macmillan Press, Ltd.

Lahav, R. (1995). 'A Conceptual Framework for Philosophical Counseling: Worldview Interpretation.' In: R. Lahav and M. da Venza Tillmanns (Eds.), *Essays on Philosophical Counseling*. New York: University Press of America, Inc.

Lahav, R. and M. da Venza Tillmanns (Eds.). (1995). *Essays on Philosophical Counseling*. New York: University Press of America, Inc.

Lyons, W. (1990). 'You Asked About Philosophy?' In S. M. Engel (Ed.), *The Study of Philosophy* (3rd ed.). San Diego: Collegiate Press.

Marinoff, L. (1995). 'On the Emergence of Ethical Counseling: Considerations and Two Case Studies.' In: R. Lahav and M. da Venza Tillmanns (Eds.), *Essays on Philosophical Counseling* (pp. 171-191). New York: University Press of America, Inc.

Rawls, J. (1971). *A Theory of Justice*. Cambridge: Harvard University Press.

Sahakian, W. S. (1979). 'Philosophical Psychotherapy.' In: W. S. Sahakian (Ed.), *Psychopathology Today: The Current Status of Abnormal Psychology* (2nd ed.) (p. 460). Itasca, Illinois: F. E. Peacock Publishers, Inc.

Schneider, A. (1998). 'More Professors Are Working Part Time, and More Teach at 2-Year Colleges.' *Chronicle of Higher Education*, March 13, 1998, p. A 14.

Schuster, S. (1999). *Philosophy Practice: An Alternative to Counseling and Psychotherapy*. Westport, Connecticut: Praeger Publishers.

Scruton, R. (1997). 'The Return of the Sophists.' *The Times* (London, August 11, 1997).

Segal, S. (1998). 'Philosophy as a Therapeutic Activity.' In K. Cust (Ed.), *Inquiry: Critical Thinking Across the Disciplines* XVII, 3, pp. 36-47.

Sharkey, J. (1998). 'I Bill, Therefore I Am—Philosophers Ponder a Therapy Gold Mine.' *New York Times*, March 8, 1998, pp. 4-1, 4-4.

van der Vlist, W. (Ed.). (1997). *Perspectives in Philosophical Practice: Collected Lectures Held at the Second International Conference on Philosophical Practice*. Groningen, Holland: Vereniging voor Filosofische Praktijk.

Wexler, L. (1997). 'Thinking Not Shrinking: I Took My Problem to a Philosopher.' *Utne Reader* (February), pp. 50-51.

Zeig, J. K. and Munion, W. M. (Eds.). (1990). *What is Psychotherapy?* San Francisco: Jossey-Bass Publishers. p. 423.

Zijlstra, B. (1997). 'The Philosophical Counsellor [sic] as an Equilibrist.' In W. van der Vlist (Ed.) *Perspectives in Philosophical Practice: Collected Lectures Held at the Second International Conference on Philosophical Practice* (p. 35). Groningen, Holland: Vereniging voor Filosofische Praktijk.

CHAPTER 13

STEPHEN WEAR

ETHICAL EXPERTISE IN THE CLINICAL SETTING

At first glance, the idea that there is "ethical expertise" in any area sounds preposterous. Particularly in America, with our deep-set notions of liberty, diversity, and individual autonomy, the idea that some people are more qualified than others to decide what is right or wrong (which seems to be what ethical expertise would amount to) just rings false. More specifically, if we move to the specific forum of the clinical setting where some of the most controversial issues of our time get played out between different parties, the notion of an ethics expert somehow assisting in the resolution of disputes and conflicts from some privileged ground sounds even less credible, if that is possible. For the resolution would be offered to people who often do not know one another, do not share common world views, faith communities and so forth, and may well have different goods and values that are variously at stake. There appears to be little if any common ground to stand on, and certainly, it would seem, no one with privileged access to it.

One might, of course, attempt to defend the idea of ethical expertise in the clinical setting by emphasizing its procedural aspects, e.g., the ability to perform values clarification and dispute mediation, or build consensus. Often such activities are at the heart of a "successful" clinical ethics consultation. But this seems like rather thin gruel, and if this is all clinical ethics expertise amounts to, then what many would think is the core sense of expertise is not being claimed. That is: go to any standard dictionary, and the definition of expertise includes not just some sort of skill(s), but also *knowledge*, and again, what knowledge can one person claim in the area of ethics, at the bedside or elsewhere?

My own view is that there is, in fact, ethical expertise worthy of the name available in the clinical setting and that it lies not just in various procedural/facilitative skills, but also involves substantial knowledge in various areas. I also believe that there are right and wrong answers to certain ethical issues, as well as better and worse ways of responding to ethical conflicts and disputes, and that certain people are much more capable of formulating such answers or responses.

I intend to justify these assertions by first proceeding in the mode of *descriptive* ethics, i.e., by describing the skills and knowledge that are ingredient in clinical ethics expertise, and then presume to observe that this description goes far toward providing sufficient proof of the normative claim that it exists and is worthy of the

L. M. Rasmussen (Ed.), Ethics Expertise: History, Contemporary Perspectives, and Applications, 243-258.

name. In effect, once one sees what is needed, and available, for clinical ethics consultation, one should see that such skills and knowledge do legitimately claim to be a form of expertise in its fuller, thick-gruel sense. How full a sense will be a pivotal issue of this paper. In the end we will have to get clearer not simply about the scope of clinical ethics expertise, but its limits. Equally, we will also have to investigate who may legitimately claim to have it, and on what grounds.

Fortunately, we are not obliged to fashion a description of the elements of clinical ethics expertise out of whole cloth. We can begin by retailing a comprehensive account of it from a single source, i.e., a consensus statement by leaders in the field, and can straightforwardly do so first by briefly documenting the weaker claim, viz., that certain types of procedural or facilitative skills are needed and available for clinical ethics consultation.

1. PROCEDURAL/FACILITATIVE EXPERTISE IN CLINICAL ETHICS

The American Society for Bioethics and Humanities (hereafter ASBH) report "Core Competencies for Health Care Ethics Consultation" should be seen as the seminal and primary document in this area (ASBH). Generated by a blue ribbon task force of the primary American bioethics umbrella organization, it went through various drafts, was extensively circulated within the field and commented on, and represents the most mature, detailed document in the field. It claims that a wide variety of procedural skills are needed and available for clinical ethics consultation. The skills claimed fall into two broad categories, i.e., skills of "ethical assessment" and "process" skills.

More specifically, the ASBH report asserts:

> Ethical Assessment skills include being able to discern and gather relevant data (e.g., clinical, psychosocial); assess the social and interpersonal dynamics of the case (e.g., power relations, hierarchical structures, manipulation, cultural differences); distinguish the ethical dimensions of the case from other, often overlapping, dimensions (e.g., legal, medical, psychiatric); identify various assumptions which concerned parties bring to the case (e.g., regarding quality of life, risk taking, unarticulated agendas); and identify relevant values of involved parties.

The ASBH report goes on to identify various "process and interpersonal" skills that are "also important for ethical assessment," including the ability to

> identify key decision-makers and relevant concerned parties and involve them in discussion; set ground rules for formal meetings (e.g., regarding length, participation, purpose, and structure of such meetings); express and stay within the limits of ethics consultants' role as ethics consultant within the meeting; and create an atmosphere of trust that respects privacy and confidentiality and that allows parties to feel free to express their concerns (e.g., skill in addressing anger, suspicion, fear or resentment; skill in addressing intimidation and disruption due to power and/or role differentials) (ASBH, p. 14).

Such "process skills also include", the ASBH report continues,

> the ability to build moral consensus. Ethics consultants must be able to help individuals critically analyze the values underlying their assumptions, their decision and the possible consequences of that decision; negotiate between competing moral views; engage in creative problem solving; the ability to utilize institutional structures and

> resources to facilitate the implementation of the chosen option; lastly process skills demand the ability to document consults and elicit feedback regarding the process of consultation so that the process can be evaluated (ASBH, p. 14).

Finally, the ASBH report asserts that

> interpersonal skills are critical to nearly every aspect of ethics consultation in individual patient cases. Interpersonal skills include the ability to: listen well and to communicate interest, respect, support, and empathy to concerned parties; educate involved parties regarding the ethical dimensions of the case; elicit the moral views of various parties; represent the views of various parties to others; enable involved parties to communicate effectively and be heard by other parties; and recognize and attend to various relational barriers to communication (ASBH, p. 14).

The preceding is certainly a mouthful, and one might well wonder how much any of it occurs in any given case, or the extent to which any individual "ethics expert" fully realizes such skills in practice at any given point, or regularly. Fair enough, but for the sake of argument, I will presume to suggest that even with a cursory reflection on the skills listed above, and the sorts of cases and conflicts that are well known to occur regularly in the clinical realm, the need for such procedural "expertise" should be apparent. As to its embodiment in any particular ethics consultant, I will simply presume to suggest that what we have is an ideal that is legitimately stated as such, may be more or less asymptotically approached, and in certain individuals, may be seen as rising to the level of true expertise, at least in a skills or procedural sense. This is the weaker "expertise" claim, and I submit it is more easily made, and sufficiently so above. To the stronger, more difficult claim that clinical ethics expertise also must and does embody knowledge in this area, I now turn.

2. SUBSTANTIVE KNOWLEDGE EXPERTISE IN CLINICAL ETHICS

Continuing with the ASBH report, it also describes one further set of skills beyond those listed above, and in doing so, addresses the issue of the knowledge that clinical ethics expertise must also bring to bear. These further skills include the ability to: "access the relevant knowledge (e.g., bioethics, law, institutional policy, professional codes, and religious teachings); clarify relevant concepts (e.g., confidentiality, privacy, informed consent, best interest,); and critically evaluate and use knowledge of bioethics, law (without giving legal advice), institutional policy (e.g., guidelines on withdrawing and withholding life-sustaining treatment), and professional codes in the case" (ASBH, p. 13). Such knowledge claims for clinical ethics expertise are categorized and extensively detailed further in a later section of the ASBH report; the types of knowledge claimed include:

(1) "Moral reasoning and ethical theory (e.g., consequentialism and non-consequentialist approaches (e.g., utilitarian approaches, deontological approaches such as Kantian, natural law, rights theories; theological/religious approaches; virtue, narrative, literary and feminist approaches; principle-based and casuistic (case-based) approaches:;related theories of justice, with particular

attention to their relevance to resource allocation, triage, and rights to health care" (ASBH, pp. 16-17).

(2) "Common bioethical issues and concepts." Here a very extensive list of about every bioethics issue is retailed, with extensive exemplification, including the issues of "informed consent, confidentiality, disclosure and deception; advance care planning, surrogate decision-making, end of life decision-making, beginning of life decision making, genetic testing and counseling, medical research, organ donation and transplantation, and resource allocation" (ASBH, p. 17).

(3) "Clinical context: knowledge of the clinical context includes: terms of basic human anatomy and those used in diagnosis, treatment, and prognosis for common medical problems; awareness of the natural history of common illnesses; awareness of the grieving process and psychological responses to illness; familiarity with current and emerging technologies that affect health care decisions; knowledge of different health care providers, their roles, relationships, and expertise; basic understanding of how care is provided on various services such as intensive care, rehabilitation, long-term care, palliative and hospice care, primary care, and emergency trauma care" (ASBH, p. 18).

(4) "The local health care institution." This will include knowledge of "the institution's mission statement; structure, including departmental, organizational, and committee structure; range of services and sites of delivery, such as outpatient clinic sites; ethics consultation resources, including financial, legal, risk management, human resources, chaplain, patient representatives; medical research, including the role of the IRB; and medical records, including location and access to patient records" (p. 19).

(5) "The local health care facility's policies," including those dealing with "informed consent; withholding and withdrawing life-sustaining treatment; advance directives, surrogate decision-making, health care agents, durable power of attorney, and guardianship, do-not-resuscitate orders; medical futility; confidentially and privacy policy; organ donation and procurement; human experimentation; conflicts of interest; admissions, discharge and transfer; impaired providers; conscientious objection; reproductive technology" (ASBH, p. 19).

(6) "Relevant codes of ethics and professional conduct and guidelines of accrediting organizations" including "codes of conduct from relevant professional organizations (e.g., medicine, nursing); local health care facility's code of professional conduct; other important professional and consensus ethics guidelines and statements (e.g., presidential commission statements); patient's bill of rights and responsibilities; relevant standards of JCAHO and other

accrediting bodies (e.g., patient rights and organizational ethics standards)" (ASBH, p. 21).

(7) "Relevant Health Law", including "both federal and state constitutional, statutory and case law" dealing with such topics as "end of life issues such as advance directives (including living wills and proxy appointment documents), nutrition and hydration, determination of death; surrogate decision-making, including determination of competence, appointment of surrogates; decision making for incompetent patients and minors; informed consent; organ donation and procurement; reproductive issues; confidentiality, privacy and release of information; reporting requirements, including child, spouse, elder abuse and communicable diseases" (ASBH, p. 21).

3. THE EMBODIMENT OF CLINICAL ETHICS EXPERTISE

Now if the earlier section describing the "procedural/facilitative" expertise of the clinical ethics expert was a mouthful, this last section describing the knowledge such expertise must include appears to call for a full belly on its way to gross obesity. That is: it would seem like we would need to fill a large room with people to get even a respectable fraction of such knowledge present and available for use. We would need various clinicians, lawyers, philosophers, clergy, and administrators to approximate such skills and knowledge. How on earth might this occur in a realistic and credible sense?

As is well known, the solution to this question is taken by many to lie in the establishment of an ethics committee, an organizational entity that most health care institutions in the U.S.A. (especially hospitals, but many nursing homes also), have in one form or another, in no small part because various governmental and accrediting agencies, e.g., the Joint Commission for the Accreditation of Hospitals (JCAHO) require it. And this seems to be a credible claim. Certainly at some point, a committee composed of clinicians, lawyers, philosophers, clergy, and administrators, should be able to incorporate many of the skills and knowledge identified in the ASBH report.

Perhaps. But a host of questions, as well as objections, will quickly assail us about such ethics committees, including the number, training and experience of the various individuals that comprise it, and whether they actually embody some respectable threshold amount of the skills and knowledge previously identified. But most crucially for our purposes: how can such a large committee (one should think that the category of clinician would necessarily include at least one physician, a nurse and a social worker, for example; as a practical matter given the type and source of common consults, one might well also hope for a geriatrician, a psychologist or psychiatrist, and a neurologist) make itself physically present at the bedside, and do so in a timely matter (many ethical issues occurring "emergently", as well as "off-shift")? The answer would seem to be that only a subset of any such committee can be present; often only an individual representative of the committee. But have we not, by this answer, largely diluted the expertise, i.e., skills and knowledge, which the committee was supposed to make available?

The preceding questions raise basic issues as to how clinical ethics expertise might actually be embodied in the clinical realm. On the one hand, one might emphasize, accurately, that much of the activity of clinical ethics expertise occurs in response to individual case dilemmas and disputes, and thus it is to the individual ethics consultant, or a sub-group of the ethics committee, that one must look for the reality of clinical ethics expertise. Such an emphasis, however, would obviously result in a substantial watering down of the presence of all the skills and knowledge that the ASBH report calls for. But, conversely, consultation by the fuller committee would be unwieldy and inadequately responsive. There appears to be a basic tension here that, either way one goes, as much is lost as is gained.

In my own experience, which I believe is echoed nationally, the responses to this tension are myriad. As a founding and ongoing member of over a dozen ethics committees, I can report that each has its own character, and each one has made different choices in responding to this tension. Some require that case consultation be done only at the full committee level; others have instituted some sort of sub-group which is empowered to speak for the whole committee; and some have authorized a single individual to field and respond to the bulk of the ethics consult requests. In each institution, where a sub-group or specific individual is given authority, retrospective monitoring occurs, and there are more or less clear senses of when certain types of cases should still go to full committee review. But the who, when and how here is quite variable from institution to institution.

To the issue of who actually embodies clinical ethics expertise, I thus submit our answer must be essentially two-fold, i.e., one form of it is embodied in some form of ethics committee, another in individual consultants, either a specific individual or an empowered sub-group of the ethics committee. The necessity of this claim, and the reasons why it can not be one or the other separately, can be gained from a fuller sense of the types of activities that ethics expertise must regularly perform in the clinical realm. Specifically, it is crucial to note, ethics expertise in the clinical realm has numerous different avenues of activity; it does not just occur in individual case consultations. As I have suggested elsewhere, these activities include:

> (1) case conferences of whatever sort, whether specifically designed to address bioethical issues and specific cases, or where such issues can be identified and addressed in passing, e.g., residents' morning report, or mortality and morbidity conferences; (2) case consultations where one is approached regarding the resolution of some specific case….; (3) hospital policy statement development, where the guiding principles of the institution are identified, and tactics for addressing, advocating for and protecting those principles are articulated….; (4) in-service presentations that generally are case-based, but more fundamentally seek to connect hospital policy to clinical behavior, and are thus only incidentally case-based; and (5) various ad hoc activities, such as quality assurance and incident reviews, or the preparation for Joint Commission on the Accreditation of Healthcare Organizations (JCAHO) surveys, where some sort of reports back to staff, as well as changes in institutional policy and procedure, often occur (Wear, 2002, pp. 435-436).

It should be apparent, in sum, that only some of the above activities can be credibly accomplished by an individual consultant or ethics committee sub-group, however much it should also be apparent that timely and effective response to case consultation requires such individuals or sub-groups. Most importantly, hospital policy statements merit wide consideration, surely beyond even the ethics committee itself, and many institutions have detailed policies on most of the bioethics issues mentioned above, policies which often include detailed tactical guidelines as to how cases will be resolved, how such resolutions will be documented, and so forth. Thus the credibility of the individual consultant, or sub-group, will rest largely on the quality of the ethics committee that fashions the policies of the institution, retrospectively monitors the case consultations performed, and sustains an ongoing mutual feedback between consultant(s) and committee toward enhancing policy to the extent recurrent problems suggest this.

The "expertise" of the two types of groups will clearly differ. Wide-ranging reflection on policy will be the committee's strong suit; no individual, however skillful, knowledgeable and experienced, could possibly possess the types of knowledge the ASBH report clearly instructs us needs to be present for this. The ethics consultant(s), conversely, can be uniquely present 24/7 for timely response to consultation requests and it should be clear that the effective presence of many of the "procedural/facilitative" skills the ASBH report highlights, e.g., dispute meditation or consensus building, will not occur in the committee format. Having witnessed the attempt many times, I will presume to report that families, and staff, who are forced to try to work out their ethical differences in front of a large committee, are more likely to just feel daunted and overwhelmed, rather than assisted.

In sum, clinical ethics expertise is embodied in the dual form of ethics committees and ethics consultant(s), and each are dependent on the other for their quality. As to the scope and limits of such expertise, especially regarding the knowledge it brings to bear, we need to take a closer look at this, beyond the ASBH report's extensive list. What sort of knowledge is being claimed here, how far does it go toward being a true expertise, and what limits, if any, does it have?

4. THE SCOPE AND LIMITS OF CLINICAL ETHICS KNOWLEDGE

It is probably sufficient to be a red-blooded American, but for those of us who have actually undergone academic study in ethics, the idea of someone having expertise in it just sounds, as I mentioned above, like an oxymoron at best, confused or plain stupid at worst. Particularly for those of us who have had substantial study in ethics proper, for example in graduate philosophy departments, it should be quite apparent that there are numerous ethical theories, generic and specific, none of which enjoy broad, primary acceptance. Further, for all that Beauchamp and Childress want to suggest that certain primary theories end up saying the same thing in the clinical setting (Beauchamp and Childress, pp. 21-22), those of us who have attempted to use the "Georgetown Mantra" of principles to resolve clinical ethics disputes know that, though the principles may assist at better appreciating the nature of the conflicts, they do not help in resolving them, primarily because the principles do not come with a differential weighting as to which one to pick when two or more of them conflict (Engelhardt, 1982).

If we move from the lecture hall to the clinical situation, however, I submit that the picture changes drastically, and a form of substantive knowledge expertise emerges. A number of basic features of the realm of clinical practice merit highlighting in this regard: first, in the lecture hall or the seminar room, debate over a bioethics issue or case can proceed interminably and without conclusion. The aim will usually be to assist the students to see both sides of an issue; not seldom, the teacher may act to resist closure on an issue, perhaps by paying devil's advocate for a position that he or she may not personally agree with.

Not so in the clinical realm. In most cases, something will or will not be done, and everyone knows this, so the best thinking available must be brought to bear at that point in time; leaving it until later is usually not an option. And believing that there is no one right answer does not match the result; a specific course of action will be chosen and instituted over alternative possibilities. So somehow we have moved from the "it's all relative" realm of theoretical ethics to the "what is the best thing we can do in this situation now, all things considered?" realm of practice.

Second, those who seek to describe the nature of clinical ethics expertise often tend to include the need for ethics experts to be knowledgeable in ethical theory, as the ASBH report does (ASBH, pp. 16-17). My experience as an ethics consultant tends to suggest, however, that ethical theory has little if any place in the clinical realm and for good reason: as no ethical theory enjoys a consensus among ethical theorists, then there is no privileged place for any of them, whether we are talking of deontology, consequentialism, natural law, situation ethics, causuistry or what have you. So the "ethics expert" who proceeds in terms of one preferred approach is either begging a crucial question, or must retail each type of theoretical approach and assist other parties to the discussion to pick which approach they each wish to utilize (an enterprise for which the clinical realm clearly offers insufficient time). My point here is simply that, in 20-plus years of ethics consulting, I have never seen a case or dilemma decided by choosing some theoretical approach over others. Rarely have I even seen such theories discussed, certainly not at the bedside, and only rarely and superficially within the ethics committee forum. And the further point is that however relative, or undetermined, or up-for-grabs, we think ethical theory is, or lacking in consensus and therefore without any privileged position, it is not the stuff of ethical reflection in the practical realm anyway. Whatever is the stuff of clinical ethics knowledge may then be subject to expertise, and the oxymoronic character of ethical expertise may well evaporate.

Thus far, the basic point is that we are in the realm of practical reason and the controversies and difference of opinion regarding theoretical ethics do not necessarily affect the knowledge claims that ethical expertise might make. Sifting the previous claims of the ASBH report, we may further point out that both ethics committees and consultant(s) clearly need and can claim knowledge in a number of areas, including the law, institutional policy, the characteristics of the clinical context, and various codes and standard statements of agreed upon concepts.

Is this all? Perhaps. One might maintain that all such knowledge comes clearly to bear on such things as hospital policy statements, but once we move to the meat and potatoes of ethics consulting, viz., the individual case consultation, then it is the consensus of the involved parties, however facilitated by the ethics consultant(s), informed by policy, etc., that is decisive for the determination of what is the right way to proceed (Moreno, 1995). And here, one might argue, is where ethical expertise meets its limit. It does not in fact have some privileged position for saying what is right or wrong, but is merely facilitative of the process of determining this. It can legitimately claim expertise, in a full-blown sense, on the basis of the skills *and* knowledge it brings to the table, but it cannot, by itself, speak with authority regarding the final resolution of any given case.

I believe, however, that though the preceding paragraph states what many ethics consultants and committee members would want to maintain, it is much too humble, and misses much of what actually occurs in individual case consultation, as well as outside this sphere of activity. I submit, in fact, that a specific "canon" of clinical ethics principles, guidelines and tactics has been evolved that, in practice, goes way beyond simply facilitating and informing the development of consensus at the bedside.

Consider: however much we value consensus, do not the knowledge claims made above clearly indicate that a great deal has already been decided before one attempts to gain consensus from interested parties? Absolutely and, as I have pointed out elsewhere in describing clinical ethics teaching, this is where we leave truly lecture hall ethics far behind:

> Consult any introductory text in bioethics and you will find discussions of basic bioethical issues, e.g., informed consent, truth telling, confidentiality, the right to refuse treatment, abortion, genetic engineering, access to and rationing of care, etc., with pro and con pieces to assist students to think such issues through for themselves. How different from the approach at (or near) the bedside, where much of what is presented as up for grabs in the lecture hall is simply accepted as correct and binding, at least in broad outline. Whether it be in the "Patients Bill of Rights" (American Hospital Association, 2001) which is hung on the walls all over the institution, or in well known legal or reviewing body (i.e., JCAHO) statements that many staff can parrot, or in the policies and procedures of the institution, many bioethical issues appear in the form of established truths. The right of competent patients to informed consent and confidentiality, to be told the truth, to refuse any and all treatment, and so forth, are seen as guiding principles for everyone. They are seen as no more up for grabs, intellectually or morally, than the clinical guidelines for managing diabetes, or the proper methods for assessing and responding to multi-infarct dementia.
>
> Teaching within such a framework of accepted moral truths goes into considerable clinical and ethical detail before any true controversy arises. Explaining how competence, i.e., decision making capacity, should be evaluated, the rank-order of surrogates for incompetent patients, what sort of interventions require informed consent and what the elements of any such disclosures should be, and so on, are clearly delineated for staff who, in the main, want to know how to proceed in the usual case. One often spends time "talking tactics," whether this regards how (not whether) to tell bad news to patients and families, when and how to encourage patients to designate a surrogate or generate advance statements regarding extraordinary scenarios, or how to document what one has discussed or determined so it will be available and useful during subsequent care.
>
> The further point here is that, however much actual bioethics teaching at (or near) the bedside may incidentally key to, or be triggered by, actual controversy or disputes, a very broad and complex background of what is accepted or assumed guides most such discussion. Often these moral truths completely control what is then done, as when a patient's specific prior statements are held, per hospital policy, to overrule contrary wishes or views of family members, or when "self-destructive" patients who are deemed competent are allowed to continue those behaviors, however much staff and family would like somehow to prevent them from occurring. Even when policy does not clearly stipulate the proper ethical course, tactics aimed at dispute mediation—attempting to restore staff-patient/family interaction—or simply talking it out in the hopes that consensus might be reached, are utilized to resolve problems as

> often as anything one would recognize as ethical reasoning. That such latter tactics as converting dispute into consensus often work suggests that the accepted truths mentioned above are not only those of the institution and its staff, but are often shared equally by patients and their families. The diversity of ethical beliefs and values itself appears to be more a creature of the lecture hall, than a presence at the bedside (Wear, 2002, pp. 436-438).

The sense of clinical ethics expertise described here has gotten pretty robust. If the above is an accurate portrayal of what individual ethics case consultation amounts to, and I believe it is, particularly that much of its character and parameters are predetermined, then one might wonder if we have, in fact, produced the true super-expert, and in an area where one might have least expected to find one. To sharpen the point: if all these issues are predetermined, from not just the law, but also by institutional policy, then one might even wonder if the search for consensus is often not mainly a political or "customer satisfaction" commodity, not an ethical one. The ethics consultant(s) may well be individually clear as to the proper result or resolution, and will intervene against any consensus that violates such guidelines and procedures. This seems surely to be the privileged ground for deciding right and wrong that we were so skeptical about initially. Is this the result we should be affirming?

In large part, I believe we must. Much has already been decided, and various other perspectives or resolutions have been disqualified. The family who wants staff to lie to an inquisitive, competent patient will be instructed that this is not permitted, and the patient informed over their objections. Similarly, those who believe a quick lethal injection is preferable to whatever "comfort measures" management of end-of-life illness that staff is proposing will be informed that an active killing will not occur. Or the family member who insists that a dying patient be resuscitated, when the patient had previously made a competent and knowledgeable decision to have a "do-not-resuscitate" order, will be informed that the patient's clear and specific wishes will be honored, not theirs. Less dramatically, anyone witnessing the attempt at gaining consensus could easily note that a good deal of this involves stipulating what is fair game and what is not: e.g., how to think about competence and the staff's primary authority to determine its presence or absence in any given case, or what is to be considered a sufficiently specific advance directive that must be honored.

I will thus admit, in my own individual consulting, that often the search for consensus is more political than ethical, and seeks to generate agreement to a way of proceeding that is already essentially clear to me as the best and only feasible solution. This is not so much because I feel wiser, or somehow better, than the other interested parties to the dispute, but rather because the weight of institutional policy, the law, the canon of clinical ethics, etc., simply, and overwhelmingly, may call for some specific resolution over any other. Hopefully I enter a family/staff conference prepared to listen closely to alternative views, and change my own views if indicated. But often this just doesn't occur and the process is much more one of getting everyone present to see the proper way to proceed. I am thus attempting to facilitate consensus and acceptance, but not decision making in its basic sense.

Having said this much, the preceding merits substantial qualification. Many clinical ethics consults or case conferences involve dilemmas that occur at the margins or gray areas of such guidelines, principles and procedures, and are thus true ethical dilemmas that neither the law, institutional policy, or the expertise of the ethics consultant speak to, at least clearly and decisively. The discussion will still assume and be guided by a vast background of accepted or assumed moral truth. But it will not be determined by it. For example, the issue of whether, per his families' wish, one should not inform a clinically depressed, somewhat confused, clearly dying patient of Italian extraction that he has an end-stage malignancy for which only palliation is possible will surely be discussed amidst the background of the institution's guiding principle of being truthful and forthcoming with patients. The family's request will have to overcome all those strong and cogent reasons that ground this guiding principle. The issue will not be, in other words, whether truth telling to the sick and dying is ethically appropriate, as in the lecture hall, but whether the present case provides a valid exception to the rule. Similarly, if one is attempting to identify and assess a patient's prior statements about aggressive end-of-life care, the principle that specific prior statements must be honored will remain operative, but the statements identified may well leave much room for interpretation, and here the status of the views of the family are at least on a par with the consultant, and the staff. In such grey areas, the limits of ethical expertise are to be found; however much the bedside reflection will be guided and constrained by the canon, the resolution of the issue(s) at hand will have much more to do with the possibilities for consensus among the major parties, and their relative authority.

In sum, there is a canon of clinical ethics, variously expressed in the law, codes and rights statements, and institutional policy, that is often decisive regarding specific case dilemmas, and surely predetermines the range of tolerable outcomes in many others. Sometimes, of course, the issue is such that only a consensus of interested parties, including the ethics expert, resolves the issue; there is no higher gold standard. But, as often, the canon is more or less operative and determining of what will actually be done. To say otherwise is to ignore what happens everyday in hospitals throughout the country on a day to day basis.

5. THE CHARACTER OF THE CLINICAL ETHICS EXPERT

A further basic point merits making here, one which I believe makes all the difference. As indicated above, I believe clinical ethics expertise resides in both ethics committees, and ethics consultant(s) for the reasons given above, and see the relationship between the two as mutually supportive and informing in the most essential of ways. For our issue of clinical ethics expertise, however, this dualism misses a further, final quality that I believe makes or breaks clinical ethics consultation. That is: in the end the eagles gather around the approach to individual case consultations as to how they are individually handled, as well as how they are carried back to the monitoring committee for further reflection, as well as policy development and improvement. In this sense, the activity of the individual consultant(s) is crucial. I say this because I have seen repeated instances where the above skills and knowledge abounded in a particular committee, but the results, at the bedside, and in the life of the institution, repeatedly fell far short of the mark.

Something else was missing. This final ingredient regards the character of the individual consultant(s) and, I believe, indicates that what we finally need to be talking about here is not just expertise, but the activity of professionalism. By professionalism, I am specifically thinking of what makes a merely skillful and knowledgeable person into someone who assumes and satisfies fiduciary duties to others. To return one more time to the ASBH report, which I believe puts it just right, I submit this lies in the character of the individual consultant(s):

> We conclude that all ethics consultants should strive to exhibit and possess these traits: *Tolerance, patience and compassion* that would enable the consultant to 'listen well and communicate interest, respect, support and sympathy'…help people with difficult problems, those who are emotionally distraught, or those who have minority views, so that these people can be fully and respectively heard.
>
> *Honesty, forthrightness and self-knowledge* will help prevent the manipulative use of information and help 'create an atmosphere of trust' necessary to facilitate formal and informal meetings. Consultants must be honest about their own limitations, their need for more knowledge, how their agendas and values are shaping a consultation, and the uncertainty of proposed solutions. *Courage* is sometimes needed to enable various parties, especially the politically less powerful, to communicate effectively and be heard by other parties. It is also sometimes required to take positions that are unpopular or contrary to the interests of one's employer or other powerful individuals. *Prudence and Humility* can inform behavior when rash or novel courses of action are being considered and enable consultants not to overstep the bounds of their role in consultation. These character traits can help consultants acknowledge possible areas of conflict between their personal moral views and their role in doing consultation. *Integrity* would enable consultants to pursue the option or range of options ethically required in the case even when it might be convenient to do otherwise. Integrity should inform all behavior of consultants as they strive to fulfill the trust placed in them by health care providers, patients, and families who seek help resolving ethical issues (ASBH, 1998, pp. 22-23).

6. IS THERE AN ETHICS EXPERT IN THE HOUSE?

I submit that the preceding sufficiently establishes that there is, in fact, clinical ethics expertise that is truly worthy of the designation, both in that it clearly incorporates substantial skills and knowledge components, and in that it should not be hard to imagine that many people can legitimately claim to possess such skills and knowledge, however much their possession must be seen as an asymptotic approach to a complex ideal, never fully realized in the flesh.

Moreover, just as with the expertise of the engineer, the surgeon, the military commander or the psychiatric counselor, much of what constitutes clinical ethics expertise can be taught. All the skills enumerated above can be identified, elaborated, rehearsed, critiqued and enhanced; and knowledge of institutional policy, relevant legal cases and statutes, and so forth, can similarly be conveyed and improved.

The issue gets murkier here if we move to consider how one determines which individuals actually have such expertise. This is so as it seems correct to say that expertise is not just the possession of certain skills and knowledge, but is most truly found to be present, or lacking, in the activity itself. And surely many well-tutored individuals may just not "have it" in this last requisite sense. Experience at such consultation is surely a necessary but not a sufficient source of such expertise; bad judgments, poor timing, flawed character, etc., can certainly render worse than useless the activity of any "expert" on the firing line, whether it be at the bedside, in a counseling session, in the operating room, or on the field of battle. So how does one determine if a given individual has it?

Interestingly, the ASBH report begins its first section with a firm rejection of the enterprise of actually picking out which individuals actually have the expertise the report then goes on to detail. The report notes: "The Task Force considered three models for using its report: formal certification of individuals or groups, accreditation of educational programs, and voluntary guidelines. *We reject the certification and accreditation models and endorse the voluntary model*" (ASBH, p. 2). The report then goes on to list various reasons for this rejection, under the heading of "certification for ethics consultation." Some of their reasons for this rejection are "political", as in: (1) "certification could undermine disciplinary diversity if it fell under the hegemony of a particular discipline (e.g., clinical medicine or academic philosophy) and was widely adopted," (2) "certification could have politically divisive ramifications for the bioethics community," and (3) certification might "disenfranchise large numbers of individuals currently active in the field who, for reasons unrelated to their competence to perform ethics consultation, may be unable to satisfy the rigor of a certification process" (ASBH, p. 2).

But to the heart of our "is there an ethics expert in the house?" issue, the ASBH report goes further:

> Certification increases the risk of displacing clinicians and patients as the primary moral problem-solvers at the bedside because it can give the impression that certified individuals are substantive moral experts with special standing to offer substantive positions on moral issues. [And] if certification of individuals or groups was based on standardized testing, the tests would have to be shown to measure the competencies in question. It is unlikely that a sufficiently reliable test could be developed to measure the competencies required for ethics consultation (ASBH, pp. 2-3).

So: the ASBH report, it turns out, rejects the idea that such clinical ethics experts are "substantive moral experts with special standing to offer substantive positions on moral issues" and goes on to admit that it is "unlikely" that the "competencies in question" can be tested for. In effect, it seems, our ASBH colleagues have rejected the ethics expertise mantle, and whatever sort of individuals they are talking about, they seem to be indicating that we are unlikely to find a way of identifying them.

This is, in part, a punt, not a position. As I have documented (mainly via the ASBH report itself), as well as argued, there are substantial skill and knowledge components to what constitutes clinical ethics competence, the presence or absence of such *identified* skills and knowledge can surely be tested for, and all this certainly comports with the standard understanding of what it means to have expertise in an area of activity.

As to the rejection that there are "substantive moral experts with special standing to offer substantive positions on moral issues," the basic distinction is already available from the preceding. That is: in grey area cases, it may well be that the final resolution must be based on consensus, with the clinical ethics consultant being merely a facilitator of this. But even in such cases, one would also need to attend to what sorts of guidelines or limits the consultant would be placing on such a resolution. As noted, the canon of clinical ethics has numerous guiding principles, e.g., patient autonomy, truth-telling, etc., that would still be operative however much the individual case was raising the possibility of an exception to the rule. That is surely a "substantive position on moral issues" however much an exception might be being countenanced. Beyond this, I submit, numerous cases will, once all the facts are known, be legitimately seen as admitting of only one resolution, and it is the clinical ethics expert who may well end up declaring this to be so. As, for example, when a clearly competent patient decides, in consultation with his or her health care provider, that he or she wishes to be covered by a DNR order; later objections to the DNR order by the patient's family will end up being rejected.

In effect, on ASBH's own terms, and given now common practice, clinical ethics experts are legitimately seen as having special standing to frame, guide and, at times, declare the end result of an ethics dispute at the bedside. And much, if not all, of what grounds this authority is known, teachable and testable.

To this section's key question, viz., "is there an ethics expert in the house?", our answer thus far is: "there may well be". But, for all of the preceding, we seem to be entitled to say, at most, that we could test for the presence of an ethics expert and probably find certain claimants wanting, given either a lack of sufficient skill or knowledge. We might well also be able to identify certain individuals who clearly embody a very robust version of the ideal, and be hard-pressed to do other than admit them to this exclusive club. But this still does not mean that we could *decisively* pick one out in the usual case; that would require a threshold conception of this expertise that just is not available, unless we do it by some sort of consensus of various "grandfathers" in the field. But then one would wonder how we pick the grandfathers.

Absent the provision of some decisive threshold concept for picking out or rejecting clinical ethics experts, which space here does not allow, I suppose that our answer to whether there is an ethics expert in the house will vary between "Yes", "No" and "Maybe", the latter being by far the most populous category. In a certain sense, no matter how impressive a given ethics consultants' credentials, the proof will always be in the pudding, i.e., in his or her activity at the bedside. In another sense, however, if we stick to well-received notions of what constitutes expertise, it is simply the possession of certain skills and knowledge, which the preceding shows to be credibly available, and perhaps the fact that the given individual is or has been substantially engaged in clinical ethics consultation. Arguably, then, the answer to our question, in any particular house, will come via a call to the switchboard to see if an individual who offers such skills and knowledge is on staff.

7. BUT IS THIS TRULY "ETHICS" EXPERTISE?

By way of conclusion, however, we need to return full circle to our initial thoughts that ethical expertise is an oxymoron, and that the idea of anyone having special standing to decide ethical issues is preposterous. The previous section arguably justifies the claim that there are "ethics experts" around, emphasis on "experts". But is this really expertise in ethics proper or something else? Can we get the word "ethics" out of the scare quotes or not?

The easy path here is to "just say no!" in company with Nancy Reagan. This is so for a number of basic reasons. First, all this talk of a bioethics "canon" may well be an accurate description of the current state of affairs in the developed West, but none of it is universally held, much is subject to dispute. During a recent visit to a Chinese society, for example, I found that informed consent to a procedure can be performed, legally and ethically, by either the patient or a family member, with a marked preference among most parties for the latter. So, in effect, the heart of the canon, i.e., patient autonomy, informed consent and truth-telling, is out the door elsewhere. One could certainly reply that universal agreement exists in no area and is thus hardly a barrier to the claim of expertise. But given that we are talking about ethics, not who is an expert at performing a coronary artery bypass graft, this surely represents a fly in the ointment. The fly might, of course, be shrunk in size by pointing out that even in more technical areas of expertise, such as surgery, counseling or architecture, the claim to expertise is not defeated by a lack of global unanimity, but this just keeps the scare quotes off the expertise part of the claim, not the ethics part.

The second and more basic reason to reply in the negative here is that there is a clear and ancient sense of "theoretical responsibility", if you will, for anyone making claims about ethics. "Is does not imply ought" as the basic ethical principle reminds us, and wherever one starts in ethics, it is hard to see that one's job is done until a theoretical foundation is supplied to justify one's ethical judgments and principles. And the history and contemporary situation seems to clearly instruct us that such a justification has not been supplied, nor is it likely to be.

Perhaps, although I will presume to close this essay with a statement of why I think the easy answer here may well not be the correct one, and suggest how a more affirmative answer might be developed.

First, if one attends to the canon of clinical ethics, as found for example in Patient's Bill of Rights statements, or institutional policies, one might note that there is a largely procedural quality to much of it. That is: protection of the autonomy and privacy of the individual patient, and deference to individual patients' present or prior views, values and statements, are surely core elements of the canon. Further, it is not as if there has not been a robust defense of such a procedural canon, e.g., in the work of thinkers like Tris Engelhardt (1982). Perhaps the typical graduate student/lecture hall view of the history and contemporary situation of ethics is just immature, and support for the canon has already been substantially supplied? And for all our contemporary obeisance to diversity, etc., reasonable and well-informed people might well be obliged to agree that there are certain core ethical truths that must be affirmed, and any society, or individual, that does not affirm and practice them is unethical.

Second and finally, there seems to be another approach here that does not presume to provide a full-blown theoretical grounding, but argues more specifically within the context of the specific bioethical issue. Take the issue of truth-telling, for example. Surely truth-telling is a part of the canon we are talking about, as well as a principle that does not enjoy unanimity of either agreement or practice around the world. For my part, I would presume to suggest that the arguments for truth-telling being a guiding principle (here I think Sissela Bok's classic work on lying captures these quite robustly) (Bok, 1978) are sufficiently strong to establish it as ethically justified in any society, whatever that societies' history, culture or preferences. And to appeal to and defer to the latter, it seems to me, is just to beg a question that has already been well answered. In my experience in a Chinese society, the sort of cases being presented often were like the following: a young woman with primary breast cancer is not told of this diagnosis, is not offered probably curative surgery, and is taken back to the provinces for herbal therapy at the insistence of her family; the case was then presented to me at the point she returned four months later, metastatic and no longer eminently treatable as she had been. Interestingly, the reasons why my Chinese hosts were "challenged" by such cases sounded remarkably like they also had the canon in their hearts and knew it was being profoundly violated. Or take the insistence that human subjects must themselves consent to submit themselves to research; do we really want to say that this principle may be violated on the basis of some other culture's history, culture or preferences, whether it be that of southern China, or Nazi Germany perhaps…..? I do not think so. I think the easy answer is a cop-out, and believe further attention to foundational theories, such as Engelhardt's, and/or attention to the full range of reasons for certain core principles of the canon, is the more appropriate path here. And we might just find that the scare quotes around ethics in the phrase "ethics expert" fall away to a large and specific extent.

REFERENCES

American Society for Bioethics and Humanities (ASBH) (1998). *Core Competencies for Health Care Ethics Consultation*. Glenview, IL: American Society for Bioethics and Humanities.

Beauchamp, T.L. & J.F. Childress (1979). *Principles of Biomedical Ethics.* New York: Oxford University Press.

Bok, S. (1978). *Lying: Moral Choice in Public and Private Life*. New York: Vintage Books.

Engelhardt, H.T. (1982). *The Foundations of Bioethics*. New York: Oxford University Press.

Moreno, J.D. (1995). *Deciding Together: Bioethics and Moral Consensus*. New York: Oxford University Press.

Wear, S.E. (2002). 'Teaching Bioethics at (or Near) the Bedside.' *Journal of Medicine and Philosophy* 27(4), 433-445.

CHAPTER 14

ANA SMITH ILTIS

BIOETHICAL EXPERTISE IN HEALTH CARE ORGANIZATIONS

The definition of expertise in bioethics is largely dependent upon the context in which the expertise is sought or is to be applied and the particular role the bioethicist is expected to play. Among the roles bioethicists play today are serving as hospital ethics consultants, public policy analysts, and expert witnesses in courts. Not all bioethicists are experts in all areas, and expertise in one area does not imply competence in another. To be a bioethics expert, an individual need not be able to perform all tasks associated with bioethicists well. Specialization within the field is legitimate.

The focus of this paper is one area of bioethics expertise, hospital ethics consultation. Expertise in hospital consultation is particularly complex and context-dependent. To be a bioethics expert in hospital consultation is not to have a purchase on any particular body of moral knowledge or consultation skills but rather on one or more of a number of approaches to moral reasoning and models of ethics consultation. The general skills and body of knowledge a person must have to be an expert depend not only on the general context within which bioethics expertise is sought (i.e., the hospital rather than the courts or the White House) but also on the specific institutions or individuals seeking the expertise (e.g., religious versus secular hospitals). Thus the knowledge and skills bioethicists need to be experts in the context of hospital ethics consultation vary depending on the health care organization within which they are to consult. There is not a single body of knowledge all bioethics consultants must understand, nor is there one model of ethics consultation they must master to be considered experts in bioethics consultation. Thus even within the particular role of ethics consultant, what constitutes expertise may vary depending on the institution, group, or individual seeking the expertise. Who counts as an expert consultant and what the consultant must do to be perceived as successful by his employer depends in large part on why the institution retains the bioethicist. I argue first for the legitimacy of diverse conceptions of expertise within bioethics. Second, I argue that in the case of hospital ethics consultation, different skills and bodies of knowledge constitute expertise, depending on the goals of the institution seeking the expertise. Third, I explore the relationship between accreditation and expertise in bioethics.

L. M. Rasmussen (Ed.), Ethics Expertise: History, Contemporary Perspectives, and Applications, 259-267.

1. BIOETHICS EXPERTISE: MULTIPLE ROLES AND THE LEGITIMACY OF SPECIALIZATION

There are multiple accounts of the history of bioethics (e.g. Jonsen, 1998, and Stevens, 2000), and they do not identify a common starting point for the field. It is plausible that many or all of these accounts reveal some truth and that bioethics truly did spring forward from many different seeds over the course of several decades during the early to middle part of the 20th century. The diverse-beginnings-theory is even more plausible when we examine the different roles bioethicists play. Bioethicists are involved in numerous segments of society where they are expected to provide different services. They may serve as educators, consultants, policy analysts and writers, and expert witnesses, among other things. Simply because these activities all fall under the heading "bioethics", there is no reason to believe that any one bioethicist is or should be an expert in all areas. Just as physicians have a general knowledge of medicine and many specialize in a particular area, there is room, and perhaps a need, for specialization among bioethicists. Specialization within bioethics may be a sign that the field is maturing and must be recognized as legitimate. It is implausible to claim (1) that only those with expertise in all areas are bioethics experts or (2) that expertise in one particular area, such as consultation, is necessary to be considered an expert in the field.

There is no reason to believe that bioethicists, any more than lawyers, physicians, chefs, and other professionals, must be well-versed in all areas of their profession to count as experts. In medicine we recognize the importance of specialization in part because of the vastness of medical knowledge; it is not feasible for any one physician to be an expert in all areas of medicine. Similarly, in law we recognize tax lawyers, criminal defense attorneys, malpractice defense attorneys and so on. We recognize as chefs individuals who specialize in French cuisine, Italian cooking, or pastry-making. Although there are not as many areas of specialization in bioethics as there are in medicine and law, there are distinct areas of expertise and we should not expect that all bioethicists are experts in all areas. Moreover, just as we would not say that if a physician is not an expert in a particular area of medicine she is not a physician, and that a chef who does not make a particular kind of food is not a chef, we should not hold that only those with expertise in a particular area of bioethics are bioethicists or are bioethics experts.

In addition to recognizing different areas of bioethics expertise based on the roles bioethicists play, bioethics expertise can also be categorized by content area. Thus we think of bioethicists as experts in research ethics, ethical issues in genetics, public health ethics, ethical issues at the end of life, and so on. Specialization in bioethics is occurring along both axes (role and content area) such that some have become well-known as experts in genetic research policy ethics, for example.

Bioethicists, regardless of the context within which they work and the kinds of expertise they have, require a general knowledge of moral reasoning (both religious and secular) and health law and policy to be literate in the field. Without this understanding, they cannot engage in a meaningful dialogue on bioethical issues with other members of the field. This suggests that there may be a body of general knowledge all bioethicsts must grasp as a necessary but not necessarily sufficient condition for being considered bioethics experts. To say that this is the case in principle is different from articulating what that body of knowledge is. As is evidenced by different requirements in graduate bioethics education, there is no universal agreement on what counts as basic bioethics knowledge. Some, for example, may hold that it is important to understand the ways different religions approach issues in bioethics even if one subscribes to no religion. Others would argue that such knowledge is irrelevant. We may agree that bioethicists need to understand different modes of moral reasoning, but which ones they need to understand and which they need to understand well enough to apply is a contentious issue. Is casuistry or principlism a better approach, for example? Regardless of what minimum conditions for bioethical knowledge we establish, they are unlikely to constitute or result in expertise by themselves. They may be necessary but not sufficient conditions for expertise. Whether the goal is to be a consultant or policy analyst, to be an expert requires more than basic knowledge.

2. BIOETHICS EXPERTISE: VARIATION IN KNOWLEDGE AMONG EXPERT CONSULTANTS

The kind of knowledge bioethicists must possess to be considered expert consultants will vary in large part depending on the goals they have or that have been established for them. This means that to be a hospital bioethics consultant does not necessarily require adherence to a specific form of moral reasoning, skills in applying a specific form of moral reasoning, or knowledge of a specific understanding of morality. The knowledge and skills necessary for expertise in this area vary because there is legitimate diversity in the goals and context of hospital consultation. If, for example, a hospital is seeking a bioethicist to ensure compliance, then this person will need to have strong knowledge of the rules and regulations that govern health care. These may include the Joint Commission on the Accreditation of Healthcare Organizations (JCAHO) standards, the Office of Human Research Protections (OHRP) regulations governing research, and local and state health care regulations. Here the bioethicist essentially serves as a compliance officer. There is a clear body of knowledge which the bioethicist must grasp and interpret, and his role is to make sure that organizational actions comport with the pertinent rules and regulations. Whether compliance is actually part of bioethics per se is open to debate, but I will not address this issue here.

If however a hospital is a religious institution with a thick set of moral commitments, it may want to hire a person trained in that religion's moral doctrine. Here the expert will have to be seen not as an impartial mediator but as a defender of organizational values. To do her job successfully she will have to be familiar with the organization's values and the implications they have for health care, and she will have to defend those positions. A secular hospital is unlikely to seek such an expert. Secular health care organizations, or religious ones that do not have a thick moral vision they wish to sustain, will want a different kind of bioethics consultant. They may want someone, for example, who has an understanding of the many moral positions patients and clinicians may hold and who is skilled at mediating between parties that disagree. The body of knowledge and the set of skills that constitute a bioethics expert may, therefore, depend on the context in which the person is to work. These are only some of many plausible accounts of what a bioethics consultant may know and do. Which one applies to a situation will be context-dependent such that there is no one uniform standard of bioethics expertise in the area of hospital ethics consultation.

The expert is the one who applies the right understanding of morality in the right context. There is not a single thick understanding of morality available in our society, and there is widespread disagreement on some of the issues that are at the heart of bioethics. Thus bioethical expertise cannot necessarily refer to knowledge of the content and application of a universally or even widely held moral code.

In addition to the many understandings of morality that flourish in our society, no standard model of consultation serves as the gold-standard in hospital ethics consultation. To be an expert consultant is not to understand and be able to apply the right model of ethics consulting, for there is not a shared understanding of what constitutes a good consultation. Different models of bioethics consultation have been recommended or developed (e.g., Rubin and Zoloth, 1994, 1996; Zoloth and Rubin, 1997; Ackerman, 1987; Johnson and Siegler, 1998; LaPuma and Schiedermayer, 1994), and there is no overlapping consensus on which one or ones should be followed.[1] Perhaps the fact that there is no accreditation for bioethicists is the most obvious sign of diverse understandings of expertise in bioethics. It is not clear what one would test for and how individuals would be evaluated or graded on an accreditation exam. So to be an expert bioethics consultant is not to understand or share one particular moral vision or model of consultation. What constitutes expertise in bioethics, I argue, depends in large part on who is requesting the expertise and where it is being applied. In the case of bioethics consultation in health care institutions, what kind of bioethicist the institution seeks will determine what knowledge and skills are necessary to be considered an expert.

Can a health care organization legitimately set any goals for a bioethics consultant, or are there limits? The work of bioethics consultants generally should involve moral reasoning about issues in health care. If a bioethicist is asked merely to "rubber stamp" institutional policies and actions, he is a mercenary, not a bioethics expert. "Rubber stamping" is not recognized anywhere, to my knowledge, as a justifiable activity of bioethicists. Bioethicists also should not be asked by institutions to act in ways that violate the law. These are not arbitrary distinctions; what holds the field of bioethics together as a field despite its interdisciplinary nature and composition is a focus on moral reasoning in health care.[2] When bioethicists are asked not to reason, as in the rubber stamping case, or when they are asked to act as experts in areas in which they do not have expertise, they are being asked to act inappropriately. It is, however, permissible for institutions to define the scope of the moral reasoning they engage in as consultants. In a religious hospital, for example, it is permissible for the institution to require that the moral reasoning that grounds ethics consultations be rooted in the religious tradition that shapes and guides the hospital. In a Roman Catholic hospital, for example, it is permissible for the role of the consultant to be interpreting and applying Roman Catholic moral doctrine and the *Ethical and Religious Directives* (1994 and 2001) in the hospital. This should be disclosed prior to employment, and the hospital should seek someone with expertise in Roman Catholic bioethics. Similarly, in a secular hospital, it is permissible for the institution to insist that Roman Catholic moral doctrine not routinely be the basis of ethics consultation. There may be situations in which it is appropriate to use religious moral reasoning because of the parties involved in a case, but in general, secular hospitals do not wish to be quasi-religious institutions. It is within their purview to give shape to institutional activities in light of the context of institutional values.

The claim that health care institutions can restrict or direct the texture of moral reasoning in bioethics consultation within their institution may give serious pause to some. This may seem like an infringement on professional autonomy. Can bioethics consultants be considered professionals if their goals are set by institutions and not within the profession or field? This understanding of expertise as context-dependent does not imply that bioethicists are not professionals. Legitimate institutional restrictions on the practices of bioethics consultation are those that are grounded in an effort to sustain the institution's mission and integrity.

Health care organizations often adopt particular goals or commitments – missions. Among the commitments a health care organization might adopt are religious values, and Roman Catholic hospitals are perhaps the most well-known examples of health care organizations with religious commitments. But a health care organization's commitments need not be religious in nature to constitute a mission. Any set of values or commitments that are intended to shape its activities constitutes an organization's mission. A hospital could be committed to providing basic health care services to the poor community surrounding the hospital, or to providing state-of-the-art cancer treatment and engaging in cutting-edge cancer research, for example. An organization's mission is its deep moral character; it is the set of fundamental commitments the organization adopts and that are intended to direct its life. Organizations, like individuals, do not always succeed in having their missions direct their lives and they do not always maintain their integrity. Only those organizations that are able both to articulate their fundamental commitments and live them out such that their actions reflect their deep moral characters have integrity. In order to maintain their integrity, that is, to articulate and act on their deep moral characters, health care organizations must shape and direct the activities of persons who act on their behalf, including physicians, nurses, public relations personnel, and bioethics consultants. Institutions that do not do this well may lose their integrity, a circumstance that can have serious consequences for individuals in our society.

There are a number of reasons for which it is important to allow for the presence in our society of health care organizations with a range of moral commitments and to encourage them to maintain their integrity. This observation should not be confused with the claim that integrity is an absolute value, a view I do not hold. Two of the most important reasons for allowing and perhaps even encouraging organizations to adopt missions and maintain their integrity concern the protection such integrity affords individuals. First, having health care organizations with a wide range of commitments that maintain their integrity can protect personal freedom and promote personal integrity. We live in a society in which persons can choose whether or not to participate in religious activities and, if so, which ones. To a large extent, we also can choose what to teach our children and who will teach them. Thus we recognize the importance of preserving personal freedom to pursue a meaningful life, insofar as one's pursuit does not violate others' basic rights of forbearance. Allowing for a range of educational and religious institutions is necessary for the preservation of individual liberty. Given the seemingly endless and irresolvable disputes regarding medical-moral matters, e.g., abortion, euthanasia, and embryonic stem cell research, we can see that in health care there also is a need to allow for a range of institutions, ones with different commitments, values and goals, in order to help preserve personal freedom. Individuals who cannot provide and receive health care in organizations whose activities are consistent with their personal values may find their personal integrity compromised. For example, assuming some form of physician assisted suicide (PAS) were legalized, a physician who holds that it is contrary to her moral and professional commitments to participate in PAS may not want to practice medicine in a health care organization that values patient autonomy above all else and requires physicians to respond to patients' requests for help in dying. Similarly, a patient whose conception of personal dignity requires control over the time and circumstances of death whenever possible might not want to receive health care in an organization that does not permit PAS.

Second, no individual or organization ever could pursue all that is good. Agents must make choices regarding which goods they will pursue. Once they have made those choices, it is reasonable to desire that they act consistently and fulfill their commitments, all things being equal. If nothing else, this gives others a reliable set of expectations regarding organizational behavior and allows others to assess the extent to which they might be willing to interact with particular organizations.

The discussion above demonstrated the legitimacy and importance of (1) allowing for the presence of health care organizations with a range of commitments and goals in our society and (2) generally speaking, allowing or even encouraging them to maintain their integrity. One implication of this argument is that insofar as bioethics consultants' activities affect institutional integrity, it is necessary for their activities to be guided by organizations' missions. This means, for example, that in a religious health care organization whose religious commitments drive the organization, it will be important that bioethics consultations be performed within this framework.

This does not imply, as some might think, that "anything goes", and that bioethics expertise is nothing more than what an institution says it is. Bioethics expertise remains focused on moral reasoning in health care. If an institution asks a bioethicist to encourage particular outcomes not for moral reasons but for financial or political ones, for example, then the bioethicist is being asked to act outside her range of expertise. Here the bioethicist must maintain her integrity by refusing to practice irresponsibly. Even if a for-profit health care institution attempted to argue that profit was central to its mission and identity and it had a moral obligation to make money for its shareholders, the bioethics consultant would not be an appropriate person to devise strategies to maximize profits or to serve as an institutional puppet or public relations star. Those are roles best played by business consultants and actors, respectively.

Bioethics consultants' activities must be focused on moral reasoning in health care. The scope of moral reasoning, however, is wide. There are multiple understandings of what constitutes good moral reasoning. Bioethics experts in the field of hospital consultation may subscribe to different "brands" of morality and models of consultation and still be considered experts. Not all experts will be compatible with all health care institutions. This, however, is not an infringement on bioethicists' professional autonomy. Instead, it reflects the reality of moral pluralism in our society.

Institutional mission and identity can justifiably frame the scope of bioethics consultation within particular institutions, just as an organization's mission and identity should shape the activities of all employees. Health care organizations that wish to maintain their integrity must not only identify and articulate their commitments, but they must live them out. To live them out requires that all individuals associated with the institution act in ways that support the organization's mission and identity. This is not a matter of professional autonomy but of professional compatibility. In a pluralistic society, individuals and institutions have diverse moral commitments, different understandings of what constitutes legitimate moral reasoning, and differing conceptions of the content of morality. The circumstance of pluralism is the context within which bioethics issues arise, and it is the context within which they must be resolved. Any conception of bioethics expertise must not only acknowledge this circumstance but must be grounded in it. There is a wide range of permissible goals and missions for health care organizations and there are important reasons for individual health care organizations to maintain their integrity. Central to establishing and maintaining their integrity is having employees, including bioethics consultants, who support the organization's mission and help it live out its commitments. Thus it is appropriate for institutions seeking bioethics consultants to establish the precise body of knowledge and skills their consultants must possess and to hire (and fire) bioethics consultants accordingly.

3. THE FLUIDITY OF BIOETHICAL EXPERTISE: ACCREDITATION AND EXPERTISE

The possibility of accrediting bioethicists has been under discussion for several years.[3] The interdisciplinary nature of the field raises numerous difficulties with the prospect of accreditation, especially because there is no single educational background that we can appeal to as a necessary (if not sufficient) condition for being a bioethicist. Our society's moral pluralism and the multiple models of consultation that follow from it add an additional layer of difficulties for accreditation. As the discussion in this paper demonstrates, there is a broad range of healthcare organizations that adopt and pursue different moral commitments. This pluralism reflects the pluralism of our society, and an important condition for the preservation of personal freedom is that healthcare organizations maintain their integrity. One aspect of this is hiring bioethics consultants with different bodies of knowledge and skill sets, and using organizations' commitments to frame bioethics consultants' activities. What would be the nature of the accreditation given this legitimately broad conception of bioethics expertise? Would accreditation be accomplished by an exam? If so, who would write the exam and what would the exam test? Where would we draw the line between general knowledge and skills all bioethicists should have, regardless of their area of focus, and knowledge only those working in particular areas need, and which would thus not be part of general accreditation standards? By raising these questions I do not mean to suggest that accreditation is impossible. These are questions any profession faces in determining what will be on its boards or exams. Those seeking to pass the Bar exam in a particular state are required to have knowledge of areas of the law in which they never intend to practice, for example. There is a sense in which professionals in a field need to be versed in areas that reach beyond their particular area of expertise or intended expertise. The same can be true in bioethics. Determining the exact nature of an exam would be difficult, but it could be done just as other professions have done. The fact that in bioethics, unlike in law and medicine, there is no standard educational background complicates matters somewhat because not everyone studies a particular kind of curriculum to become a bioethicist. Nevertheless, the absence of a uniform educational background does not make it impossible to develop general standards.

The discussion above is not intended to be an analysis of the accreditation debate in bioethics, and my goal is not to argue for or against accreditation of bioethicists. What I wish to point out is that accreditation should not be equated with expertise. Accreditation standards most likely would have to be global in scope and would measure only baseline knowledge. They would, therefore, be unable to measure expertise. For someone to be accredited in bioethics, therefore, would not tell us what expertise, if any, the person had. One could be an accredited bioethics consultant, for example, and not have the expertise necessary to serve a particular kind of health care organization. The extent of a bioethicist's expertise is unlikely to be measured by accreditation. Expertise is a much richer concept. In the area of bioethics consultation in particular, expertise is determined not only by what a person knows and can do, but also by what an institution's integrity requires. To be an expert consultant for a religious hospital may not mean that one is able to be an expert consultant for a secular hospital, and vice versa. This is a reflection of the circumstance of moral pluralism in our society. It does not suggest that bioethicists are mercenaries available to do and say whatever an institution asks of them. Rather, individuals with different understandings of morality and moral reasoning can be expert consultants; not all individuals will be compatible with all institutions, and what one institution deems expert, another may consider inappropriate. Expertise in bioethics is context-dependent.

NOTES

1. For a thorough discussion of consultation methods, see Agich (2001).
2. For further discussion of what defines the field of bioethics, see Kopelman (1998).
3. See, for example, American Society for Bioethics and Humanities (1998) and volume 10, number 1 of the *Journal of Clinical Ethics.*

REFERENCES

Ackerman, T. (1987). 'The Role of an Ethicist in Health Care.' In: G. R. Andeson and V. A. Glesnes-Anderson (Eds.), *Health Care Ethics* (pp. 308-320). Rockville: Aspen Publishing Company.

Agich, G. J. (2001). 'The Question of Method in Ethics Consultation.' *American Journal of Bioethics* 1(4), 31-41.

American Society for Bioethics and Humanities (1998). *Core Competencies for Heatlh Care Ethics Consultation: The Report of the American Society for Bioethics and Humanities*. Glenview, Illinois: ASBH.

Jonsen, A. (1998). *The Birth of Bioethics*, New York: Oxford University Press.

Jonsen, A. and M. Siegler (1998). *Clinical Ethics: A Practical Approach to Ethical Decisions in Clinical Medicine*, 4th ed. New York: McGraw-Hill.

Kopelman, L. M. (1998). 'What Makes Us One Field?' *The Journal of Medicine and Philosophy* 23(4), 356-368.

LaPuma, J. and D. Schiedermayer (1994). *Ethics Consultation: A Practical Guide*. Sudbury, MA: Jones and Bartlett.

National Conference of Catholic Bishops (1994 and 2001). *Ethical and Religious Directives for Catholic Healthcare Organizations*. Washington, D.C.: National Catholic Bishops Conference.

Rubin, S. and L. Zoloth-Dorman (1994). 'First-person Plural: Community and Method in Ethics Consultation.' *The Journal of Clinical Ethics* 5(1): 49-54.

Rubin, S. and L. Zoloth-Dorfman (1996). 'She Said/He Said: Ethics Consultation and Gendered Discourse.' *Journal of Clinical Ethics* 7(4) 321-332.

Stevens, T. (2000). *Bioethics in America: Origins and Cultural Politics*. Baltimore: Johns Hopkins University Press.

Zoloth-Dorfman, L. and S. Rubin (1997). 'Navigators and Captains: Expertise in Clinical Ethics Consultation.' *Theoretical Medicine* 18(4), 421-432.

CHAPTER 15

KENNETH KIPNIS

THE EXPERT ETHICS WITNESS AS TEACHER[1]

1. THE PROBLEM

For more than a decade I have regularly taken on responsibilities as an expert witness in ethics-related litigation, usually – but not always – in the context of health care. Trained and steadily employed as an academic philosopher, burdened with the standard university-level instructional and research responsibilities, I have often reflected both on the path that led me to this peculiar extra-mural activity and on this work's relationship to my professional education and experience. The pages that follow are an attempt to organize some of what I think I have learned about ethics expertise and my role in the judicial process. There is a connection, I will argue, between how professors ought to teach medical ethics to university students and how experts in medical ethics should testify about medical ethics in court.

In most of the cases where my expertise is sought, the medical profession's "standard of practice" is a central issue. Litigation in tort law characteristically requires a plaintiff's attorney to show a court that a physician defendant (#1) was subject to some duty, (#2) failed in the performance of that duty, and (#3) caused injury as a result of that failure. Malpractice cases typically arise out of an allegation that a clinician caused injury by acting outside of some standard of practice – e.g., failing to close up a wound properly – and so violating a duty of care. But there are ethical standards as well as technical ones, and these too can be breached. Broadly, the distinctive role of the expert ethics witness is to illuminate for the court the scope, the justification, and the import of pertinent ethical standards applicable to doctors. If asked whether alleged conduct would have been in violation, his or her professional opinion must focus narrowly on whether there would have been a violation if the facts are as others have described them. Expert ethics testimony does not extend to the truth or falsity of claims about what was actually done (#2 above) and, were a breach to have occurred, whether or not it caused injury (#3). The court has a concern about ethics – specifically any professional responsibilities of the practitioner that may have been breached. The expert witness is there to say something about the topic that will be helpful under the circumstances.

I think it is a serious question whether ethics professors are, as a class, particularly skilled in and knowledgeable about the management of ethical problems

L. M. Rasmussen (Ed.), Ethics Expertise: History, Contemporary Perspectives, and Applications, 269-278.

as these occur in practice. I personally don't think so. My philosophical education dealt almost not at all with the responsible engagement of concrete moral issues, and the doubts sometimes expressed about philosophy's ability to illuminate quandaries are certainly ones I have thought about. Accordingly, it may be helpful to describe the path this one philosopher has taken from dissertation defense to deposition. Many of the philosophical questions I have encountered along the way have prepared me for work as an expert witness.

2. THE PATH

I began the systematic study of ethics in 1962, as a sophomore at Reed College. The Sixties encompassed the Civil Rights Movement, the counterculture, the Vietnam war, the beginnings of modern feminism, and all of the controversies and questions that accompanied these. But by 1972, ten years later, with a Doctorate in philosophy and an appointment at a state university, I had appreciated that my specialization, curiously, did not in any readily apparent way equip me to disentangle and resolve concrete ethical problems occurring in what Wittgenstein called "the stream of life." I was able to trot out the main accounts of ethics and their histories. I knew the salient tenets of the chief proponents, the arguments and the counter-arguments. Given an ethical dilemma, I could set out several lines of inquiry. But it was a serious problem for me that philosophy, as a discipline, lacked consensus on a favored approach and, more importantly, did not seem to be very concerned about carrying its conversation beyond the campus, where there was work to be done. To be sure, there were a few exceptional scholars who were beginning to break out of the academy and I was not alone in believing we should be more energetic in taking on nontraditional tasks.

The phrase "applied ethics" suggests an analogy with "applied science". Engineers, for example, are often said to be in the business of applying what physicists have established. But there is nothing in philosophical ethics corresponding to the engineer's received view of physics, no broadly consistent knowledge-base about which specialists are in consensus. Thus the views I might share with others about substantive ethical issues seemed more like personal opinions than professional judgments. Nor was it clear that what I knew was helpful to those with ethical problems. (And what exactly did "helpful" mean in this context?)

These doubts were reinforced by parallel worries about the medical ethics courses I had started to teach in the early 1970's. Then as now, philosophical textbooks on medical ethics typically began by setting out some principal ethical theories. Kantian deontology, utilitarianism and virtue theory are three of the most common exemplars. Medical ethics, so the story went, was the application of these (and perhaps other abstract normative accounts) to ethical problems in health care. Approached in this way, expertise in "medical ethics" began to look alarmingly like skill in disingenuous rationalization. My best students knew how to advance Natural Law or Kantian arguments if minded to keep a dying patient alive and, just as skillfully, how to invoke libertarian or utilitarian justifications if inclined to pull the plug. As long as we philosophers taught ethics at an abstract level, this troubling problem was submerged. My medical ethics classes forced it to the surface and, for me, it was a pedagogical crisis. As a teacher of medical ethics, what should I be trying to do? And what did I need to know in order to do it?

In the face of this profound difficulty, I undertook to have sustained and systematic contact with ethically perplexed persons. I conjectured that philosophical "field work" in the Socratic tradition would help me to determine what, if anything, our "expertise" in ethics had to offer those in ethical distress. It seemed to me that practical problems were worthy of philosophical attention, if only to throw light on what it is we know. But as I began writing about these problems, I started to think about when the philosophical job is over. The most common view is that a philosophical task is complete when one publishes original results. However, I began to think that the work is over only when those to whom the arguments were addressed had been rationally persuaded by them. (Who those others were was also an issue.) Work may remain if they are unpersuaded. Like an autopsy, the careful scrutiny of "rationality failure" can be a source of philosophical advancement. It was in that sense that my excursion into "practical ethics" (a much better term) was theoretical research. To be sure, there are irrational people who can not be persuaded by flawless justifications. But, in practice, it is often unclear whether persuasion has failed because of irrationality in the hearer or because of ignorance and ineptitude in the speaker. If medicine's novices were to be knowledgeable about their responsibilities, we philosophers would have to understand the issues they would face and how to communicate with them. Fieldwork held out the hope of giving me a fresh perspective on what I had learned in my own education. It might show me what I should be teaching and how.

By the late 1970's I was trying to obtain an unpaid clinical position. Not finding success in the Midwest, I accepted an academic appointment in Hawaii – an isolated but culturally diverse area where there were no philosophers in medical ethics. By 1981, I had a grant from the Hawaii Committee for the Humanities that allowed me to work at Kapiolani-Women's and Children's Hospital. With access to the mid-Pacific's only tertiary level NICU at the height of the Baby Doe era, I was finally working with professionals who faced a bracing cascade of ethical problems: the status of intended "abortuses" showing signs of life at delivery, conflict of interest in hospital social work, the nature of brain death, withdrawal of consent in perinatal intensive care, "heroic" treatments in pediatric oncology, nontreatment decisions for severely compromised newborns, and professional burnout. I began giving talks and workshops throughout Hawaii: an average of two per week for several years.

In the years since, I have served on an institutional review board and several hospital and professional ethics committees. I have chaired and helped establish some of these bodies, drafting policies and bylaws. I have served on state commissions in Hawaii, overseeing programs for drug-abusing pregnant women, reviewing human rights violations in care homes for mentally disabled adults, recommending statewide strategies for confronting sexually transmitted diseases, and issuing recommendations to the Family Court in response to petitions for the sterilization of mentally incapacitated adult wards. I co-wrote the Hawaii statute governing the authorization of such sterilizations and have participated in the drafting of other health-related laws. I helped to develop ethics curricula for the schools of medicine and nursing at the University of Hawaii. I have done regular patient rounds in an ICU and on-the-floor case consultation. I completed a book on the ethics of the legal profession and was centrally involved in the development of a code of ethics for a 60,000-member professional association, the National Association for the Education of Young Children. During a 2001-2 sabbatical, I was a Visiting Senior Scholar/consultant at the American Medical Association's headquarters in Chicago.

What I can bring to my medical ethics courses now is a sense of the way a profession can collectively build and maintain a cogent sense of responsibility in a multi-cultural society. A medical ethics course can map the dimensions and limitations of that responsibility, point out strengths and weaknesses, unearth philosophical underpinnings and illustrate how it can be embraced, critiqued and improved. While I began my study of ethics with the question of which ethical theory provided the soundest foundation for ethical judgment, I have now moved to a much more eclectic – some would say "pragmatic" – approach. Socrates, Aristotle, Hume, Mill, Kant, Rawls, and many others have crafted a remarkable collection of intellectual tools. Ethical "practice" can benefit from the intelligent use of these strategies, provided one knows when and when not to use them. Automobile mechanics know about materials, mechanical linkages, gases and liquids, electrical circuitry, and so on. Practical ethicists should have a mastery of multiple areas, and the ability to match them to the jobs they do best.

3. EXPERT WITNESSING

I began testifying as an expert witness out of a conviction that what I had been doing in the classroom (and a dozen other venues) could be done as well in a court of law. Both activities, as I have come to understand them, draw from six key concepts.

1. The fact of ethical pluralism
2. The need for professional consensus
3. The practice of ethical justification
4. The varieties of ethical competence
5. The role of professional statespersons
6. The renegotiation of the social contract

In an earlier article (2001), I sketched the application of these ideas to teaching. Here I want to focus on their implications for the expert ethics witness.

3.1. The Fact of Ethical Pluralism

Some years ago I heard a physician complain about the difficulties he had had treating a 25-year old Samoan with a cancerous tumor in his leg. It had metastasized and death could not be prevented. It is well-known in Hawaii that Samoans can acknowledge a far greater deference to paternal authority than most Westerners can accept. As is common for Samoans, the large family had gathered in the young man's hospital room to meet with the physician. The doctor recounting the story had recommended amputation of the leg to prolong the patient's life for perhaps a few additional months. The young man's father, who was also the chief, readily agreed and the patient nodded his assent. But later on, alone in preop, the son told the surgeon that he did not want to live out his last weeks as an amputee. But even more, he did not want his father to know he had questioned his judgment. The social consequences would be severe: worse than amputation.

I have asked my medical ethics students, "What should a physician do under these circumstances?" For Samoans in the class, the question is a no-brainer: The leg comes off! Why are we talking about this? Caucasians backtrack, telling the father a story if necessary, but absolutely not doing the surgery. Asians and Pacific islanders are split, some siding with the Caucasians and others honoring the chief's judgment either because of a feeling that the patient should be treated under Samoan rules or out of a sense of filial respect. It is wonderfully clear that there is no consensus. Collectively, they do not know what a good doctor should do. My students need to make room for the probability that the moral principles that have served them and their families well over the years may be inadequate to the task of providing 21st Century physicians with sound ethical guidance. As with the technical aspects of medicine, responsible practitioners have to be open to the likelihood that what they have learned about ethical responsibility is not the final word.

What is so evident in Hawaii is on its way to becoming true in most places. Pluralism, John Rawls tells us (2001), has become a permanent feature of the human condition, unlikely to fade away. What the representative peoples in my medical ethics class need to be able to do – and this is not easy – is somehow to bracket the personal moralities they are bringing with them into the profession and learn new and different ways of understanding obligations. As a teacher and an expert witness, I have had to learn to bracket my pre-professional personal values and morality. I am not there to evangelize on behalf of some particular comprehensive religio-metaphysical view that I may have absorbed in childhood. I am not there to discuss how to be good Catholic physicians, or Jewish physicians or Jehovah's Witness physicians. For Catholic doctors may be stymied while treating a rape victim who – medically – needs to be informed about the morning-after pill: a contraceptive/abortifacient. Orthodox Jewish doctors may find it impossible to manage a patient who has withdrawn permission for ventilator support: removal of the breathing tube is morally culpable homicide. Jehovah's Witness doctors may be unable to transfuse a patient who is bleeding to death: the Bible forbids using blood in that way. The question medical ethics must ask – and that the medical profession must somehow answer – is how to be a good physician, period. Those with conflicting personal values or moralities will either have to trim the professional responsibilities they take on or choose a different career.

Since the decline of Natural Law theory, Western ethics has been struggling with the problem of how to reach responsible consensus on matters of obligation, without reliance upon sectarian premises. The tools we have crafted can be used to create an intellectual space within which reasonable people from different backgrounds can reach consensus on the ethical dimension of professional practice. I like to begin by asking what good doctors ought to care about (Kipnis, 2000). The specification and prioritization of core professional values can constitute a foundation for responsibility. There are three hallmarks for these values: (1) celebratability by the profession, (2) serious incorporation into professional education and practice, and (3) endorsement by the communities the profession serves. It is not enough that doctors (1) subjectively concur that patient benefit (for example) is highly important, and that (2) they develop and polish their therapeutic skills and enlarge their related knowledge bases. It is equally important that (3) the rest of us devoutly want our doctors to care about patient benefit. This reciprocity is at the heart of professional responsibility.

3.2. *The Need for Professional Consensus*

Dilemmas in professional ethics are either a consequence of conflicts between core values, necessitating a priority rule, or a result of ambiguity in a core value, necessitating what we philosophers call "disambiguation". In the case of the Samoan, the core values in conflict involve beneficence and respect for the family. Arising in many cases involving parents and children, these values need to be prioritized. There is also ambiguity in how we are to understand beneficence. Given the patient's outlook, are we benefiting or harming him by keeping him alive as long as possible, but as an amputee he does not want to be? And what about the meaning of "family"? Do the parents of mature offspring have the requisite standing to decide? We need to disambiguate. The core values approach mirrors in pedagogy the creative process that can give rise to professional consensus. Participants in class and within the field of medicine can articulate the profession's core values and, in the context of carefully chosen cases, disambiguate and prioritize these in the process of reaching consensus about what good physicians should do. Both the values and the prioritization are in play and wholly the product of discussion.

The dialog in a successful class can therefore mirror the ethical progress of the medical profession. For as contributions to the bioethics literature accumulate, it gradually becomes possible to distinguish between what might be called "consensus issues", about which a mature, broadly accepted, well-founded professional judgment has been formed, and "cutting edge issues", about which responsible professional judgments either fall on both sides or are sparse. It is a fact of contemporary life that the publicity surrounding dilemmas in medical ethics is greatest when they first appear and professional opinions are poorly informed. Cases of newborns with correctable but potentially fatal congenital anomalies coupled with Down syndrome (e.g., the Baby Doe cases of the 1980s) were highly publicized examples of this. Do parents have the authority to withhold effective treatment for their infant's anomaly solely because they do not want a retarded child? Now, dozens of articles and years later, there is broad and responsible consensus on the dimensions of the obligation to treat such infants. But the media have not covered the evolution of informed opinion as they earlier publicized the initial cases that precipitated the debate. The disparity between the striking initial dramas and the lethargic emergence of professional consensus has perhaps led members of the public and even some scholars to opine that ethical questions are inherently resistant to responsible answers.

But this is not so and it is possible to ascertain the conventional ethical wisdom of the profession. As an expert witness, I can call on three types of evidence. First, there are the codes and other consensus documents that regularly pop up in medicine. While these are a key source, the positions to be found in them have to be consistent with the second source: professional literature. Where a topic has been well-explored, the relevant arguments developed and assessed, it is sometimes clear that, at least for the time being, the issue is settled. In the United States, for example, there is a good deal of effective consensus on (1) the nature of and need for informed consent, (2) queuing patients for treatment during civilian disasters, (3) many areas of confidentiality, (4) most cases of withdrawal and withholding of life support, (5) medical decision-making for decisionally-incapacitated adult patients and (6) the use of patients as research subjects. Both the consensus documents and the medical ethics literature are important in characterizing the main elements of the profession's current sense of its ethical responsibility. The third type of evidence involves the arguments that have been and can be adduced in support of candidate obligations. We now turn to the element of justification.

3.3. The Practices of Ethical Justification

The broad endorsement of general ethical standards is not enough to ground obligation. The authority of the profession's standards is not a function of their bare acceptance but rather (one hopes), a function of the good reasons there are for endorsing them. For medical ethics, like all medical knowledge, must rest upon good reasons. Philosophers are characteristically skilled in the analysis and assessment of arguments and, as teachers, should be able to convey insight to juries and judges. Setting out these justifications and exposing their ethical significance is large part of what I prepare to do in classrooms as well as courtrooms.

However, I sometimes find myself setting out the consensus view, spelling out the arguments that have persuaded the profession, but then going on to criticize those same arguments. Philosophy has a rich history of challenges to the conventional wisdom. I have occasionally written that, with respect to some issue, the medical profession's reasons for consensus are unsound and that a different position, with less obviously flawed arguments (I think), is to be preferred. Of course, when the arguments call for it, I can opine that the medical profession seems to have gotten something right. Medical ethics is, as it were, always "in play": its precepts should never be carved in stone.

At the broadest level, sound principles of medical ethics are rules that, if broadly adhered to by practitioners, will enable the profession to discharge the social responsibilities it has assumed in the political process of professionalization,[2] rules that will enable the profession to secure and further its distinctive core values. These rules derive their obligatory force from the profession's publicly expressed commitment to significant social values (as in its codes of ethics) and in the reciprocal and exclusive reliance of the public upon the profession. We make it a crime to practice medicine without a license, so only doctors can properly treat us. The public relies upon professions to its detriment if its trust is misplaced.

As knowledgeable expert witnesses, we can play a role in the clarification of the legal and ethical context within which medicine is practiced. As teachers, we can help novice practitioners do responsible work in that setting. There can be progressive evolution in medical ethics (as in all reflective traditions) and professionally supported ways of bringing that about. My colleagues tell me that medicine's technical knowledge has a half life of seven years. Something similar is (or at least should be) true for medicine's ethical knowledge.

3.4. *The Varieties of Ethical Competence*

When we speak of "competence" in medicine, what typically springs to mind has mainly to do with the extent of the practitioner's knowledge-base as well as his or her skill in differential diagnosis and treatment. But there is something that can be called ethical competence. Physicians need to be able to spot potential ethical problems before they erupt. The doctor treating the 25-year-old Samoan man neglected to explore the young man's views before discussing options with his family. But, in addition, doctors should know where to look for information about ethics, they should be minimally competent to discern and assess the arguments that are set out in those materials, and they should have the skills that are needed to manage sensitive issues in the clinical setting. Medical education should nurture the development of practical ethical skills. There is something to be learned, for example, about how to conduct a family case conference when an incapacitated loved one has reached a crisis and a decision needs to be made. We know the mistakes that are possible in that setting and it is not difficult to teach students simple ways of avoiding them. Other skills involve chairing hospital ethics committees, preparing expert testimony for a legislature, approaching impaired or errant colleagues, dealing with seriously "noncompliant" patients, discussing medical errors, reviewing the bioethics literature and presenting findings in an intelligible way, and so on.

An ethically competent physician doesn't merely have an open mind, a grasp of the profession's conventional ethical wisdom and a sense of the debates that give rise to that consensus. The ethically competent physician is also capable of doing certain things. These ethics-related procedures are also part of the standard of practice.

3.5. The Role of Professional Statespersons

Most teachers tend to be focused on the individual student, and while this can be noble and praiseworthy, it can also blind us to a larger and, on my view, a much more important objective. Instructors tend to measure success in their assessment of discrete students, as if ethics in medicine were a talisman bestowed upon initiates. But it is better to conceive of the "student" as the medical profession as a whole. The job is not done when a student acquits himself or herself respectably but, rather, when the collectivity of student-doctors can model within itself the character of an ethically competent profession. Medicine is arguably one of Western civilization's most impressive traditions and, as Alasdair MacIntyre has argued (1984), to remain vital a tradition must sustain an ongoing and vigorous debate about the nature of its distinctive goods and how these are to be realized in the course of practice. Testimony by an expert witness, as an important part of that conversation, must be motivated by an understanding of what can be called an ethically competent profession. Each of the four preceding objectives is a prerequisite for this one. Unless medicine enjoys a critical mass of what we might call "statespersons", it will be unable to participate systematically and cogently in the social organization of health care, unable to chart its own course. The courtroom is a formal context within which a profession can be measured, in part, by the quality of its arguments. When voices are clear and cogent, when intellectual positions are well-informed and well-reasoned, and when open-mindedness is conjoined with critical attentiveness, there can be an expectation of progress. Ethics education in medicine can model habits of mind that can serve the abiding interests of the profession as well as individual professionals. Ethics testimony is a setting within which society takes the measure of medicine's sense of its responsibility.

3.6. The Renegotiation of the Social Contract

In the end, the medical profession exists to serve – centrally – the health needs of its community. The relationship between the two expresses something of a social contract: benefits and resources flow directly and indirectly to the profession as it ministers to the medical needs of others. Collectively, doctors enjoy a *de jure* monopoly on the delivery of their distinct services to the community and reap rewards in consequence. A community is in good order when both parties are reasonably satisfied with the goods they receive and the burdens they are obligated to shoulder. Accordingly, pertinent questions of justice, social values, ethical obligation and public policy will pervade aspects of that transaction and, ironically, physicians will almost inevitably find themselves on both sides, as patients and providers.

If, in one sense, the ultimate student of the medical ethics teacher is the medical profession itself, in another the ultimate client of the professional medical ethicist is the community as a whole. In the background of our ethical deliberations must be a sense of the proper place of medical care in a just society, the standards that should inform "the deal".

The expert ethics witness can make a similar contribution. For it is in courts of law where arguments come under the most intense public criticism and review, where the soundness of judgment meets one of its strongest challenges. For a central collective task of medicine and medical ethics is to identify and achieve a stable and just accommodation between those involved in the delivery of health care and the members of the communities they serve. The task of reconciliation is rendered more difficult by moral pluralism and the tendency to act and judge out of personal interest rather than out of professional and civic values. When they are clear, sound ethical standards help a profession to discharge the social responsibilities assumed in the political process of professionalization, exactly as citizens are entitled to expect it to.

It is curious that Western medicine and philosophy trace their origins to two Greeks who flourished 2400 years ago. While the art of Hippocrates would be easily recognizable to today's physicians, Socrates work is quite unlike what most philosophers do today, for he practiced in the marketplace, in prison, in the courts and in people's homes. It was Plato, Socrates' student, who sequestered philosophy in a specialized institution where it has historically educated the sons of the well-off while nurturing a discipline that Plato had made Academic rather than practical. Those of us who have recently come to bioethics from philosophy owe a debt to medicine. For bioethics has reminded us of the challenges to be found in mending a fraying social fabric, in healing a part of the world in a small but perhaps important way.

NOTES

1. This article draws heavily on three essays I have published earlier: Kipnis 2001; 1997; 2000.
2. This account of occupational professionalization and its relationship to ethical obligation is further developed in Kipnis, 1986.

REFERENCES

Kipnis, Kenneth (1986). *Legal Ethics.* Englewood Cliffs, NJ: Prentice-Hall.
Kipnis, Kenneth (1997). 'Confessions of an Expert Ethics Witness.' *Journal of Medicine and Philosophy* 22(5), 325-343
Kipnis, Kenneth (2000). 'Medical Ethics Education in a Problem-based Learning Curriculum.' *APA Newsletter on Philosophy and Medicine* (Fall). Accessible at: http://www.apa.udel.edu/apa/publications/newsletters/v00n1/medicine/05.asp.
Kipnis, Kenneth (2001). 'Professional Ethics and Instructional Success.' In: Audiey Kao (Ed.), *Professing Medicine.* Chicago: American Medical Association.
MacIntyre, Alasdair (1984). *After Virtue.* Notre Dame: University of Notre Dame Press.
Rawls, John (2001). *Justice as Fairness.* Cambridge: Harvard University Press.

NOTES ON CONTRIBUTORS

Kenneth Cust, Ph.D., is Associate Professor in the Department of English and Philosophy, Central Missouri State University.

Mary Ann G. Cutter, Ph.D., is Associate Professor in the Department of Philosophy, University of Colorado at Colorado Springs

Corinna Delkeskamp-Hayes, Ph.D., is Director of European Programs, International Studies in Philosophy and Medicine, Freigericht, Germany.

Ben Eggleston, Ph.D., is Assistant Professor in the Department of Philosophy, University of Kansas.

Ana Smith Iltis, Ph.D., is Assistant Professor in the Center for Health Care Ethics, Saint Louis University.

Carrie-Ann Biondi Khan, Ph.D., is Assistant Professor, Department of Art, Music and Philosophy, John Jay College of Criminal Justice, City University of New York.

Kenneth Kipnis, Ph.D., is Professor in the Department of Philosophy, University of Hawaii, Manoa.

Scott LaBarge, Ph.D., is Assistant Professor in the Department of Philosophy and the Department of Classics at the University of Santa Clara.

Dale E. Miller, Ph.D., is Assistant Professor in the Department of Philosophy and Religion, Old Dominion University

Lisa S. Parker, Ph.D., is Associate Professor and Director of Graduate Education in the Center for Bioethics and Health Law, University of Pittsburgh.

Lisa M. Rasmussen, Ph.D., is Adjunct Instructor in the Department of Philosophy, University of Alabama at Birmingham.

Christopher Tollefsen, Ph.D., is Associate Professor in the Department of Philosophy, University of South Carolina.

Griffin Trotter, M.D., Ph.D., is Associate Professor of Ethics and Associate Professor of Surgery in the Center for Health Care Ethics, Saint Louis University.

Robert Veatch, Ph.D., is Professor of Medical Ethics at the Kennedy Institute of Ethics, Georgetown University.

Stephen Wear, Ph.D., is Associate Professor in the Department of Medicine, University of Buffalo.